Peptide Self-Assembly and Engineering

Peptide Self-Assembly and Engineering

Fundamentals, Structures, and Applications

Volume 1

Edited by Xuehai Yan

WILEY VCH

Editor

Prof. Xuehai Yan
Institute of Process Engineering, CAS
Zhongguancun
Haidian District
Beijing, 100190
China

Cover Image: © Xuehai Yan

Library of Congress Card No.: applied for

British Library Cataloguing-in-Publication Data
A catalogue record for this book is available from the British Library.

Bibliographic information published by the Deutsche Nationalbibliothek
The Deutsche Nationalbibliothek lists this publication in the Deutsche Nationalbibliografie; detailed bibliographic data are available on the Internet at http://dnb.d-nb.de.

Print ISBN: 978-3-527-35195-4
ePDF ISBN: 978-3-527-84124-0
ePub ISBN: 978-3-527-84125-7
oBook ISBN: 978-3-527-84126-4

Typesetting: Straive, Chennai, India
Printing and Binding: CPI Group (UK) Ltd, Croydon, CR0 4YY

C9783527351954_240124

Contents

Volume 1

Preface *xiii*

1 **Introduction to the Concept, History, and Significance of Peptide Self-assembly and Materials** *1*
Xuehai Yan
References *3*

2 **Peptides: Molecular Basis, Secondary Structures, and Synthesis Methods** *5*
Luyang Zhao, Xiaokang Ren, and Xuehai Yan
2.1 Molecular Basis of Peptides *5*
2.2 Peptide Secondary Structures *6*
2.2.1 Helices *7*
2.2.2 Sheets *9*
2.2.3 Turns *10*
2.3 Synthesis Strategies of Peptides *10*
2.3.1 Solid-phase Peptide Synthesis *11*
2.3.2 Peptide Ligation *14*
2.3.3 Biological Manufacturing of Peptides *16*
2.4 Conclusion *18*
Acknowledgment *18*
References *18*

3 **Principles of Peptide Self-assembly and Material Design Rules** *23*
Chengqian Yuan, Shuai Cao, and Xuehai Yan
3.1 Introduction *23*
3.2 LLPS-mediated Multistep Nucleation Mechanism *24*
3.3 HOO-mediated Multistep Growth Mechanism of Peptide Self-assembly *28*
3.4 Principles of Peptide Self-assembly *31*
3.4.1 Thermodynamics and Kinetics *31*

3.4.2 The Role of Intermolecular Interactions *33*
3.4.2.1 Electrostatic Interaction *34*
3.4.2.2 Hydrogen Bonding Interaction *34*
3.4.2.3 π–π Stacking *35*
3.4.2.4 Hydrophobic Effect *36*
3.5 Design Rules for Materials Based on Peptide Self-assembly *37*
3.5.1 Designing Amphiphilic Peptide Structures with Desired Functions *37*
3.5.2 Modulation of Intermolecular Interactions *38*
3.5.3 Regulating Self-assembly Thermodynamics and Kinetics *39*
3.6 Conclusions and Outlook *41*
Acknowledgments *42*
References *42*

4 Molecular Simulations and Computational Chemistry of Peptide Self-assembly *51*
Peng Zhou, Hai Xu, and Xuehai Yan
4.1 Introduction *51*
4.2 Molecular Models and Computational Methods *52*
4.2.1 Theory *52*
4.2.1.1 Quantum Mechanics *52*
4.2.1.2 Molecular mechanics *52*
4.2.1.3 Molecular dynamics *53*
4.2.2 Molecular Model of Force Field in Molecular Simulation *54*
4.2.2.1 All-atom Force Field *54*
4.2.2.2 United-atom Force Field *54*
4.2.2.3 Coarse-grained Force Field *55*
4.2.3 Acceleration Method of Molecular Simulation *55*
4.2.3.1 Replica-exchanged Molecular Dynamics *55*
4.2.3.2 Discrete molecular dynamics *56*
4.2.4 Nonequilibrium Molecular Dynamics *57*
4.3 Analysis Object of Molecular Simulation in Peptide Self-assembly System *58*
4.3.1 Analysis of Molecular Interactions *58*
4.3.1.1 Types of Noncovalent Interaction in Peptide Self-assembly System *58*
4.3.1.2 Inter- and Intramolecular Interactions *60*
4.3.2 Molecular Conformation and Secondary Structure Analysis *61*
4.3.2.1 α-Helix Secondary Structure *61*
4.3.2.2 β-Sheet Secondary Structure *61*
4.3.2.3 α-Sheet Secondary Structure *62*
4.3.2.4 Random Coil Secondary Structure *63*
4.4 Molecular Simulation of Self-assembling Process *64*
4.4.1 Self-assembling Models of Peptide Systems *64*
4.4.1.1 Hierarchical Self-assembly Model *64*
4.4.1.2 Transition Model from Nanoribbon to Nanotube *64*
4.4.1.3 Surfactant-like Self-assembly Model *64*

4.4.1.4 Self-assembly Model Mediated by Phase Separation *65*
4.4.2 Simulation of Phase Separation *65*
4.4.2.1 Liquid–Liquid Phase Interface *66*
4.4.2.2 Aggregation and Fluidity in Liquid-like State *66*
4.4.3 Simulation of Structural Propensity and Evolution *67*
4.4.3.1 Aggregation Process *67*
4.4.3.2 Molecular Alignment *67*
4.4.4 Simulation of Gelation *68*
4.4.4.1 Structural Propensity Toward Nanofiber *68*
4.4.4.2 Formation of Nanofiber Network *70*
4.4.5 Simulation of Chirality Transfer *70*
4.4.5.1 Chirality of Single Molecule *70*
4.4.5.2 Chirality of Building Blocks *71*
4.5 Conclusion and Outlook *72*
Acknowledgments *73*
References *73*

5 Stimuli-responsive Structural Transformations of Peptide Supramolecular Gels *81*
Simona Bianco and Dave J. Adams
5.1 Introduction *81*
5.2 Designing Peptide-based Supramolecular Gelators *83*
5.3 Stimuli-responsive Changes in Peptide Supramolecular Gels *85*
5.3.1 Aging *88*
5.3.2 Light-triggered Response *89*
5.3.2.1 Photodimerization *90*
5.3.2.2 Photoisomerization *91*
5.3.2.3 Photoreduction *92*
5.3.3 Chemical-responsive Gels *93*
5.3.3.1 pH-triggered Changes *93*
5.3.3.2 Dynamic Chemical Transformations *94*
5.3.3.3 Redox Response *97*
5.3.4 Mechanical Response *98*
5.4 Conclusion and Future Perspective *100*
References *100*

6 Self-assembling Cyclic Peptide Nanotubes: Methods and Characterization *109*
Ángel L. Fuentes de Arriba and Juan R. Granja
6.1 Introduction *109*
6.2 Preparation of Self-assembling Cyclic Peptide Nanotubes *109*
6.2.1 Requirements *109*
6.2.2 Design *113*
6.2.3 Synthesis *116*
6.3 Characterization of Self-assembling Cyclic Peptide Nanotubes *117*

6.3.1 NMR *117*
6.3.2 X-ray Diffraction *118*
6.3.3 Microscopy *121*
6.3.3.1 Scanning Electron Microscopy *121*
6.3.3.2 Transmission Electron Microscopy *121*
6.3.3.3 Electron Diffraction *122*
6.3.3.4 Atomic Force Microscopy *122*
6.3.3.5 Fluorescence Microscopy *122*
6.3.4 Fourier-transform Infrared Spectroscopy *125*
6.3.5 Fluorescence Spectroscopy *126*
6.3.6 Scattering Techniques *127*
6.3.7 Size-exclusion Chromatography *128*
6.3.8 Computational Calculations *128*
6.3.8.1 Molecular Modeling *129*
6.3.8.2 Molecular Dynamics Simulations *130*
6.4 Recent Applications *131*
References *133*

7 Design and Control of Self-sorting Patterns of Supramolecular Peptide Nanofibers *143*
Keisuke Nakamura and Itaru Hamachi
7.1 Introduction *143*
7.1.1 Self-sorting in a Biological System *143*
7.1.2 Self-sorting of Supramolecular Peptide Nanofibers *144*
7.2 How to Characterize Self-sorting Patterns of Supramolecular Peptide Nanofibers *146*
7.2.1 Spectroscopy (IR/UV–Vis/CD/NMR) *146*
7.2.2 Small-angle Scattering *147*
7.2.3 Electron Microscopy/Atomic Force Microscopy *148*
7.2.4 Confocal Laser Scanning Microscopy *149*
7.3 How to Design and Control Self-sorting Patterns of Supramolecular Peptide Nanofibers *151*
7.3.1 Kinetics *151*
7.3.2 Chemical Structure *153*
7.4 Multiscale Self-sorting Pattern *154*
7.4.1 Network-level Self-sorting *154*
7.4.2 Macroscopic Segregation *156*
7.5 Functions of Self-sorted Nanofibers *157*
7.5.1 Multiple Stimuli-responsive Hydrogels *157*
7.5.2 Cascade Reaction *159*
7.5.3 Optoelectronics *159*
7.5.4 Biological Application of Self-sorting *160*
7.6 Conclusion *160*
References *162*

8 Protein-decorated Artificial Viral Capsids Self-assembled from β-Annulus Peptides *167*
Kazunori Matsuura
8.1 Introduction *167*
8.2 Artificial Viral Capsids Self-assembled from β-Annulus Peptides *168*
8.2.1 Self-assembly of β-Annulus Peptides *168*
8.2.2 Encapsulation of Guest Molecules into Artificial Viral Capsids *169*
8.3 Protein-decorated Artificial Viral Capsids *170*
8.3.1 Decoration of Functional Molecules on Artificial Viral Capsids *170*
8.3.2 Human Serum Albumin-decorated Artificial Viral Capsids *171*
8.3.3 Horseradish Peroxidase-decorated Artificial Viral Capsids *172*
8.3.4 Ribonuclease S-decorated Artificial Viral Capsids *173*
8.4 Enveloped Viral Replicas Embedded with Membrane Proteins *175*
8.4.1 Enveloped Artificial Viral Capsids *175*
8.4.2 Embedding of Connexin-43 onto Enveloped Artificial Viral Capsids *178*
8.5 Summary and Perspectives *180*
References *181*

9 Peptide Assemblies on Surfaces: A Study Using Scanning Tunneling Microscopy *185*
Lanlan Yu, Yanlian Yang, Chen Wang, and Chenxuan Wang
9.1 Introduction *185*
9.2 Structural Characterization of Individual Peptides *187*
9.2.1 Isolated Peptide Molecules Immobilized on Surfaces *187*
9.2.2 Single Peptides Immobilized by Molecular Matrix *191*
9.3 Structural Characterization of Peptide Assemblies *192*
9.4 Modulation Effects on Peptide Assemblies *197*
9.4.1 Peptide Assemblies Modulated by Mutations *197*
9.4.2 Peptide Assemblies Modulated by Post-Translational Modifications *201*
9.4.3 Peptide Assemblies Modulated by Chirality *203*
9.5 Peptide–Organic Molecule Co-Assemblies *206*
9.5.1 Peptide Assemblies Interacting with Small Molecules by Peptide Termini *206*
9.5.2 Peptide Assemblies Interacting with Small Molecules by Peptide Side Chains *208*
9.5.3 Assemblies of Peptide–Organic Conjugates *211*
9.6 Conclusions and Perspectives *212*
Acknowledgments *213*
Abbreviations *213*
References *214*

10 Molecular Assembly of Cyclic Dipeptide: Structure to Function *221*
Debasis Ghosh and Thimmaiah Govindaraju
10.1 Introduction *221*
10.2 The Synthesis of CDPs *222*

10.3 Molecular Assembly of CDP *223*
10.3.1 CDP Nanoarchitectures *223*
10.3.1.1 Amino-acid Composition and Stereochemistry *223*
10.3.1.2 The Effect of Molecular Assembly Conditions *225*
10.3.2 CDP Molecular Assembly for Various Applications *226*
10.3.2.1 CDP Amphiphiles for Drug Encapsulation and Delivery *226*
10.3.2.2 CDP-based Catalytic Architecture *229*
10.3.2.3 CDP Assembly: Optical Waveguiding *231*
10.3.2.4 CDP-based Quantum-confined Materials *232*
10.4 Biological Activities *234*
10.4.1 Neuroprotective CDPs *234*
10.4.2 Antitumor activity *235*
10.4.3 Antivirus activity *235*
10.5 Conclusions *236*
Acknowledgments *236*
References *236*

11 Minimalistic Peptide Assemblies for Sustainable Optoelectronics *241*
Yan Zhang, Jiahao Zhang, Junbai Li, Ehud Gazit, and Kai Tao
11.1 Introduction *241*
11.2 Photoluminescent Properties *242*
11.2.1 Quantum Confinement-induced Inherent Photoluminescence *242*
11.2.2 Influence of Solvents *244*
11.2.3 The Effect of Temperature *246*
11.2.4 The Effect of Chemical Conjugations *247*
11.2.5 The Effect of Doping *248*
11.3 Optical Waveguide Properties *251*
11.3.1 Intrinsic Optical Waveguiding Properties *251*
11.3.2 Oriented Optical Waveguide Devices *251*
11.3.3 Second-harmonic Generation (SHG) *253*
11.3.4 Electroluminescent Properties *254*
11.4 Conclusions and Perspectives *255*
Acknowledgment *256*
References *256*

12 Peptide-based Coacervates: A Biomimetic Protocell *263*
Manzar Abbas
12.1 Introduction *263*
12.2 Driving Forces for Coacervation and Types of Phase Separation *265*
12.3 Peptide-based Complex Coacervates *267*
12.3.1 Peptide/Nucleotide-based Coacervates *267*
12.3.2 Peptide/Peptide-based Coacervates *269*
12.3.3 Peptide/Metal-based Coacervates *271*
12.4 Peptide-based Simple Coacervates *272*

12.4.1 Designer Small Peptide Derivatives *273*
12.4.2 Mussel Foot Peptide Derivatives *274*
12.5 Biomimetic Coacervates Protocells *276*
12.5.1 Compartmentalization and Sequestration *276*
12.5.2 Catalytic Nature *277*
12.6 Conclusion and Future Perspectives *278*
Acknowledgment *279*
References *279*

13 Water-induced Peptide Self-assembly and Its Function *283*
Wei Fan, Chengqian Yuan, and Xuehai Yan
13.1 Introduction *283*
13.2 The Structure of Water *284*
13.2.1 Radial Distribution Function of Water *284*
13.2.2 Liquid and Solid Water Cluster Structures *285*
13.2.3 Interface-ordered Water *286*
13.3 Water Determines Protein Structure and Function *286*
13.3.1 Dynamic Hydration *286*
13.3.2 Water Participating in Protein Proton Transfer *289*
13.3.3 Water Inducing Protein Allosteric Effect *290*
13.3.4 Antifreeze Protein Inhibits Water Crystallization *290*
13.4 Water-induced Peptide Self-assembly and Its Function *291*
13.4.1 The Role of Water in Protein Fibrosis *291*
13.4.2 The Impact of Water Molecule Behavior on the Peptide Self-assembly Process *293*
13.4.3 Water Molecules Participating in and Stabilizing the Peptide Crystal Structure *294*
13.4.4 Water-induced Mechanical Deformation of Self-assembled Materials *295*
13.5 Conclusions *296*
Acknowledgment *296*
References *297*

Volume 2

Preface *xiii*

14 Functional peptide coatings *301*
Xingcen Liu[1] and Junbai Li

15 Self-assembling Peptide Amyloids: From Structure to Function in Nanotechnology *335*
Hao Kong and Gang Wei *335*

16 **Piezoelectric Self-assembling Peptides for Engineering Applications** *367*
Haoran Wu, Sarah Guerin, Syed A. M. Tofail, Rusen Yang, Ehud Gazit, Damien Thompson and Kai Tao

17 **Peptide-based Hydrogels for Soft Electronic Devices and Wearable Biosensors** *393*
Sen Liang, Yajie Tian, and Shuo Bai

18 **Self-assembled Peptide-based Biocatalyst** *421*
Jingjing Han, Xuehai Yan, and Juyoung Yoon

19 **Short Peptide Supramolecular Hydrogels for Antimicrobial Applications** *449*
Xuzhi Hu, Haoning Gong, Mingrui Liao, and Jian Ren Lu

20 **Supramolecular Peptide-based Nanomaterials for the Treatment of Fibrosis** *479*
Jikang Liu and Qianli Zou

21 **Self-assembly of Peptides and Chromophores for Design of Theranostic Nanodrugs and Cancer Precision Therapy** *505*
Rui Chang, Ruirui Xing, and Xuehai Yan

22 **Self-assembling Bioactive Peptides for Supramolecular Cancer Immunotherapy** *531*
Shukun Li, Xuehai Yan, and Jan van Hest

23 **Self-assembled Stimuli-responsive Nanomaterials Using Peptide Amphiphiles for Targeting Delivery of Drugs** *563*
Nayeong Jeon, Ilhyung Choi and Eunji Lee

Index *629*

Preface

It is with great pleasure that I introduce this book, which marks the 30th anniversary of the discovery of the first peptide building block capable of self-assembly. Over the past three decades, this field has undergone remarkable growth in scientific understanding and technical expertise. The materials and techniques derived from this research have not only expanded our knowledge but also provided us with innovative tools to address the challenges faced in the development of our modern society. The use of peptide-based materials with biocompatibility, biodegradability, and biorecyclability holds immense potential for promoting healthy, green and sustainable development on our planet.

In celebration of this significant milestone, I propose the publication of a comprehensive book that provides an overview of the advancements and future directions in peptide self-assembly and engineering. One particularly captivating area of exploration is the recent discovery of amino acid and peptide glasses, which exist in a thermodynamically metastable state. This finding opens up exciting avenues for the creation of dynamic, complex, and metastable glassy structures and materials that closely resemble biological systems. I firmly believe that more fascinating discoveries and groundbreaking achievements in the field of peptide self-assembly and engineering are yet to come, making this an ideal time to compile a book that presents a summary of the latest scientific findings from the past decades.

This book covers a wide range of topics related to peptide self-assembly and engineering, including self-assembly principles, assembled structures and functions, and materials engineering and applications. As such, it will cater to a diverse readership, comprising individuals with backgrounds in chemistry, materials science, nanotechnology, molecular engineering, biomedicine, and life science. Whether you come from academia, industry, or the education sector, I am confident that you will find valuable insights and information within the pages of this book.

Finally, I would like to express my deepest gratitude to my mentors, colleagues, collaborators, and students, as well as family members, whose unwavering support has been instrumental in my academic journey. I extend my heartfelt appreciation to the contributors of each chapter for accepting my invitation and sharing their expertise to make this book a reality.

Xuehai Yan
Beijing, China
August 2023

1

Introduction to the Concept, History, and Significance of Peptide Self-assembly and Materials

Xuehai Yan

State Key Laboratory of Biochemical Engineering, Institute of Process Engineering, Chinese Academy of Sciences, No.1 North 2nd Street, Zhongguancun, Haidian District, Beijing, P. R. China
University of Chinese Academy of Sciences, School of Chemical Engineering, No.1 Yanqihu East Rd, Huairou District, Beijing, P. R. China
Center for Mesoscience, Institute of Process Engineering, Chinese Academy of Sciences, No.1 North 2nd Street, Zhongguancun, Haidian District, Beijing, P. R. China

Self-assembly is a ubiquitous phenomenon in nature. Biological molecules such as phospholipid, protein, and DNA can undergo hierarchical and multiscale self-assembly to form complex and dynamic architectures and ultimately cells with life activities [1, 2]. Over the past few decades, extreme efforts have been made to design and develop self-assembling biomolecular building blocks so as to obtain biologically based materials with structures and functions similar to the natural ones.

Peptides typically comprise 2–50 amino acids and are as a structural motif of proteins. Due to their availability, programmability, biocompatibility, and biodegradability, peptides have been recognized as a kind of popular and promising molecular building blocks for fabrication of well-ordered nanostructures and functional materials [3–5]. Hence, the study of peptide self-assembly and engineering not only enriches biologically based materials with abundant structures and functions but also contributes to a broad range of applications from energy conversion to disease treatments.

The history of peptide self-assembly for design of functional materials can be traced back to the early 1990s. In 1993, Ghadiri and coworkers experimentally demonstrated the assembly of cyclic peptides containing alternating D- and L-amino acids, resulting in a new class of hollow nanotubes [6] that were originally predicted by De Santis et al. In the same year, Zhang and coworkers discovered the repetitious and ionic self-complementary peptide segment in yeast protein Zuotin [7]. Such ionic complementary peptides were found to spontaneously form ordered nanofibers through noncovalent interactions. After that, a variety of peptides with diverse secondary structures of β-sheets [8], β-hairpins [9], α-helices [10],

Peptide Self-Assembly and Engineering: Fundamentals, Structures, and Applications, First Edition.
Edited by Xuehai Yan.

and coiled coils [11] are developed as building blocks for preparation of soft materials such as supramolecular hydrogels. In 2001, Stupp and coworkers developed a family of peptide amphiphile (PA) as characterized by a hydrophilic peptide head modified with a hydrophobic alkyl tail [12, 13]. Such a design allows for the flexible incorporation of functional peptide sequence into the end of PA that usually displays biological functions. The discovery of the nanotube-forming aromatic diphenylalanine [14], the core recognition motif of Alzheimer's β-amyloid, stimulates a passion to explore short peptides consisting of 2–3 amino acids for self-assembly. With the motivation of integrating multiple functions into the peptide-based materials, a series of chromopeptides that combine peptide motifs and chromophores, similar to chromoproteins, are developed as building blocks [15, 16]. The chromopeptides can self-assemble to form a variety of architectures with both biofunctions and photofunctions.

Nowadays, peptide self-assembly has been a subject of intensive research for both fundamental and applied aims due to its customizability and superiority based on the "bottom-up" methodology. Peptide self-assembly is a process that associates with noncovalent interactions, such as hydrogen bonds, hydrophobic interactions, van der Waals forces, and electrostatic forces [17]. This process can lead to the formation of a variety of nanostructures, ranging from simple nanoparticles and nanofibers to more complex hierarchical structures such as microtubes, microwires, and capsules. Understanding the fundamental principles that guide peptide self-assembly, particularly the intermolecular interactions and multiscale dynamic processes, is critical to engineering peptide-based materials [17–19]. In recent years, there have been numerous innovative strategies developed to precisely regulate the structures and functions of peptide assemblies, thus leading to the development of peptide-based functional materials for various applications.

Molecular peptide design is a crucial factor that determines the properties and functions of peptide assemblies, and it involves a balance between structural stability and functional performance [20, 21]. To achieve the desired properties and functions, various factors such as charge, hydrophobicity, and secondary structure must be rationally considered in the design of molecular peptide building blocks. The solvent environment also plays a critical role in peptide self-assembly, and various solvent conditions such as pH, ionic strength, and polarity can affect the structure and stability of peptide assemblies. Different solvent conditions can lead to changes in the intermolecular interactions between peptides, affecting their ability to self-assemble and ultimately impacting their structures and functions. Computational simulations, such as molecular dynamics, have become essential tools for investigating the thermodynamics and kinetics of peptide self-assembly at the molecular level [22]. Using the computational methods, researchers can well optimize peptide design and solvent conditions to achieve the desired structures and properties of the resulting assemblies.

Peptide self-assembly enables the generation of nanostructured materials with unique and useful properties that have expanded their applications in fields such as biomedical engineering, nanotechnology, and photoelectronics. These materials possess unique mechanical, optical, electronic, and biological properties that make them highly suitable for developing new materials. For instance, self-assembled peptide materials have been used for tissue engineering, drug delivery, and biosensing devices due to their biocompatibility and ability to mimic natural extracellular matrices. Overall, peptide self-assembly has opened up new opportunities for developing next-generation green and eco-friendly materials and technique toward a sustainable future [23].

References

1 Whitesides, G., Mathias, J., and Seto, C. (1991). Molecular self-assembly and nanochemistry-a chemical strategy for the synthesis of nanostructures. *Science* 254: 1312–1319.

2 Lehn, J. (2013). Perspectives in chemistry-steps towards complex matter. *Angew. Chem. Int. Ed.* 52: 2836–2850.

3 Ulijn, R. and Woolfson, D. (2010). Peptide and protein based materials in 2010: from design and structure to function and application. *Chem. Soc. Rev.* 39: 3349–3350.

4 Lowik, D., Leunissen, E., van den Heuvel, M. et al. (2010). Stimulus responsive peptide based materials. *Chem. Soc. Rev.* 39: 3394–3412.

5 Levin, A., Hakala, T., Schnaider, L. et al. (2020). Biomimetic peptide self-assembly for functional materials. *Nat. Rev. Chem.* 4: 615–634.

6 Ghadiri, M., Granja, J., Milligan, R. et al. (1993). Self-assembling organic nanotubes based on a cyclic peptide architecture. *Nature* 366: 324–327.

7 Zhang, S., Holmes, T., Lockshin, C., and Rich, A. (1993). Spontaneous assembly of a self-complementary oligopeptide to form a stable macroscopic membrane. *Proc. Natl. Acad. Sci. U. S. A.* 90: 3334–3338.

8 Yokoi, H., Kinoshita, T., and Zhang, S. (2005). Dynamic reassembly of peptide RADA16 nanofiber scaffold. *Proc. Natl. Acad. Sci. U. S. A.* 102: 8414–8419.

9 Schneider, J., Pochan, D., Ozbas, B. et al. (2002). Responsive hydrogels from the intramolecular folding and self-assembly of a designed peptide. *J. Am. Chem. Soc.* 124: 15030–15037.

10 Ogihara, N., Ghirlanda, G., Bryson, J. et al. (2001). Design of three-dimensional domain-swapped dimers and fibrous oligomers. *Proc. Natl. Acad. Sci. U. S. A.* 98: 1404–1409.

11 Pandya, M., Spooner, G., Sunde, M. et al. (2000). Sticky-end assembly of a designed peptide fiber provides insight into protein fibrillogenesis. *Biochemistry* 39: 8728–8734.

12 Hartgerink, J., Beniash, E., and Stupp, S. (2001). Self-assembly and mineralization of peptide-amphiphile nanofibers. *Science* 294: 1684–1688.

13 Cui, H., Webber, M., and Stupp, S. (2010). Self-assembly of peptide amphiphiles: from molecules to nanostructures to biomaterials. *Biopolymers* 94: 1–18.

14 Reche, M. and Gazit, E. (2003). Casting metal nanowires within discrete self-assembled peptide nanotubes. *Science* 300: 625–627.

15 Zou, Q., Zhang, L., Yan, X. et al. (2014). Multi-functional porous microspheres based on peptide-porphyrin hierarchical co-assembly. *Angew. Chem. Int. Ed.* 53: 2366–2370.

16 Chang, R., Zhao, L., Xing, R. et al. (2023). Functional chromopeptide nanoarchitectonics: molecular design, self-assembly and biological applications. *Chem. Soc. Rev.* 52: 2688–2712.

17 Yuan, C., Ji, W., Xing, R. et al. (2019). Hierarchically oriented organization in supramolecular peptide crystals. *Nat. Rev. Chem.* 3: 567–588.

18 Wang, J., Liu, K., Xing, R., and Yan, X. (2016). Peptide self-assembly: thermodynamics and kinetics. *Chem. Soc. Rev.* 45: 5589–5604.

19 Yuan, C., Li, Q., Xing, R. et al. (2023). Peptide self-assembly through liquid-liquid phase separation. *Chem* https://doi.org/10.1016/j.chempr.2023.05.009.

20 Zhao, X., Pan, F., Xu, H. et al. (2010). Molecular self-assembly and applications of designer peptide amphiphiles. *Chem. Soc. Rev.* 39: 3480–3498.

21 Hamley, I. (2011). Self-assembly of amphiphilic peptides. *Soft Matter* 7: 4122–4138.

22 Ilie, I. and Caflisch, A. (2019). Simulation studies of amyloidogenic polypeptides and their aggregates. *Chem. Rev.* 119: 6956–6993.

23 Xing, R., Yuan, C., Fan, W. et al. (2023). Biomolecular glass with amino acid and peptide nanoarchitectonics. *Sci. Adv.* 9: eadd8105.

2

Peptides: Molecular Basis, Secondary Structures, and Synthesis Methods

Luyang Zhao[1], Xiaokang Ren[1,2], and Xuehai Yan[1,2,3]

[1] *State Key Laboratory of Biochemical Engineering, Institute of Process Engineering, Chinese Academy of Sciences, Beijing 100190, China*
[2] *School of Chemical Engineering, University of Chinese Academy of Sciences, Beijing 100049, China*
[3] *Center for Mesoscience, Institute of Process Engineering, Chinese Academy of Sciences, Beijing 100190, China*

2.1 Molecular Basis of Peptides

Peptides are short chains of covalently-linked amino acids. The amino acids are joined by an amide linkage, also called peptide bond. Peptides can be crudely classified as oligo- and polypeptides, depending on the number of joined amino acids. Oligopeptides are formed if a few amino acids are joined together, while polypeptides are formed if many amino acids are joined into one chain. Peptides contain only one free α-amino group and one free α-carboxylic group located on the two terminals of the peptide chain, respectively. These two terminal groups together with the polar groups of the amino-acid residues amid the peptide chain may ionize to charged groups in aqueous media. This leads to the acid–base behavior with characteristic titration curves and characteristic isoelectric pH (pI). Of course, the ionization makes peptides pH-responsive and property-variable in certain conditions.

Naturally occurring peptides are widely distributed in living organisms with diverse functions. The amino-acid residues of natural peptides range in quantities from two to many thousands. It is interesting that even the shortest peptides may have important biological effects. A typical example is the aspartylphenylalanine methyl ester, a commercially available dipeptide (ester derivative) known as aspartame that is used as a sugar substitute in foods and beverages. Many peptides represent their functions at very low concentrations. For example, a number of vertebrate hormones include oxytocin (a nonapeptide), which is released by the posterior pituitary and stimulates uterine contractions. Thyrotropin-releasing factor, a tripeptide that is formed in the hypothalamus, stimulates the release of thyrotropin, another

Peptide Self-Assembly and Engineering: Fundamentals, Structures, and Applications, First Edition.
Edited by Xuehai Yan.

hormone from the anterior pituitary gland. Some highly toxic mushroom poisons, such as amanitin, are also small peptides like many antibiotics.

Inspired by the great values of peptides, scientists designed peptide sequences that are not found in nature. These are classified as artificial peptides. Artificial peptides differ from naturally occurring peptides in both structure and functionality. Artificial peptides may also include nonnatural amino-acid residues or derived peptide backbone. It is well known that there are 20 common amino acids, while nonnatural amino acids that have been reported thus far exceed more than 300 [1–3]. Besides, the basic peptide backbone skeleton may be arbitrarily modified or inserted by external groups. Therefore, the space for peptide and peptide derivative construction is dimensionless. The artificial peptides are applied not only on biological processes as substitutes with optimized properties to natural peptides but also expanded to therapeutics, drug delivery, catalysis, and so forth [4–7].

2.2 Peptide Secondary Structures

The key difference between peptides and proteins is that the former do not possess tertiary structure. However, peptides own secondary structures, and this is usually crucial for peptide assembly and many other properties. Naturally occurring peptide secondary structures may be divided into three kinds: helices, sheets, and turns. Each kind includes several subtypes, depending on the peptide backbone skeleton. As shown in Figure 2.1, other than the planar conformation of amide bond, the backbone orientation around the C_α atom of each amino-acid residue of the peptide has rotational freedom and leads to the flexible peptide chain. The orientation is described by two dihedral angles concerning the C_α atom, namely $\phi(C_{i-1}-N_i-C_{\alpha,i}-C_i)$ and $\psi(N_i-C_{\alpha,i}-C_i-N_{i+1})$. The combination of these two angles gives rise to a certain peptide secondary structure and is feasibly revealed by the Ramachandran plot [8].

The secondary structure of peptides is partially influenced by intra- or intermolecular interactions including electrostatic interaction, hydrogen-bonding (HB) interaction, hydrophobic interaction, and van de Waals (VdW) interaction. Although these interactions are not exclusive to one another, for instance HB interaction contains electrostatic interaction because the electronegative atoms and hydrogen among one HB group are

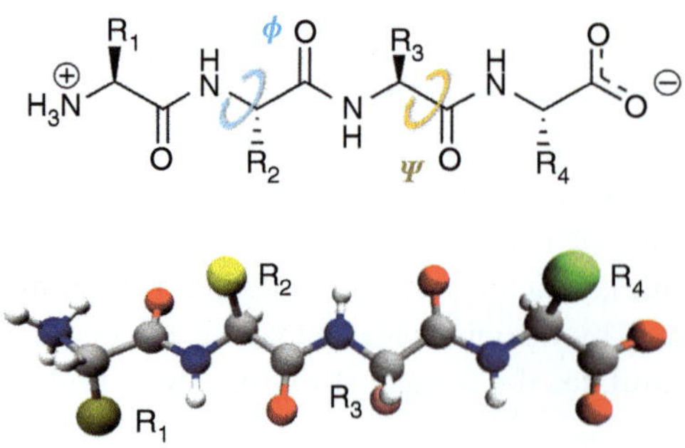

Figure 2.1 Peptide structure and stereo representation.

charged [9], they well depict the peptide structure variation in most cases, which will be detailed below. On the other hand, the secondary structure is also affected by the amino-acid residues, especially their spatial arrangement neighboring to each other. If these steric effects are repetitively arranged (and this occurs in many cases), periodic helices and sheets will be formed [10]. Especially, the adjacent C_α atoms with identical or similar dihedral angles play a vital role in the structural periodicity.

Nowadays, it is widely believed that the peptide and protein structures are determined solely by the primary amino-acid sequence. This is because the sequentially connected amino acids provide two approaches of influences with the backbone: (i) the steric effects of side chains require that the number of amide H-bonds be reduced, and (ii) the side chains have various weak interactions with backbone or other side-chain groups [11, 12]. Following the hypothesis, much effort was paid to predict the secondary or higher level structures of peptides or proteins via the primary sequence. However, a few works denoted that the secondary structure can be tuned by thermodynamic conditions such as temperature, solvent pH, and ionic strength [13–15], which means that the evolution from a randomized structure is not unique and may lead to different secondary structures along with different pathways [16–18]. The multiple pathways appended more complexity to the peptide secondary structure and are extremely important for peptide assembly.

2.2.1 Helices

Helices are the most common secondary structures distinguished by the intramolecular hydrogen bonding patterns. α-Helix is featured by the hydrogen-bonding interaction between the CO and NH groups of the amino acids indexed by i and $i+4$, respectively, on the backbone, which is briefly denoted as $CO_i \leftarrow HN_{i+4}$. This means four residues apart in the sequence [19]. Such an intra-backbone H-bonding pattern gives rise to a helical turn with 3.6 amino acids per turn, and the amino acids in a linear sequence are spatially far apart in spatial proximity. In nature, due to the chiral L-configuration of amino acids, clockwise arrangement is energetically superior to counterclockwise arrangement in a supramolecular assembly structure with low steric hindrance.

3_{10}-Helix is featured by the H-bonds between the amino-acid residues indexed by i and $i+3$ and denoted as $CO_i \leftarrow HN_{i+3}$, which leads to three residues per turn [20]. However, 3_{10}-helix owns a more stretched structure than that of the α-helix (Figure 2.2a) and has 10 atoms in the turn formed by making the H-bond. Each amino acid in the 3_{10} helix corresponds to a 120° turn and a translation of 2.0 Å along the helical axis. Residues in long 3_{10}-helices adopt (ϕ, ψ) dihedral angles near (−49°, −26°), but the value is usually deviated because 3_{10}-helices are normally shorter than α-helices.

The third type is π-type helix, which is featured by a $CO_i \leftarrow HN_{i+5}$ H-bond pattern and 4.4 residues per turn (Figure 2.2a) [21]. The standard π-helix is right-handed with 4.1 residues per turn, corresponding to an 87° turn for each amino acid and a translation of 1.15 Å along the helical axis. It is much smaller than the other two types of helices, usually containing no more than seven residues in length. Consequently, its regular repeats have much fewer dihedral angles than the other two types of helices.

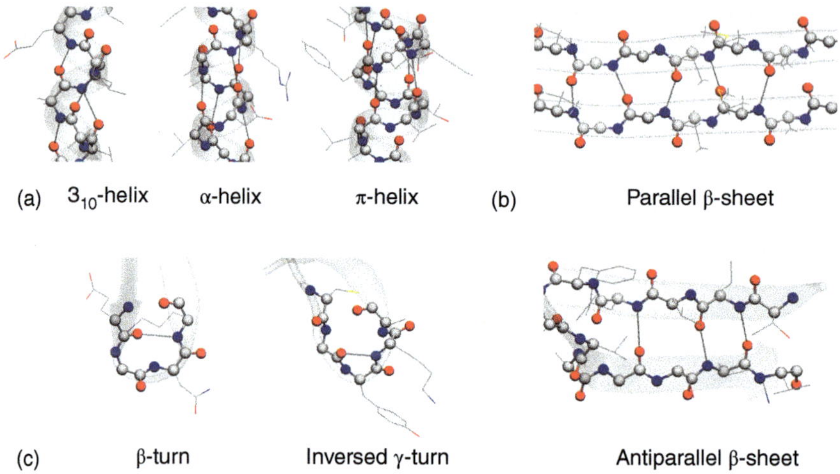

Figure 2.2 Main peptide secondary structure models. (a) Helices, including 3_{10}-helix (PDB ID: 7QDI), α-helix (PDB ID: 111L), and π-helix (PDB ID: 1X38); (b) sheets, including parallel β-sheet (PDB ID: 2BEG) and antiparallel β-sheet (PDB ID: 2TRX); (c) turns, including β-turn (PDB ID: 1N09) and inversed γ-turn (PDB ID: 2EW4). Source: Adapted with permission from Ref. [10]. Copyright 2020 American Chemical Society.

These naturally occurring helical structures stimulated researchers to explore hitherto unseen structures with spectacular functions. Nam et al. identified a pentapeptide sequenced by YFCFY that can fold to helix via the formation of disulfide bond and further assemble into macroscopic flat sheets at the air/water interface [22]. The pentapeptide can be prolonged to a heptapeptide with YYACAYY sequence and the mutation of F to A [23], which exhibited unambiguous α-helix according to single-crystal characterization. The formation of α-helix also underwent the formation of disulfide bond which is oxidized by silver ion. This peptide template is expected to facilitate efficient electron transfer to improve the electrochemical properties.

In many cases, α-helical peptides have considerably long sequence. Researchers deciphered the sequence code to construct self-assembling α-helical peptides. The key strategy is to construct a helix based on the heptad repeat sequence *abcdefg*, with the a, d, e, and g positions responsible for directing the generation of interhelical dimers or multimers (Figure 2.3) [26]. The dimers are combined via hydrophobic interactions between the apolar amino-acid residues from each monomer at the

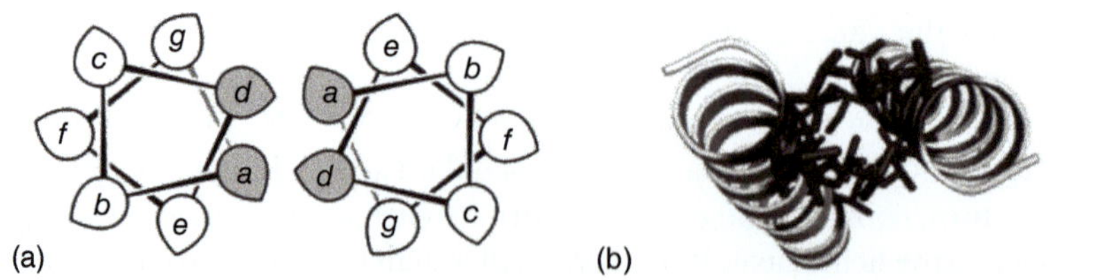

Figure 2.3 (a) Helical wheel representation of the repeated heptad of coiled-coil sequence. The helical dimers are formed due to the hydrophobic interactions on *a* and *d* sites. (b) Top view of an coiled-coil dimer crystal structure (PDB ID: 4DZM) [24]. Source: Reprinted with permission Ref. [25]. Copyright 2019, John Wiley & Sons.

specific sites. On this basis, the self-assembling behavior of peptides could be regulated by mutating the amino-acid residues. Adams et al. for the first time reported self-assembling hydrogels based on the purely α-helical linear peptides [27]. They designed a series of peptides encoded by IXXLKXK IXXLKXE IXXLEXE NXXLEX, where X is either A or Q. The peptides formed α-helical fibril arrays and interacted with each other to form hydrogels supporting >99% water content. They showed an intriguing property that gelled only by mixing, so that their formation or assembly could be rigorously controlled. Such a property rendered the α-helical peptides to be applied as substrates in cell cultivation, which was found to promote the growth and differentiation of pheochromocytoma cells.

In addition to the fibrils, Woolfson and Thomson et al. developed a geometrical framework for constructing α-helical barrels based on α-helical peptides and their complex assembly pattern, known as coiled-coil packing [28]. The unit of these *de novo* designed barrels showed near-parallel oligomeric structures, which expand the rationally designed coiled-coil assemblies from commonly prepared dimer to heptamer. Such coiled-coil oligomers could be applied as fluorescence label of proteins and to mimic diverse biological enzymatic reactions such as peptide ligation, transfer reactions, and redox reactions [25].

2.2.2 Sheets

β-Sheets describe the secondary structure that the peptide amide bonds are arranged in-plane like a pleated sheet [29]. In the β-sheet structure, the stretched polypeptide chains, called β-strands, are extended into a zigzag pattern by a series of sequential interstrand H-bonds rather than helical structure. The adjacent β-strands that form a β-sheet may be either parallel or antiparallel (Figure 2.2b). They have a similar structure, with a repeat period of 6.5 Å for the parallel conformation and a slightly longer period of 7.0 Å for the antiparallel conformation. They also have the same dihedral angles ϕ and ψ. The difference lies on their hydrogen-bonding patterns, where the hydrogen bonds of the antiparallel conformation adopt an alternating sequence of 10- and 14-membered rings. In contrast, the hydrogen bonds in the parallel conformation always consist of 12-membered rings. These similarities lead to the possibility that multistrand β-sheet structures may contain either purely parallel or antiparallel features or a mixture of both.

Naturally occurring β-sheet structures are like native silk protein [30] or immunoglobulins [31]. β-Sheet structures are also observed in certain diseases such as neurodegenerative diseases [32]. Inspired by the naturally existing peptide β-sheets, researchers explored various protocols for constructing artificial ones. One of the successfully developed design is that of the ionic-complementary peptides, which are characterized by the alternating arrangement of negatively and positively charged amino-acid residues. The ionic-complementary peptides contain the sequence segments typically such as –AEAEAKAK– and –RIQIRIQI [33, 34]. Their hydrophilic and -phobic regions are arranged alternately and divided into two ordered regions. The hydrophobic amino-acid residues are folded to shield water molecules, and the hydrophilic region has regular and orderly positive and negative charges to attract each other. The formation of intermolecular hydrogen bonds accelerated the self-assembly of peptides, and the mutual chimerism of ionic bonds increased the strength of self-assembled structures. It is worth noting that β-sheets are not always flat, and twisted β-sheets in a right-handed fashion are found in proteins [35]. Especially, a slight structural deviation from flat β-sheet may lead to drastically different self-assembly structure [36].

In addition, β-sheets may form a β-sandwich when stacked together through hydrophobic face-to-face interactions. β-Sandwiches are widely found in protein domains and can assemble and lead to unique quaternary structure and function. β-Sheets may also stack to form enclosed structures without exposing hydrogen-bonding edges, called β-barrels. Small β-barrels are structurally similar to β-sandwiches, with the neighboring layer edges being hydrogen-bonded. Comparably, large β-barrels typically form cavities accommodating with additional molecule or molecular groups [37]. Typically, green fluorescent protein (GFP) has a β-barrel structure with 11 strands, in the middle of which is located a fluorescent chromophore [38]. The chromophore is protected by the barrel against interaction with external species, thus enabling a considerable emission efficiency.

Notably, β-barrel may participate in the formation of small amyloid oligomers, which are often considered key neurotoxicants in Alzheimer's disease [39]. Another interesting structure associated with these diseases is the β-helix, which is composed of parallel β-strands and presents a helical protein structure [40]. β-Helix is mainly stabilized by hydrogen-bonding interaction, and sometimes electrostatic and hydrophobic effects also make a contribution. Remarkably, Schmuck et al. reported an octapeptide derivative with substituted *N*-terminal group self-assembles into a supramolecular β-helix via head-to-tail intermolecular interactions, based on charge interactions between the individual peptides [41]. The peptide demonstrated intriguing transitions from small particles to helical fibers in a time-variant process. And fibers may be switched to vesicles by modifying the pH value.

2.2.3 Turns

Turns are broadly defined as the regions of a peptide (and usually polypeptide) where a direction change in the backbone occurs [42]. They connect defined secondary structure regions in a protein and are thus necessary for protein compact. They generally take the forms of 16- (π-turn) [43], 13- (α-turn) [44], 10- (β-turns) [45], and 7-membered (γ-turn) [46] H-bonding rings (Figure 2.2c). Turns exhibit the characteristic dihedral angles depending on their types, and changing their sign usually converts them to their inverse form. Frequently occurring turn structures are γ-turn and β-turn, and β-turn can participate in ring formation. These loops can induce orientation changes that occur within the β-strand, which can lead to intramolecular folding and eventually an antiparallel β-sheet structure.

2.3 Synthesis Strategies of Peptides

Many peptides have unique functions and commercial values. The investigation of peptide cannot leave peptide synthesis. Nowadays, there are majorly two ways to obtain a peptide: (i) direct chemical synthesis and (ii) biological engineering.

Chemical synthesis of peptides underwent a long history since 1901, when Fischer and Fourneau created the term "peptide" [47]. After that, Bergmann and Zervas explored reversible N^{α}-protecting group, the carbobenzoxy (Cbz) group for peptide synthesis in 1932. The early classical strategies for peptide synthesis were then widely applied, such that DuVigneaud successfully constructed an oxytocin-like active peptide. A milestone event was the synthesis of the world's first artificial bovine insulin by Chinese research teams in 1965 [48, 49]. However, the synthetic difficulties soared along with the length of peptide sequence, and so did the large-scale manufacturing. Methods with higher efficiency and economy are required.

2.3.1 Solid-phase Peptide Synthesis

A new avenue was opened till the invention of solid-phase peptide synthesis (SPPS) by Merrifield in 1963, who used insoluble cross-linked polystyrene resin as peptide carrier and N^{α}-protected amino acid for synthesis [50]. The major procedures for SPPS lie as follows: (i) cleavage of the N^{α}-protecting group; (ii) washing; (iii) coupling of a protected amino acid; and (iv) washing. These steps are initiated from (amino-acid preloaded) resin and recycled accompanied by the peptide chain growing until the last amino-acid residue. Then, the peptide chain is cleaved from the resin followed with purification steps (Figure 2.4). Nowadays, microwave-assisted

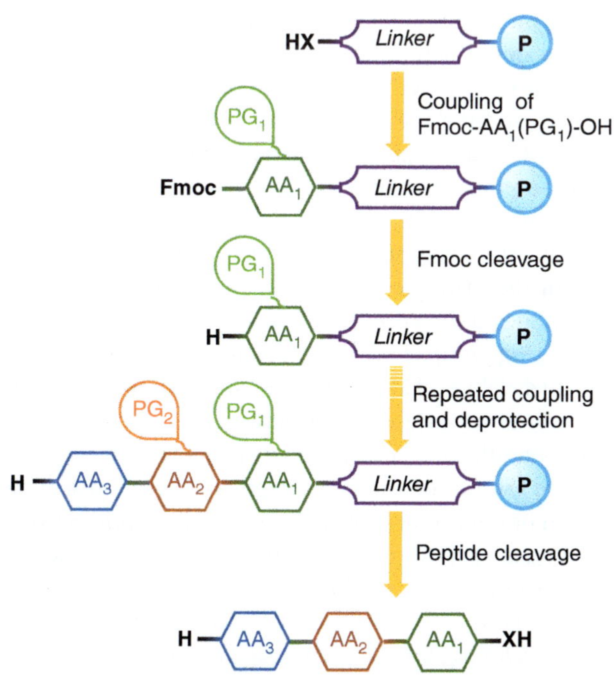

Figure 2.4 Scheme of general solid-phase peptide synthetic procedures. X = O or NH, P refers to the polymer resin, AA_n refers to the *n*th coupled amino acid, and PG_n the protected group of the *n*th coupled amino-acid residue side chain that will be cleaved at the final step, orthogonal to the cleavage of the Fmoc group. Note that not all the amino acids require a protecting group. Source: Adapted with permission from Ref. [51].

Fmoc-SPPS, where Fmoc refers to 9-fluorenylmethyloxycarbonyl which is utilized as the *N*-protecting group, has been developed to be a standard method for peptides with flexible length, various terminal groups, and abundant topologies (i.e. linear, cyclic, and branched). It is not difficult to prepare a decapeptide with crude purity above 90% using this technique.

SPPS shows exceeding efficiency for linear peptide synthesis. In terms of cyclic peptides, however, SPPS also makes sense but requires additional procedures. Cyclic peptides have a more rigid conformation than linear peptides and possess potentially diverse biological activities. Natural cyclic peptides are formed via reactions involving lactamization, lactonization, thiolactonization, or disulfide bridge formation. Correspondingly, the reactions can be classified into different types, depending on the reaction site (head, tail, or side) [4]. For those peptides that are cyclized by merely amide bonds, they can be usually synthesized on resin via the following route: Anchor the side chain of an amino acid to the resin, and the C-terminal carboxylate could be cyclized after orthogonal deprotection [52]. Sometimes the anchoring site and cyclization terminals require delicate design. In addition to such "direct" cyclic amidation, nonpeptide groups may be employed for selective synthesis of cyclic peptides, for example azide and alkyne, that cycloadded to a triazole linkage [53, 54]. The introduction of various cycloadditive groups tremendously expanded the cyclopeptide family [55, 56].

The peptide cyclization will be more complicated if there are disulfide bonds. Disulfide bonds that come from the direct linkage of two intra- or intermolecular cysteines play an important role in the peptide structure and biological activities [57, 58]. The cysteine and disulfide bond is active to environmental conditions such as redox agents and pH value. In addition, several disulfide bonds in one peptide may form cross-linked networks that are topologically much more complex than amidation cyclization [59]. These properties make the synthesis of peptides with disulfide bond (network) troubling. Currently, an effective synthetic protocol for constructing disulfide network is orthogonal stepwise linkage during an SPPS procedure, which means to control the orthogonal reactivity of the thiol protecting groups, so that it enables sequential and site-specific disulfide synthesis (Figure 2.5). Fortunately, a few tens of protecting groups have been developed for thiol groups, and they are classified as reductively labile, oxidatively labile, and "safety catch" [60]. This tool box facilitates the study of the relationships between peptide secondary structure and biological activity and the advances for novel peptides into clinics as well as therapies. By the way, the multidimensional orthogonal protection scheme may also be applied on branched or other topologically complicated peptides [61].

In addition to the efforts paid on the peptide topology, endeavors are also made on improving the SPPS efficiency. The key step in SPPS is the amino-acid coupling, where a coupling reagent must be used for the formation of amide bond. The problem is that, the "window" for developing the coupling reagent is narrow, because underactive reagents lead to insufficient activation of carboxyl group, and on the contrary, excessively active reagents may result in the notorious racemization of α-chiral center and other side reactions. Currently widely used and effective condensing agents mainly include carbodiimides, phosphoniums, and uroniums such

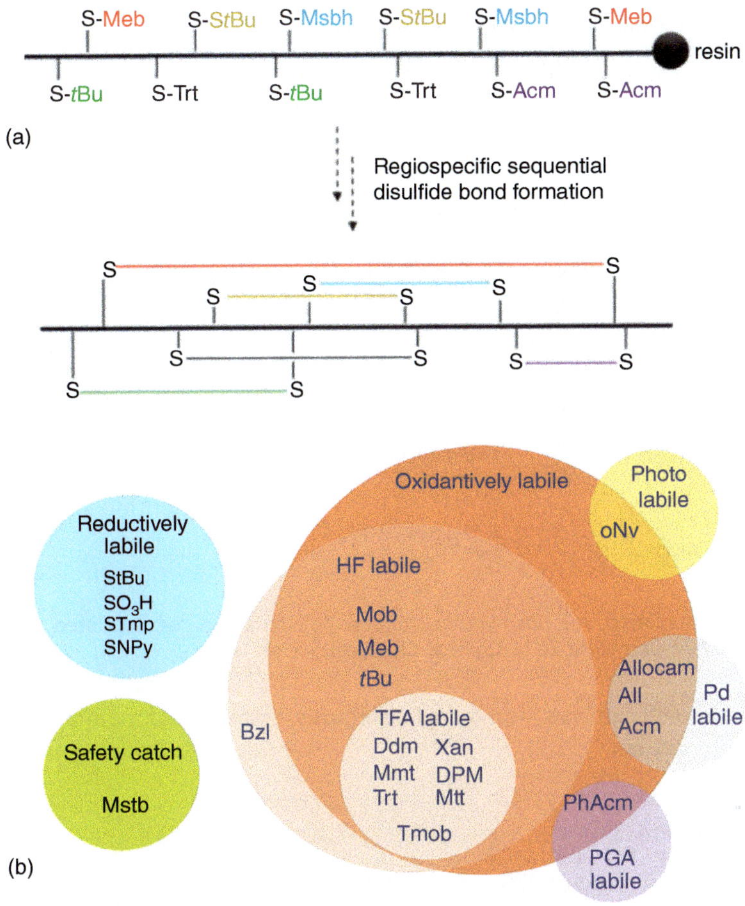

Figure 2.5 (a) Representation of orthogonal stepwise construction of disulfide bond network based on SPPS; (b) classification of orthogonal cleaving reagents. Source: Reprinted with permission Ref. [60]. Copyright 2019 Wiley-VCH.

as DIC [62], PyBOP [63], and HATU [64] with high condensation efficiency and low-level racemization. However, they suffer low step economy and atom economy [65]. Severely, the synthesis efficiency reduced along with the peptide chain elongation. Very recently, a potential new route was opened by Zhao et al., who intelligently developed a ynamide reagent for efficient amino-acid coupling [66]. The resonance structure of ynamide is analogous to the popular diimide. They screened many possible structures and discovered that tosyl- or mesyl-substituted ynamide, MYTsA, or MYMsA had a well-balanced activity and selectivity to the coupling reaction. That is, the chirality of the amino acid can be kept intact, and the target product can be obtained in almost quantitative yield (Figure 2.6). At the same time, these ynamides were universally available for most common amino acids. Compared with the current popular condensation technology, ynamide coupling reagents have the following advantages: mild reaction conditions, simple operation,

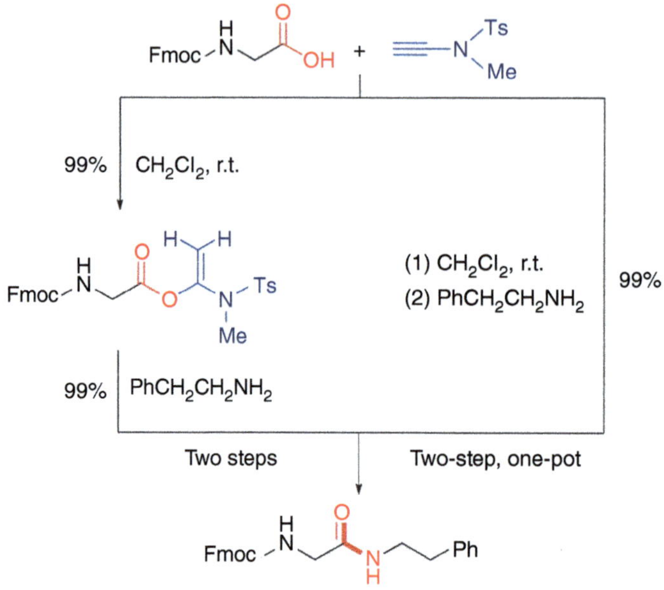

Figure 2.6 Ynamide-mediated "one-pot" synthesis of peptide amides. Source: Reprinted with permission Ref. [66]. Copyright 2021, Chinese Chemical Society.

no need for additives and catalysts, and more importantly, chirality was maintained during the coupling process.

2.3.2 Peptide Ligation

Despite the power for efficiency, SPPS suffers limitation for peptide length because small impurity in one coupling step accumulates with the peptide chain growth. For peptides with 50 amino acids or more, Kent et al. discovered an ingenious strategy named native chemical ligation (NCL) by which two peptide segments obtained by SPPS, respectively, are ligated via transthioesterification between a thioester on the C-terminal and a cysteine on the N-terminal. The thioester intermediate was then converted to a peptide bond through S-to-N acyl transfer [67]. The NCL was further developed to many more efficient strategies, including an extension of the ligation site to the other amino acids [68, 69]. The SPPS and ligation methods are competent for peptides with fewer than 100 residues. They also succeed on synthesizing longer peptides or proteins containing more than 200 amino acids, and in such cases, sequential ligation of multiple peptide segments in an organized manner is needed [70]. To achieve efficient synthesis, multisegment ligation methods such as sequential ligation, convergent ligation, and and one-pot ligation strategies are developed [71–73].

On the other hand, the combination of solid- and liquid-phase (ligation step) synthesis of peptide suffers accumulation of by-products and cumbersome purification. Overuse of protection and deprotection reagents also reduces the overall synthetic yield and process economy. Alternatively, chemoenzymatic peptide synthesis (CEPS), which is to ligate chemically synthesized peptide segments by enzyme, shows great potential for peptides with 20–50 amino acids with superior cost-efficiency [74]. CEPS is advantageous in the less requirement for side-chain protection, inhibition of racemization, and environmental friendliness.

The key for CEPS is the ligating enzyme, wherein proteases in the reverse reaction direction are applied conventionally [75, 76]. However, protease-mediated peptide ligation is thermodynamically unfavorable in aqueous media, while other natural enzymes such as sortase and butelase as the substitutes of protease require specific sequence motifs of each ligating peptide segment. To achieve a high efficiency of CEPS, researchers constructed proteins with higher activity and adaptivity via protein engineering. Remarkably, Janssen and Wu et al. reconstructed an organic cosolvent-stable enzyme from a stable calcium-independent of *Bacillus amyloliquefaciens* subtilisin BPN' with double mutations for peptide coupling reactions (Figure 2.7) [74]. The enzyme, named peptiligase, showed exceptionally catalytic efficiency on peptide coupling in water with synthesis over hydrolysis (S/H) ratio higher than 100 (which means a >99% conversion) for a 13-mer peptide. Furthermore, the peptiligase is robust and broadly applicable for peptide motifs, potentially enabling industrial synthesis of diverse pharmaceutical peptides.

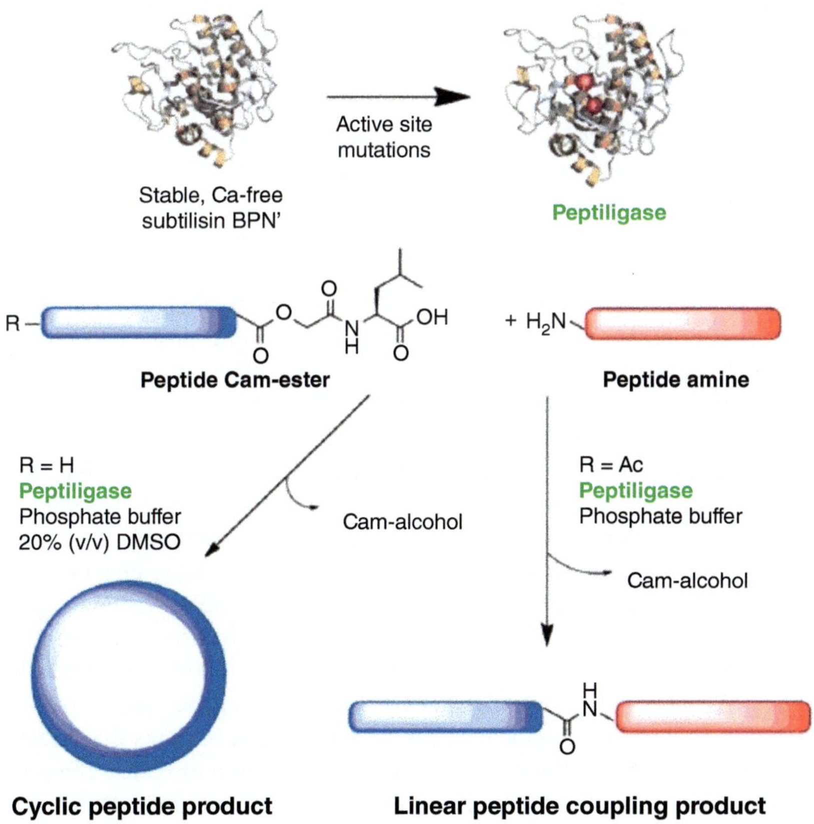

Figure 2.7 Construction of peptiligase and its catalytic mechanism. Source: Reprinted with permission Ref. [74]. Copyright 2016 Wiley-VCH.

2.3.3 Biological Manufacturing of Peptides

While chemical synthetic techniques provide almost limitless opportunities for novel peptide creation, obstacles still remained for peptides with complex structures, such as long-chain peptides and peptide derivatives. The limitations of peptide chemical synthesis are as follows: (i) the amount of raw materials such as protected amino acids, reagents, and organic solvents for use is huge, which resulted in high cost and low atomic economy; (ii) the synthetic route is complicated, which results in a low yield; and (iii) racemization may occur in the synthetic steps, which resulted in a low purity. In contrast, biomanufacturing technology has great advantages. Biomanufacturing refers to the process of using biological systems (such as enzymes or cells) to produce and process target products. It is characterized by (i) mild reaction conditions, strong demand adaptability, and environmental friendliness; (ii) that artificial processes can be constructed to match biological production processes with high efficiency; and (iii) strict regional, stereo, and chemical selectivity. Biomanufacturing is especially suitable for peptide compounds and derivatives that have complex structures, that are difficult to achieve by traditional chemical synthesis methods, and that require large-scale preparation.

The features of chemical synthesis and biological fermentation of dipeptides are outstanding in the following comparison. Many dipeptides have biological activities, such that Ala-Gln and Gly-Tyr are often used as the main nutritional components of injections for postoperative recovery of patients. Chemical synthesis of dipeptide (protocols introduced in Section 2.3.1) provides a high yield with controllable process, but the toxic reagents used may suffer potential biosafety risk. Alternatively, the bioproduction of dipeptides is carried out using nonribosomal peptide synthetase (NRPS). NRPS is a large enzyme system with a molecular weight of 100–1700 kDa. It is composed of many modules with independent catalytic ability, and each module contains a series of structural domains to bind an amino acid as a substrate to the nascent peptide chain. Doeckel and Marahiel designed a dipeptide biosynthetic organism by linking part of the *bac* gene (encoding the NRPS for bacitracin biosynthesis in *Bacillus licheniformis*) and the *tyc* gene (encoding the NRPS for brevibacterium tyrosin biosynthesis in *B. licheniformis*) and successfully expressed it in *Escherichia coli*. Among the artificial enzyme, the first A region, Bac A1, can recognize and activate isoleucine and use it with the C-PCP-A2 (recognize and activate leucine) region contained in Tyc C. As a total consequence, by inputting the starting materials of isoleucine, leucine, and ATP with the enzyme, the desired dipeptide Ile-Leu was produced. Further improvement to the enzymatic synthesis lies on two aspects: (i) the selectivity of the C region, which is crucial to control the sequence of dipeptide synthesis and (ii) the activity of the artificially synthesized NRPS, which is still not high enough to promise a good yield.

Another biomanufacturing method is direct fermentation. It contains the process of coupling amino-acid fermentation and Lal catalytic reaction to realize the production of dipeptides using the basic substrates glucose and ammonia. Direct fermentation requires more delicate design of gene encoding because the bacteria themselves could use the dipeptide as a nitrogen source and lead to trivial dipeptide production [77]. The problem can be overcome by destroying some related protease genes and dipeptide uptake enzyme genes, such as knocking out pep A, pep B, pep D, or pep N gene in several *E. coli* encoding dipeptidase or aminopeptidase and weakening the dipeptide transporter gene dpp. By this way the dipeptide degradation by the bacteria itself could be effectively inhibited. Tabata and Hashimoto used this strategy to construct a genetically engineered *E. coli* that produces Ala-Gln. After 47 hours of fermentation, the yield of Ala-Gln reached 100 mmol l^{-1} (25 g l^{-1}), the yield of by-product Ala-Ala was 26 mmol l^{-1} (3.6 g l^{-1}), and the conversion rate of glucose reached 11.5% [78].

For manufacturing peptides or peptide derivatives with complex structure, chemical and biological synthetic protocols are combined. Anidulafungin, a typical peptide antibiotics, consists of cyclic hexapeptide and fatty side chains, which functions by resisting deep fungal infections. Anidulafungin is synthesized by stepwise microbial fermentation, biotransformation, and chemical modification. A precursor echinocandin B (ECB) was prepared by fermentation of microorganisms (*Aspergillus nidulans*), then biotransformed (ECB deacylase) to prepare ECB core, and finally acylated with *p*-pentyloxyterphenyl-4-carboxylic acid to obtain anidulafungin (Figure 2.8).

The biological preparation processes with complex metabolic pathway include many factors that could be further optimized for higher production efficiency. Taking the synthesis of echinocandin as an example, the fermentation unit, deacylase activity, and the biotransformation efficiency are the key factors limiting the production yield. Zheng et al. showed that high-yielding strains may be selected and bred to improve the biological fermentation process [80, 81]. Experimentally, multiple rounds of ultraviolet mutagenesis, chemical mutagenesis, and plasma compound mutagenesis were used to obtain the ECB high-yielding mutant strain with the highest fermentation level, and the fermentation level increased from the initial 500 mg l^{-1} to more than 3500 mg l^{-1}. Besides, a metabolic engineering strain able to utilize cheap starch as the carbon source instead of fructose was screened out, which reduced the cost of ECB production dramatically by around 45% [81]. Moreover, the enzyme-catalyzed conversion process was optimized by gene cloning and gene mining technology. An enzyme library and strain library of ECB deacylase were established through the construction of recombinant engineering bacteria [79]. On this basis, the structural characteristics of the active pocket of ECB deacylase were studied, and a G287Q/R527V mutant whose activity was increased by 30 times was obtained by screening. As a total result, the biological engineering of anidulafungin was optimized more efficiently and cost-economically.

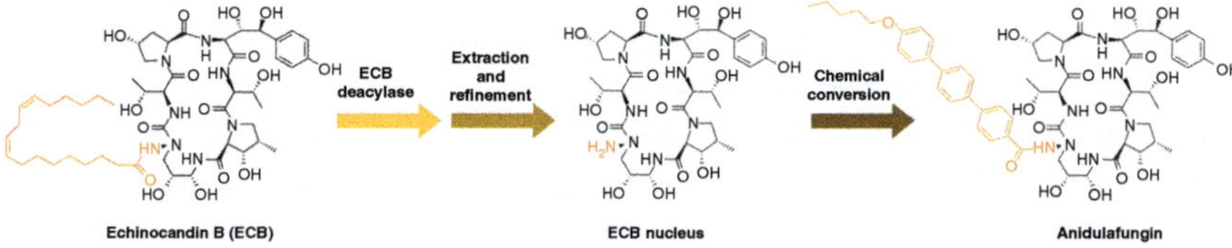

Figure 2.8 Biomanufacturing preparation procedures of an antibiotic peptide, anidulafungin. Source: Adapted with permission Ref. [79]. Copyright 2021 Elsevier.

2.4 Conclusion

This chapter briefly introduced the basic properties, secondary structure, and synthetic methodologies of peptides. The secondary structure of peptides is extremely important for their self-assembly property as well as chemical and biological functions. Synthetic methodology provides a possible way to obtain peptides with specific structures. In the future, with the continuous development of synthesis technology, more and more intriguing new peptide compounds will be discovered. They will have novel structures and physiochemical properties and will also present more interesting and important chemical and biological functions and applications.

Acknowledgment

This work was financially supported by the National Natural Science Foundation of China (project nos. 22077122, 22232006, and 22025207).

References

1 Kessler, J.L., Kang, G., Qin, Z. et al. (2021). Peptoid residues make diverse, Hyperstable collagen triple-helices. *J. Am. Chem. Soc.* 143: 10910–10919.

2 González, S.S., Ad, O., Shah, B. et al. (2021). Genetic code expansion in the engineered organism V_{max} X2: high yield and exceptional fidelity. *ACS Cent. Sci.* 7: 1500–1507.

3 Hu, B., Song, N., Cao, Y. et al. (2021). Noncanonical amino acids for hypoxia-responsive peptide self-assembly and fluorescence. *J. Am. Chem. Soc.* 143: 13854–13864.

4 Chow, H.Y., Zhang, Y., Matheson, E., and Li, X. (2019). Ligation technologies for the synthesis of cyclic peptides. *Chem. Rev.* 119: 9971–10001.

5 Hamley, I.W. (2017). Small bioactive peptides for biomaterials design and therapeutics. *Chem. Rev.* 117: 14015–14041.

6 Sheehan, F., Sementa, D., Jain, A. et al. (2021). Peptide-based supramolecular systems chemistry. *Chem. Rev.* 121: 13869–13914.

7 Tang, J.D., Mura, C., and Lampe, K.J. (2019). Stimuli-responsive, pentapeptide, nanofiber hydrogel for tissue engineering. *J. Am. Chem. Soc.* 141: 4886–4899.

8 Perczel, A., Angyan, J.G., Kajtar, M. et al. (1991). Peptide models. 1. Topology of selected peptide conformational potential energy surfaces (glycine and alanine derivatives). *J. Am. Chem. Soc.* 113: 6256–6265.

9 Izgorodina, E.I. and MacFarlane, D.R. (2011). Nature of hydrogen bonding in charged hydrogen-bonded complexes and imidazolium-based ionic liquids. *J. Phys. Chem. B* 115: 14659–14667.

10 Gloaguen, E., Mons, M., Schwing, K., and Gerhards, M. (2020). Neutral peptides in the gas phase: conformation and aggregation issues. *Chem. Rev.* 120: 12490–12562.

11 Gazit, E. (2002). A possible role for π-stacking in the self-assembly of amyloid fibrils. *FASEB J.* 16: 77–83.

12 Tatko, C.D. and Waters, M.L. (2002). Selective aromatic interactions in β-hairpin peptides. *J. Am. Chem. Soc.* 124: 9372–9373.

13 Zhao, L., Ren, X., and Yan, X. (2021). Assembly-induced superlarge redshifted absorption: a burgeoning way for organic near-infrared materials. *CCS Chem.* 3: 678–693.

14 Wang, J., Liu, K., Xing, R., and Yan, X. (2016). Peptide self-assembly: thermodynamics and kinetics. *Chem. Soc. Rev.* 45: 5589–5604.

15 van Esch, J.H., Klajn, R., and Otto, S. (2017). Chemical systems out of equilibrium. *Chem. Soc. Rev.* 46: 5474–5475.

16 Sorrenti, A., Leira-iglesias, J., Sato, A., and Hermans, T.M. (2017). Non-equilibrium steady states in supramolecular polymerization. *Nat. Commun.* 8: 15899.

17 Shi, J., Fichman, G., and Schneider, J.P. (2018). Enzymatic control of the conformational landscape of self- assembling peptides. *Angew. Chem. Int. Ed.* 21702: 11188–11192.

18 Tantakitti, F., Boekhoven, J., Wang, X. et al. (2016). Energy landscapes and functions of supramolecular systems. *Nat. Mater.* 15: 469–476.

19 Pauling, L., Corey, R.B., and Branson, H.R. (1951). The structure of proteins: two hydrogen-bonded helical configurations of the polypeptide chain. *Proc. Natl. Acad. Sci.* 37: 205–211.

20 Tonlolo, C. and Benedetti, E. (1991). The polypeptide 310-helix. *Trends Biochem. Sci.* 16: 350–353.

21 Fodje, M.N. and Al-Karadaghi, S. (2002). Occurrence, conformational features and amino acid propensities for the π-helix. *Protein Eng. Des. Sel.* 15: 353–358.

22 Nam, K.T. (2016). Water-floating giant nanosheets from helical peptide pentamers. *ACS Nano* 10: 8263–8270.

23 Nanosheets, T.S., Lee, H., Jang, H. et al. (2020). Redox-active tyrosine-mediated peptide template for large-scale single-crystalline two-dimensional silver nanosheets. *ACS Nano*. 14: 1738–1744.

24 Fletcher, J.M., Boyle, A.L., Bruning, M. et al. (2012). A basis set of de novo coiled-coil peptide oligomers for rational protein design and synthetic biology. *ACS Synth. Biol.* 1: 240–250.

25 Rink, W.M. and De Thomas, F. (2019). Novo designed α-helical coiled-coil peptides as scaffolds for chemical reactions. *Chem. Eur. J.* 25: 1665–1677.

26 Pandya, M.J., Spooner, G.M., Sunde, M. et al. (2000). Sticky-end assembly of a designed peptide fiber provides insight into protein fibrillogenesis. *Biochemistry* 39: 8728–8734.

27 Banwell, E.F., Abelardo, E.S., Adams, D.J. et al. (2009). Rational design and application of responsive α -helical peptide hydrogels. *Nat. Mater.* 8: 596–600.

28 Thomson, A.R., Wood, C.W., Burton, A.J. et al. (2014). Computational design of water-soluble α-helical barrels. *Science* 346: 485–488.

29 Pauling, L. and Corey, R.B. (1951). The pleated sheet, a new layer configuration of polypeptide chains. *Proc. Natl. Acad. Sci.* 37: 251–256.

30 Lefèvre, T., Rousseau, M.-E., and Pézolet, M. (2007). Protein secondary structure and orientation in silk as revealed by Raman spectromicroscopy. *Biophys. J.* 92: 2885–2895.

31 Buijs, J., Norde, W., and Lichtenbelt, J.W.T. (1996). Changes in the secondary structure of adsorbed IgG and F(Ab')2 studied by FTIR spectroscopy. *Langmuir* 12: 1605–1613.

32 LaFerla, F.M., Green, K.N., and Oddo, S. (2007). Intracellular amyloid-β in Alzheimer's disease. *Nat. Rev. Neurosci.* 8: 499–509.

33 Goeden-Wood, N.L., Conticello, V.P., Muller, S.J., and Keasling, J.D. (2002). Improved assembly of multimeric genes for the biosynthetic production of protein polymers. *Biomacromolecules* 3: 874–879.

34 Zhang, H., Park, J., Jiang, Y., and Woodrow, K.A. (2017). Rational design of charged peptides that self-assemble into robust nanofibers as immune-functional scaffolds. *Acta Biomater.* 55: 183–193.

35 Ho, B.K. and Curmi, P.M.G. (2002). Twist and shear in β-sheets and β-ribbons 11 Edited by J. Thornton. *J. Mol. Biol.* 317: 291–308.

36 Clover, T.M., O'neill, C.L., Appavu, R. et al. (2020). Self-assembly of block heterochiral peptides into helical tapes. *J. Am. Chem. Soc.* 142.

37 Cheng, P.-N., Pham, J.D., and Nowick, J.S. (2013). The supramolecular chemistry of β-sheets. *J. Am. Chem. Soc.* 135: 5477–5492.

38 Verkhusha, V.V. and Lukyanov, K.A. (2004). The molecular properties and applications of Anthozoa fluorescent proteins and chromoproteins. *Nat. Biotechnol.* 22: 289–296.

39 Laganowsky, A., Liu, C., Sawaya, M.R. et al. (2012). Atomic view of a toxic amyloid small oligomer. *Science* 335: 1228–1231.

40 Fitzpatrick, A.W.P., Falcon, B., He, S. et al. (2017). Cryo-EM structures of tau filaments from Alzheimer's disease. *Nature* 547: 185–190.

41 Li, M., Radić Stojković, M., Ehlers, M. et al. (2016). Use of an octapeptide–guanidiniocarbonylpyrrole conjugate for the formation of a supramolecular β-Helix that self-assembles into PH-responsive fibers. *Angew. Chem. Int. Ed.* 55: 13015–13018.

42 Marcelino, A.M.C. and Gierasch, L.M. (2008). Roles of β-turns in protein folding: from peptide models to protein engineering. *Biopolymers* 89: 380–391.

43 Rajashankar, K.R. and Ramakumar, S. (1996). π-Turns in proteins and peptides: classification, conformation, occurrence, hydration and sequence. *Protein Sci.* 5: 932–946.

44 Pavone, V., Gaeta, G., Lombardi, A. et al. (1996). Discovering protein secondary structures: classification and description of isolated α-turns. *Biopolymers* 38: 705–721.

45 Wilmot, C.M. and Thornton, J.M. (1988). Analysis and prediction of the different types of β-turn in proteins. *J. Mol. Biol.* 203: 221–232.

46 Milner-White, E.J. (1990). Situations of gamma-turns in proteins: their relation to alpha-helices, beta-sheets and ligand binding sites. *J. Mol. Biol.* 216: 385–397.

47 Fischer, E. and Fourneau, E. (1901). Über einige derivate des glykocolls. *Berichte der deutschen chemischen Gesellschaft* 34: 2868–2877.

48 Institute of Biochemistry, Chinese Academy of Sciences; Institute of Organic Chemistry, Chinese Academy of Sciences; Department of Chemistry, Peking University. V (1966). Total synthesis of crystalline insulin (translation). *Chinese Sci. Bull.* 17: 266–271.

49 Gong, Y., Du, Y., Huang, W. et al. (1965). Total synthesis of crystalline bovine insulin. (translation). *Chinese Sci. Bull.* 16: 941–945.

50 Merrifield, R.B. (1963). Solid phase peptide synthesis. I. The synthesis of a tetrapeptide. *J. Am. Chem. Soc.* 85: 2149–2154.

51 Tips and tricks for solid phase peptide synthesis from the experts at Bachem. https://www.bachem.com

52 Isied, S.S., Kuehn, C.G., Lyon, J.M., and Merrifield, R.B. (1982). Specific peptide sequences for metal ion coordination. 1. Solid-phase synthesis of cyclo-(Gly-His)3. *J. Am. Chem. Soc.* 104: 2632–2634.

53 Rostovtsev, V.V., Green, L.G., Fokin, V.V., and Sharpless, K.B. (2002). A stepwise Huisgen cycloaddition process: copper(I)-catalyzed regioselective "ligation" of azides and terminal alkynes. *Angew. Chem. Int. Ed.* 41: 2596–2599.

54 Tornøe, C.W., Christensen, C., and Meldal, M. (2002). Peptidotriazoles on solid phase: [1,2,3]-triazoles by regiospecific copper(I)-catalyzed 1,3-dipolar cycloadditions of terminal alkynes to azides. *J. Org. Chem.* 67: 3057–3064.

55 Ulrich, S., Boturyn, D., Marra, A. et al. (2014). Oxime ligation: a chemoselective click-type reaction for accessing multifunctional biomolecular constructs. *Chem. – A Eur. J.* 20: 34–41.

56 Botti, P., Pallin, T.D., and Tam, J.P. (1996). Cyclic peptides from linear unprotected peptide precursors through thiazolidine formation. *J. Am. Chem. Soc.* 118: 10018–10024.

57 Fass, D. and Thorpe, C. (2018). Chemistry and enzymology of disulfide cross-linking in proteins. *Chem. Rev.* 118: 1169–1198.

58 Landeta, C., Boyd, D., and Beckwith, J. (2018). Disulfide bond formation in prokaryotes. *Nat. Microbiol.* 3: 270–280.

59 Ahangarzadeh, S., Kanafi, M.M., Hosseinzadeh, S. et al. (2019). Bicyclic peptides: types, synthesis and applications. *Drug Discov. Today* 24: 1311–1319.

60 He, R., Pan, J., Mayer, J.P., and Liu, F. (2020). Stepwise construction of disulfides in peptides. *ChemBioChem* 21: 1101–1111.

61 Itoh, H. and Inoue, M. (2019). Full solid-phase total synthesis of macrocyclic natural peptides using four-dimensionally orthogonal protective groups. *Org. Biomol. Chem.* 17: 6519–6527.

62 Fattahi, N., Ayubi, M., and Ramazani, A. (2018). Amidation and esterification of carboxylic acids with amines and phenols by N,N′-diisopropylcarbodiimide: a new approach for amide and ester bond formation in water. *Tetrahedron* 74: 4351–4356.

63 Coste, J., Le-Nguyen, D., and Castro, B. (1990). PyBOP®: a new peptide coupling reagent devoid of toxic by-product. *Tetrahedron Lett.* 31: 205–208.

64 Carpino, L.A., El-Faham, A., and Albericio, F. (1994). Racemization studies during solid-phase peptide Synthesis using azabenzotriazole-based coupling reagents. *Tetrahedron Lett.* 35: 2279–2282.

65 Constable, D.J.C., Dunn, P.J., Hayler, J.D. et al. (2007). Key green chemistry research areas—a perspective from pharmaceutical manufacturers. *Green Chem.* 9: 411–420.

66 Liu, T., Xu, S., and Zhao, J. (2021). Recent advances in ynamide coupling reagent. *Chinese J. Org. Chem.* 41: 873–887.

67 Dawson, P.E., Muir, T.W., Clark-Lewis, I., and Kent, S.B.H. (1994). Synthesis of proteins by native chemical ligation. *Science* 266: 776–779.

68 Jin, K., Li, T., Chow, H.Y. et al. (2017). P–B desulfurization: an enabling method for protein chemical synthesis and site-specific deuteration. *Angew. Chem. Int. Ed.* 56: 14607–14611.

69 He, Q., Li, J., Qi, Y. et al. (2017). Chemical synthesis of histone H2A with methylation at Gln104. *Sci. China Chem.* 60: 621–627.

70 Zhang, B., Li, Y., Shi, W. et al. (2020). Chemical synthesis of proteins containing 300 amino acids. *Chem. Res. Chinese Univ.* 36: 733–747.

71 Zuo, C., Zhang, B., Yan, B., and Zheng, J.-S. (2019). One-pot multi-segment condensation strategies for chemical protein synthesis. *Org. Biomol. Chem.* 17: 727–744.

72 Bang, D. and Kent, S.B.H. (2004). A one-pot total synthesis of crambin. *Angew. Chem. Int. Ed.* 43: 2534–2538.

73 Erlich, L.A., Kumar, K.S.A., Haj-Yahya, M. et al. (2010). N-methylcysteine-mediated total chemical synthesis of ubiquitin thioester. *Org. Biomol. Chem.* 8: 2392–2396.

74 Toplak, A., Nuijens, T., Quaedflieg, P.J.L.M. et al. (2016). Peptiligase, an enzyme for efficient chemoenzymatic peptide synthesis and cyclization in water. *Adv. Synth. Catal.* 358: 2140–2147.

75 Bordusa, F. (2002). Proteases in organic synthesis. *Chem. Rev.* 102: 4817–4868.

76 Yokozeki, K. and Hara, S. (2005). A novel and efficient enzymatic method for the production of peptides from unprotected starting materials. *J. Biotechnol.* 115: 211–220.

77 Plasschaert, R.N. and Bartolomei, M.S. (2014). Genomic imprinting in development, growth, behavior and stem cells. *Development* 141: 1805–1813.

78 Kazuhiko, T. and Shin-ichi, H. (2007). Fermentative production of L-alanyl-l-glutamine by a metabolically engineered *Escherichia coli* strain expressing l-amino acid α-ligase. *Appl. Environ. Microbiol.* 73: 6378–6385.

79 Zou, S.-P., Han, X., Zhu, H.-Y. et al. (2021). Functional expression of an echinocandin B deacylase from actinoplanes utahensis in *Escherichia coli*. *Int. J. Biol. Macromol.* 187: 850–857.

80 Zou, S.-P., Zhong, W., Xia, C.-J. et al. (2015). Mutagenesis breeding of high echinocandin B producing strain and further titer improvement with culture medium optimization. *Bioprocess Biosyst. Eng.* 38: 1845–1854.

81 Hu, Z.-C., Peng, L.-Y., and Zheng, Y.-G. (2016). Enhancement of echinocandin B production by a UV- and microwave-induced mutant of *Aspergillus nidulans* with precursor- and biotin-supplying strategy. *Appl. Biochem. Biotechnol.* 179: 1213–1226.

3

Principles of Peptide Self-assembly and Material Design Rules

Chengqian Yuan[1], Shuai Cao[1], and Xuehai Yan[1,2,3]

[1] *State Key Laboratory of Biochemical Engineering, Institute of Process Engineering, Chinese Academy of Sciences, Beijing, P. R. China*
[2] *School of Chemical Engineering, University of Chinese Academy of Sciences, Beijing, P. R. China*
[3] *Center for Mesoscience, Institute of Process Engineering, Chinese Academy of Sciences, Beijing, P. R. China*

3.1 Introduction

Self-assembly is a universal phenomenon in nature, especially in biological systems [1–5]. Protein filaments, with their significance in regulating the mechanical characteristics, motility, and differentiation of live cells, are prevalent examples of supramolecular polymers that are formed through protein self-assembly [6–9]. Biomolecular self-assembly has proven to be not only a successful method for producing a variety of useful materials but also an excellent study model for comprehending biological phenomena, including the origin of life [5, 10, 11]. Peptides consisting of several amino acids, due to their remarkable benefits of structural, mechanical, and functional diversity as well as excellent biocompatibility and biodegradability, have evolved into a form of exquisite building blocks [4, 12]. By adjusting environmental factors during self-assembly, numerous peptide self-assemblies have been accomplished in recent years [13, 14]. In addition to the functional role that proteins play in manufacturing the essential cellular machinery of life, aberrant biomolecular self-assembly has been proven to be associated with the occurrence and progression of numerous neurological diseases, including Alzheimer's disease [15–17]. Therefore, achieving controlled self-assembly of peptides becomes a challenging yet crucial task. To address this problem, it is highly needed and extremely crucial to investigate the peptide self-assembly mechanism. The clarification of molecular mechanisms underlying peptide self-assembly is crucial not only to the understanding of biological function and malfunction but also to advancing the development of novel approaches for the prevention of some diseases associated with aberrant self-assembly as well as a wide range of potential applications ranging from nanotechnology to biomimetic self-assembly.

Peptide Self-Assembly and Engineering: Fundamentals, Structures, and Applications, First Edition.
Edited by Xuehai Yan.

In general, peptide self-assembly follows the nucleation and growth mechanism [18–20], which determines the structures and properties of final assemblies. Given the crucial significance of nucleation and growth in the structural transition of peptides from soluble form into mature assemblies, increasing research focus has switched from the preparation of peptide self-assemblies, with various structures based on the trial-and-error method, to the self-assembly process [21–25]. Tremendous efforts have been made to investigate the self-assembly process to understand the underlying molecular mechanisms [23, 26, 27]. For example, the whole structural transformation process, particularly the crucial steps preceding the nucleation of peptide self-assemblies, has been tracked experimentally, indicating that peptide self-assembly follows a multistep nucleation-growth process via multiple metastable phases [27]. However, the metastable phases have been reported as different intermediates ranging from oligomers and liquid droplets to amorphous aggregates [28]. Therefore, a general principle of peptide self-assembly is still lacking.

In this chapter, we focus on the peptide self-assembly mechanism and materials design rules. First, we illustrate the multistep self-assembly mechanism of peptides, especially the multistep nucleation through liquid–liquid phase separation (LLPS) and growth through hierarchically oriented organization (HOO). Then, we highlight the principle of peptide self-assembly based on the analysis of thermodynamics and kinetics as well as the role of intermolecular interaction. Finally, we propose the design rules of supramolecular materials based on peptide self-assembly. We aim to provide a thorough knowledge of peptide multistep self-assembly to pave the way for the exploration and development of smart and multifunctional peptide materials.

3.2 LLPS-mediated Multistep Nucleation Mechanism

The formation of peptide supramolecular polymers is a nucleation-dependent self-assembly process [18–20]. Previously, the classical nucleation theory (CNT) had been widely accepted to describe nucleation, where the ordered nucleus generates from the supersaturated systems directly. However, increasing evidence has suggested that CNT fails to explain all the aspects of nucleation from solution due to its oversimplified assumptions [29]. For instance, the capillary approximation presupposes that a cluster is a sufficiently big spherical entity with uniform density and a stiff border, and that the surface energy of the cluster can be represented in terms of the plain layer interfacial tension. Thus, this assumption is invalid for tiny clusters because the idea of macroscopic interfacial tension loses significance. With technical advancements, several theoretical and modeling investigations suggested that nuclei can generate via a metastable phase-mediated two-step nucleation process [30–32].

A prominent example based on *N*-(*t*-butoxycarbonyl)-L-Phe-L-Phe-COOH (Boc-FF) clearly illustrates the multistep nucleation of peptide self-assembly [23]. It was found that the self-assembly of Boc-FF follows a multistep phase transition from nanospheres to nanofibers and finally to the nanotubes. The results suggest that nanospheres and nanofibers, generally viewed as final peptide self-assemblies in certain conditions, can also serve as precursors for the formation of more stable self-assemblies. Further results of thermodynamic study demonstrate that Boc-FF self-assembly is a multistep process governed by Ostwald's rule of stages.

Although the multistep self-assembly process is discovered in a simple dipeptide model, such phenomena are also observed in the self-assembly of other peptides with longer sequences such as Aβ peptide of Alzheimer's disease [24, 33, 34]. The previous study on the formation of amyloid nanofibers based on the core sequence of Aβ peptide, Ac-KLVFFAE-NH_2 (Aβ [16–22]) also clearly displays similar multistep evolution processes. At the early stage of self-assembly, unstructured aggregates formed, followed by the emergence of amyloid assemblies from them [24]. Similar multistep phase transition of self-assembly has also been observed in amino-acid amphiphile self-assembly, where a progressive phase transition from spherical structures to nanofibril hydrogel and finally to crystals occurred [26, 35]. Notably, similar multistep structural evolution has also been discovered in the dynamic peptide network system [34].

The above examples have demonstrated the multistep pathway of peptide self-assembly, while the fundamental molecular events underlying the initial stages prior to nucleation are less well understood. This can be ascribed to two main reasons. Firstly, peptide self-assembly is a multiscale hierarchical self-assembly process spanning multiple time and length scales, which involves the formation and transformation of multiple transient intermediates. Secondly, the weak noncovalent interactions and flexible molecular conformations are susceptible to dynamic microenvironment changes. Therefore, the inherently dynamic and multiscale hierarchical features limit the tracking of the self-assembly process [34, 36, 37]. To address such challenges, an effective approach is to improve the stability of self-assembly intermediates by selecting relatively rigid building blocks to slow down the self-assembly kinetics. Another approach is to monitor the self-assembly process through using advanced techniques with high spatiotemporal resolutions. Amphiphilic peptides, such as diphenylalanine modified with a carboxybenzyl group (Z-FF), become ideal building blocks as they possess both hydrophilic and hydrophobic moieties, improving the stability of intermediates through the cooperative interplay of hydrophobic effects, hydrogen bonding, and π–π stacking interactions. A pioneering work was conducted to monitor the structural evolution of Z-FF nanofibers using time-lapse bright field microscopy, where the coalescence of spherical structures can be observed (Figure 3.1a) [27]. This indicates the liquid nature of such structures and confirms the formation of liquid droplet through LLPS at the early stage of self-assembly. The stronger fluorescence of the liquid droplets reveals that the droplet is a peptide-rich phase (Figure 3.1b). Moreover, time-dependent cryogenic transmission electron microscopy (cryo-TEM) was used to monitor the structural evolution process of the Z-FF self-assembly from the initial liquid droplets to the final nanofibrils (Figure 3.1c). With the progression of self-assembly, both the intermediates and nanofibrils were detected, where the number of intermediates decreased with time. Finally, it was found that flexible nanofibrils became the predominant phase in the system at the cost of the intermediate aggregates. Therefore, the liquid intermediates served as metastable nucleation precursors for the nanofibrils. Notably, such a phenomenon is general and can occur for other

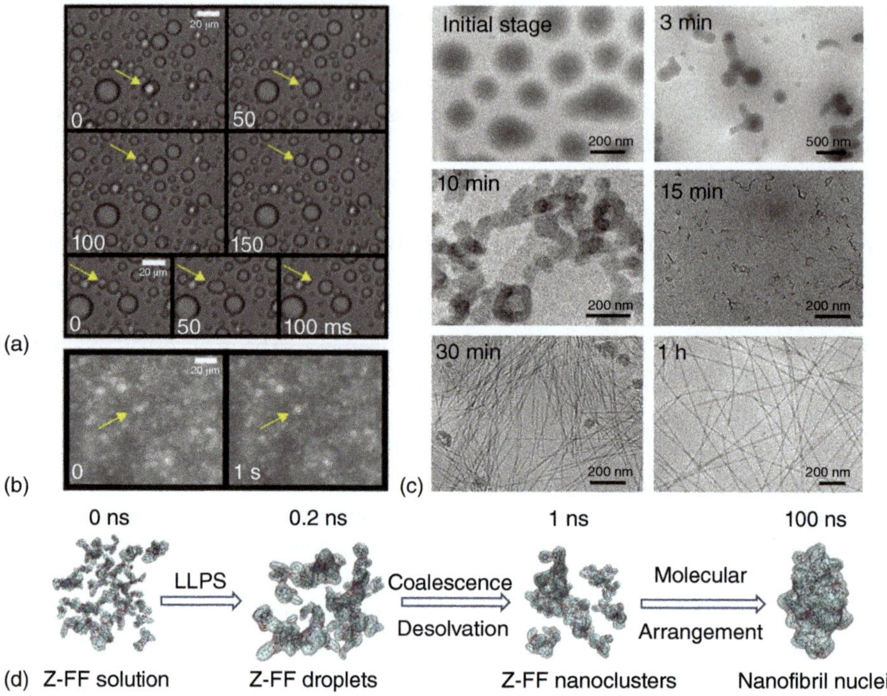

Figure 3.1 LLPS-mediated peptide self-assembly. (a–c) Formation of Z-FF self-assembling nanofibers through LLPS-mediated multistep pathway. (d) Schematic illustration of LLPS-mediated multistep nucleation mechanism. Source: Adapted with permission from Ref. [27]; © 2019, John Wiley & Sons.

peptides, even for more simple biomolecules such as individual amino acids, as well as the co-assembly system containing peptides [25, 37–41]. Further, combined with the all-atom molecular dynamics simulations results, an LLPS-mediated multistep nucleation mechanism of peptide self-assembly has been proposed. The LLPS-formed dense liquid droplets can act as metastable precursors of peptide supramolecular assemblies (Figure 3.1d). The presence of LLPS can effectively reduce the nucleation barrier through a metastable liquid phase, making such a nucleation mechanism of peptide self-assembly significantly distinguishable from the classical nucleation mechanism. Moreover, the LLPS-mediated nucleation mechanism provides the possibility to achieve precise self-assembly of peptides and other biomolecules toward functional materials in the future.

Regarding the peptide self-assembly mechanism mediated by LLPS, the occurrence of LLPS that leads to the generation of peptide-rich droplets from a monomeric peptide solution is an entropy-driven process. Solvent release from bound peptides to the bulk mainly contributes to this process. Subsequently, directional and enhanced intermolecular interactions facilitate nucleation within the droplets [27]. These findings suggest that a series of desolvation processes are necessary to make the solvation shell be completely depleted for peptide self-assembly. For the traditional peptide self-assembly mechanism described by CNT, the formation of ordered nucleus is a single-step process from the supersaturated peptide solution. This implies that a single-step desolvation process with large barrier should occur in the traditional context. As such, it is assumed that desolvation dynamics are of significance for peptide self-assembly, at least to some extent. To investigate the desolvation mechanism and how it mediates peptide self-assembly, the thorough desolvation process can be monitored by spectral tracking at different stages, and a succession of intermediates can be captured by cryo-TEM imaging with high spatiotemporal resolution.

In this context, we investigate the molecular events that trigger LLPS-mediated self-assembly using a designed amphiphilic peptide and find that multistep desolvation is crucial in this process (Figure 3.2) [42]. Specifically, solvated peptide molecules undergo a sequence of desolvation during the formation and transition of peptide-rich liquid droplets, resulting in the nucleation and growth of peptide self-assembly. We identified a multistep desolvation mechanism that drives LLPS-mediated peptide self-assembly. The first step involves the generation of peptides with partially exchanged solvents, followed by the selective desolvation of hydrophobic groups, which triggers LLPS and gives rise to the formation of peptide-rich liquid droplets. Further desolvation of these droplets induces the subsequent nucleation and growth. Our study also demonstrates that altering the desolvation degree at different self-assembly periods can be a useful tactic for regulating the pathways

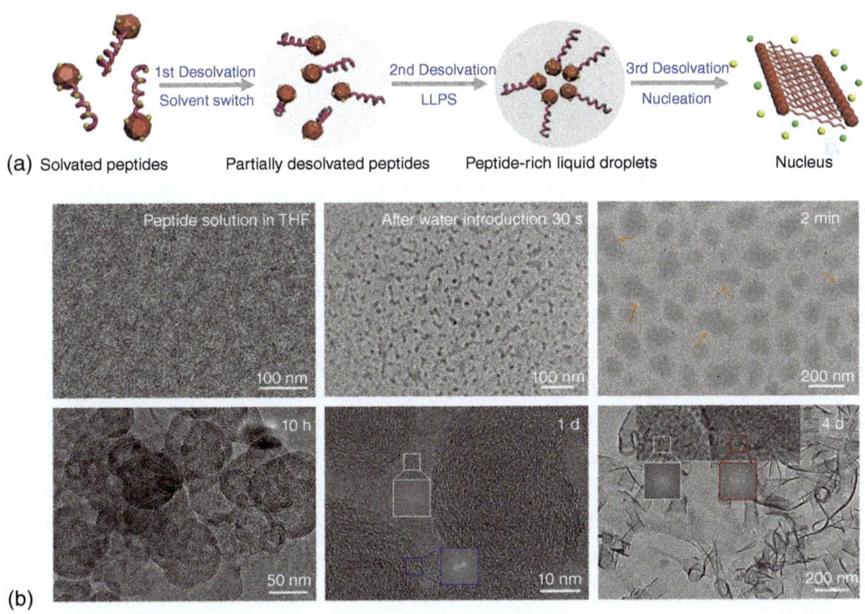

Figure 3.2 LLPS-mediated peptide self-assembly governed by multistep desolvation. (a) Schematic illustration of the multistep desolvation. (b) Time-lapse Cryo-TEM showing the multistep desolvation during TPPGFF self-assembly. Source: Adapted with permission from Ref. [42]; © 2023, CCS.

by which peptides self-assemble, which results in the formation of self-assembly polymorphs. This finding provides insights into the fundamental distinction between multistep self-assembly mediated by LLPS and conventional one-step self-assembly from the standpoint of desolvation, which opens the door for precise biomolecular self-assembly.

3.3 HOO-mediated Multistep Growth Mechanism of Peptide Self-assembly

It is widely recognized that ordered nuclei typically grow through monomer-by-monomer addition to form final self-assemblies. However, it is difficult to explain the formation of complex structures with long-range ordering, which are ubiquitous in biological systems and play a vital role in regulating structures and biological functions [43]. Therefore, designing and creating sophisticated functional systems with long-range ordering requires a thorough understanding of the basic principles behind the creation and organization of multiscale structures. In this section, we summarize examples of delicate structures formed through peptide hierarchical self-assembly and propose the HOO-mediated growth mechanism.

To gain an in-depth understanding of growth pathway of peptide self-assembly, a simple model possessing diverse self-assembly capability is urgently needed. Linear short peptides based on diphenylalanine (FF) have proven to be an ideal model due to their remarkable self-assembling ability. FF and its derivatives have been reported to form various self-assemblies ranging from nanotubes, nanofibers, nanosphere, nanorods to nanosheets [44]. Therefore, we first systemically analyze the supramolecular nanostructures assembled from FF and its derivatives to unravel the underlying mechanism. The hexagonal FF microtubes prepared through solvent thermal annealing are an excellent example for investigating the self-assembly mechanism [45]. Interestingly, disassembly and debundling of these microtubes lead to the formation of nanotubes by decreasing the FF concentration (Figure 3.3a). This indicates that the crystalline hexagonal microtubes are assembled from the hierarchically oriented nanotubes. Furthermore, FF microtubes formed through capillary-driven assembly also display similar hierarchical organization [48]. It should be noted that 3D peony-flower-like FF mesocrystals were also found to form through the oriented organization of the primary flakes with lamellar structures [49]. More interestingly, 3D organization of cyclo-FF crystalline platelets consisted of hierarchically oriented nanobelts (Figure 3.3b) [46]. Moreover, such crystalline platelets obtained through solvothermal treatment exhibited curved optical waveguiding, which advances the design and development of advanced devices involved in bio-optics and optoelectronics [50].

In addition to simple dipeptides, crystalline nanostructures that self-assembled from other short peptides and derivatives, such as YFD, DYF, PFF, A6K, KLVFF, Boc-FF, Fmoc-GG, Ac-IVE, Ac-LLE-NH2, and Ac-LVE, have also demonstrated their HOO of primary assemblies [51–58]. For example, macroscopic lamellar structures of acetylated KLVFF, namely Aβ(16-20), can rapidly form within seconds

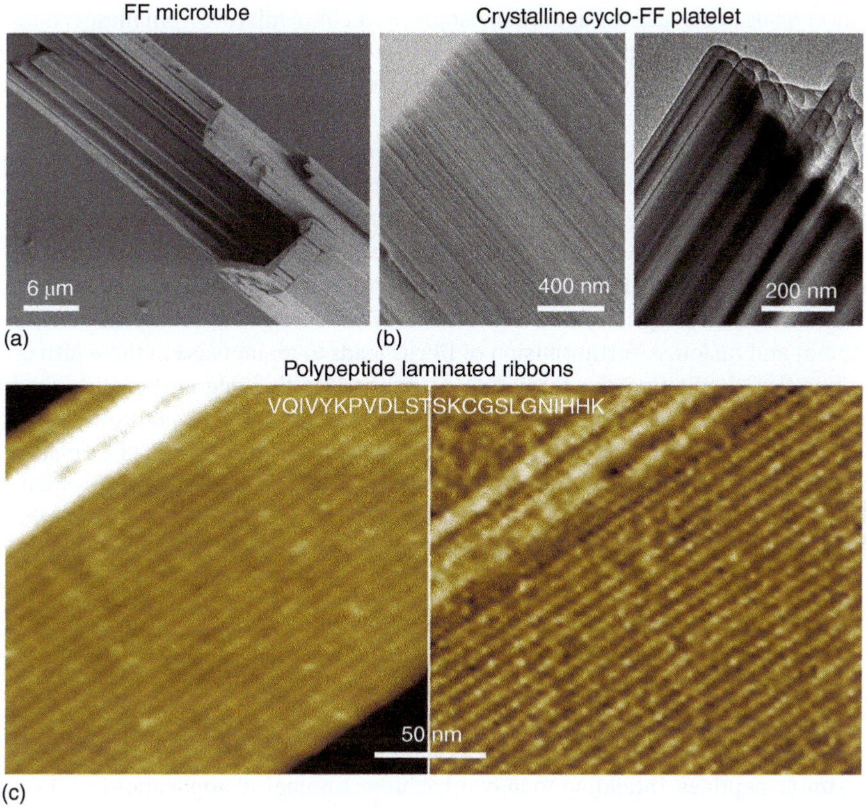

Figure 3.3 HOO-mediated growth mechanism of peptide self-assembly. (a) FF microtubes consisting of oriented nanotubes. (b) Crystalline cyclo-FF platelet consisting of oriented nanobelts. (c) Ac-KLVFF microcrystal assembled from stacked lamella. (c) Polypeptide ribbons formed through lateral association of protofilaments. (e) Hydrogel-crystal phase transition through fibril–fibril lamination. Source: (a) Reproduced with permission from Ref. [43]/John Wiley & Sons. (b) Adapted with permission from Ref. [46]; © 2011, John Wiley & Sons. (c) Adapted with permission from Ref. [47]; © 2016, John Wiley & Sons.

through lamellar stacking of microcrystals with a thickness in the nanoscale. Driven by interlamella interactions, face-to-face lamellae attach, resulting in the formation of a more compact conformation [54]. It should be noted that a lamellar structure formed via lateral-association-formed fiber bundles is widely existing in the self-assemblies of short peptides [55]. Intriguingly, the hierarchically oriented organization is a universal phenomenon in the self-assembly of polypeptides and even amino acids [59–63]. One typical example of a polypeptide is illustrated with a Tau-derived polypeptide that laminated ribbons with a lateral arrangement of protofilaments (Figure 3.3c) [47]. The prominent example of amino acids illustrated the formation of a compact single crystal of DL-alanine via the fusion of mesocrystal nanoparticles. Such a transformation process was tracked by the time-resolved small-angle neutron scattering [59]. A further study investigated the formation of well-defined hollow crystals of DL-alanine in a water/isopropyl alcohol mixture from mesocrystal comprising 3D nanocrystals [60]. Thus, it can be inferred that the hierarchically oriented organization may represent a kind of nonclassical growth mechanism of peptide self-assembly, in addition to the traditional growth mechanism.

Notably, a similar process was also found during the structural evolution from peptide hydrogel to crystals [64–68]. A prominent example based on a phenylalanine derivative clearly demonstrated the formation of crystals from the hydrogel fibers, as supported by the time-lapse TEM images. Once the hydrogel was formed, primary fibrils with a diameter of 11.9 ± 2.0 nm and a length of several micrometers were formed. Following this, thicker fibril bundles were generated by the lateral alignment and fusion. A further fusion of fibrils leads to an increase in the width of fibril bundles, facilitating the formation of nanocrystalline aggregates with rigid geometry and pronounced curvature. Finally, hollow crystalline microtubes were generated and existed as the dominant form. This process proves that the crystals formed through a fibril-based self-assembly, rather than the rule of Ostwald ripening. Thus, it is expected that hydrogel fibrils and final crystals should possess the same molecular arrangements. This further verifies the validity of the HOO for the hydrogel-crystal transition [67].

The above examples suggest that hierarchical multiscale structures can form through the HOO-mediated growth of primary peptide assemblies. Therefore, HOO-mediated growth has been proposed as a general mechanism for peptide self-assembly (Figure 3.4) [43]. Such a nonclassical growth mechanism provides critical clues for exploiting an effective strategy to achieve complex superstructures arising from simple peptides, intending to maximize biotechnological applications.

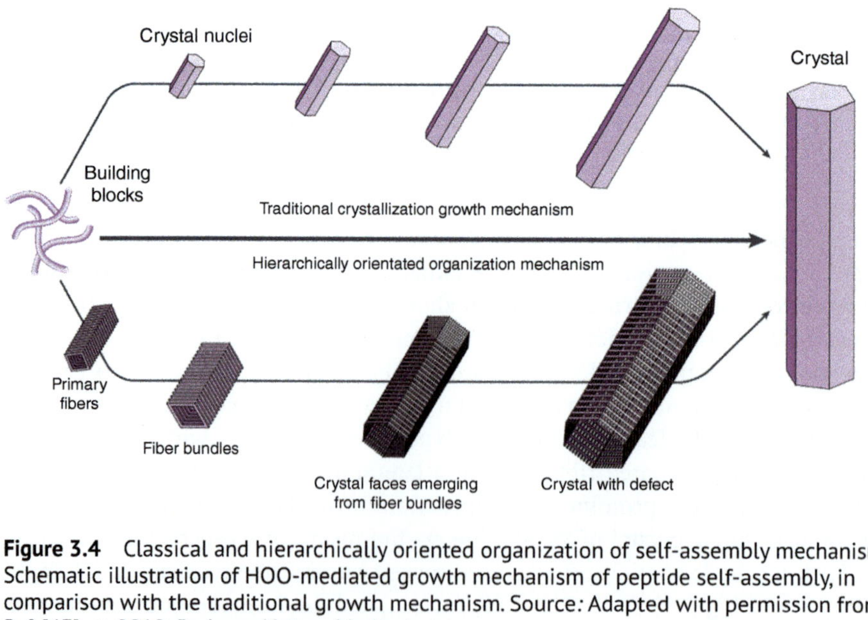

Figure 3.4 Classical and hierarchically oriented organization of self-assembly mechanism. Schematic illustration of HOO-mediated growth mechanism of peptide self-assembly, in comparison with the traditional growth mechanism. Source: Adapted with permission from Ref. [43]; © 2019, Springer Nature Limited.

3.4 Principles of Peptide Self-assembly

Based on the examples presented above, the LLPS-mediated multistep nucleation and the HOO-mediated growth mechanisms of peptide self-assembly have been proposed. To gain a deeper understanding of these mechanisms and their underlying principles, we discuss two major aspects: the thermodynamics and kinetics and the role of intermolecular interactions. In this section, we also compare nonclassical and classical peptide self-assembly pathways to unravel the theoretical origin behind their differences.

3.4.1 Thermodynamics and Kinetics

The examples given demonstrate the generality of the nonclassical peptide self-assembly mechanism, which involves LLPS-mediated multistep nucleation and hierarchically oriented organization of multistep growth. This suggests that peptide self-assembly undergoes a multiscale process that includes the generation and transition of different intermediates (Figure 3.5). The relative stability of these intermediate phases and the energy barriers between them determine the self-assembly pathway [69]. In general, the thermodynamic preference of intermediates at different self-assembly stages is determined by the free-energy landscape, while the diffusion and relaxation of intermediates and monomers affect the energy barrier for structural evolution. Therefore, we can understand the principles underlying multistep pathways of peptide self-assembly from the perspective of the free-energy landscape and self-assembly dynamics.

Various self-assembly intermediates are mapped on the free-energy landscape, where some local minima and energy barriers are displayed. The nature and distribution probability of intermediates depend on the scale of the energy barrier to the

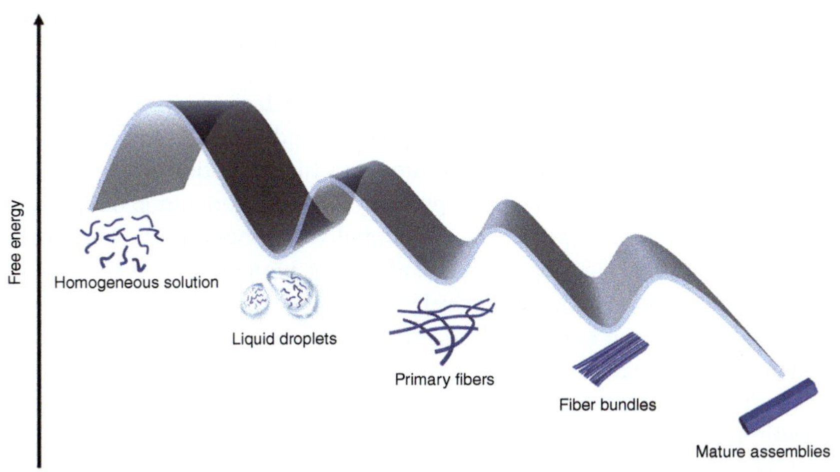

Figure 3.5 Free-energy landscape along peptide multistep self-assembly. Primary fibers and fiber bundles are the schematic intermediates for the formation of mature assemblies.

thermal energy, K_BT. The peptide self-assembly pathway varies with the shape and magnitude change of the energy barrier [70]. Regarding the free-energy landscape without local minima, the nucleation energy barrier is relatively large at a low supersaturation, which makes the formation of a critical nucleus a rare event. In this context, the monomer-by-monomer pathway dominates the nucleation and growth of peptide self-assembly, which is consistent with the CNT [71]. However, with increasing supersaturation, the energy barrier decreases, significantly improving the nucleation probability.

Regarding the free-energy landscape with local minima, it represents a self-assembly pathway involving free-energy minimum corresponding to the thermodynamically metastable intermediates that are subsequently replaced by more stable phases. During the nucleation period, two types of intermediates have been observed in this kind of pathway. One typical example is the metastable solid phase generated prior to nucleation, such as the amorphous solid phases found in the self-assembly process of peptides such as Boc-FF, Fmoc-DOPA, and amyloid peptides [23, 24, 26, 34]. Another prominent metastable precursor involves the emergence of peptide-rich liquid droplets through LLPS [27]. Actually, the solid amorphous phase can be considered as the next metastable intermediate of liquid droplets. Whether the metastable phases are solid or liquid in nature, their presence can significantly decrease the nucleation barrier compared to the direct nucleation from the monomer solution. Ordered nuclei would form at the interface or within the interior of intermediates [27, 34]. After nucleation, the growth of peptide self-assembly occurs in the HOO manner, which involves a particle-based self-assembly. This means that different intermediates reside in the local minima regions during the growth stage. It should be noted that kinetically trapped (i.e. long-lived metastable) materials could be achieved when the species' internal relaxation is slow enough. A typical example of this case is the emergence of self-assembling peptide hydrogel prior to the formation of a stable crystal [26, 64–67].

In addition to the free-energy landscape showing the thermodynamic preferences of self-assembly intermediates and mature assemblies, deciphering the entropic and enthalpic contributions during peptide self-assembly, especially in the nucleation and growth stages, could provide deeper insights into the crucial factors triggering self-assembly. The thermodynamic study on the LLPS-mediated multistep nucleation and growth demonstrates that the formation of metastable liquid droplet phases is an entropy-driven process, while the generation of primary self-assemblies from liquid droplets is an enthalpy-driven process [27]. The favorable entropy contribution is mainly associated with large amounts of solvent release from peptide surfaces, which is governed by hydrophobic interactions. The entropy-driven process enables the sudden aggregation of peptides into liquid droplets, increasing the local peptide concentration in the dense liquid phase. This results in a significant decrease in the nucleation barrier, making the LLPS-mediated multistep nucleation a thermodynamically favorable pathway compared to the classical nucleation pathway with a large nucleation barrier. The dominant enthalpy contribution is associated with the enhanced and directional intermolecular interactions such as hydrogen bonding and highly ordered π–π stacking interactions [27]. Such a trend was also observed in amorphous aggregates-mediated self-assembly [23], which further confirms that liquid droplets and amorphous aggregates are on the same pathway.

Moreover, it was found that the hierarchically oriented organization of primary assemblies is driven by entropy maximization [72]. This can be explained from the geometrical perspective [73]. The conventional theoretical explanation for this phenomenon is Onsager's argument, which highlights the crucial role of entropy in ordering. Based on Onsager's argument, the oriented ordering of primary assemblies will result in the entropy loss of them, which will induce the entropy gain of the system due to the free volume increase at sufficiently high concentrations [74]. This can be ascribed to the fact that parallel arrangements occupy less space than random patterns, hence leaving more space for solvent molecules to move in to maximize the entropy and minimize the free energy of the system.

3.4.2 The Role of Intermolecular Interactions

Self-assembly is a hierarchical, multiscale process that occurs under specific thermodynamics and kinetics conditions and spans multiple length and time scales. Different intermolecular interactions play distinct roles at different scales [75–77]. Furthermore, the weaker strength of noncovalent intermolecular interactions compared to covalent interactions makes the self-assembly process highly sensitive to its flexible microenvironment, leading to the generation of intermediates with polydispersity, polymorph, and transient features. To design and develop innovative nanostructures with desired functionality based on peptide self-assembly, it is crucial to understand the nature of the various intermolecular noncovalent interactions and exploit their interplay to the fullest extent. In this section, we first introduce the main types of intermolecular noncovalent interactions involved in peptide self-assembly (Figure 3.6) and then highlight the roles that different interactions play in the different stages of self-assembly.

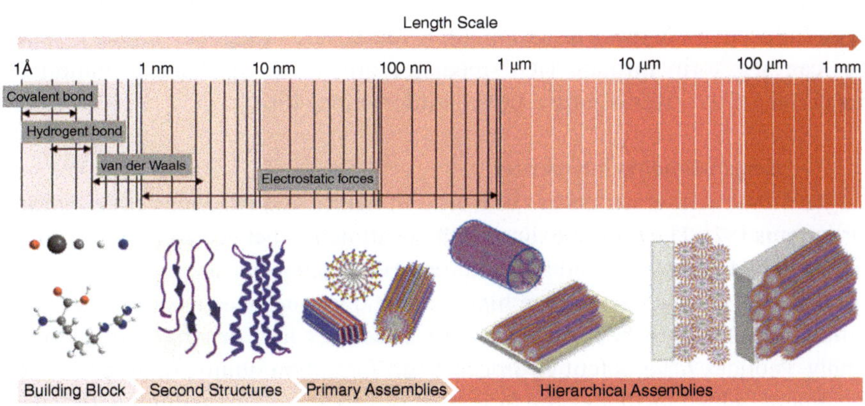

Figure 3.6 Intermolecular interactions responsible for peptide self-assembly. Source: Adapted with permission from Ref. [77]; © 2013, Wiley-VCH.

3.4.2.1 Electrostatic Interaction

Electrostatic interactions are a critical type of nonspecific and nondirectional interactions that occur between charged moieties of molecules. They can drive peptide self-assembly in aqueous environment through electrostatic attraction and electrostatic repulsion interactions [78, 79]. However, their strength and effectiveness are dependent on the solvent polarity, ionic strength, and pH. In highly polar water media, the electrostatic interactions can be drastically screened, resulting in a decrease in attractions between charged groups [80]. Conversely, a hydrophobic microenvironment can reduce the interaction of charged groups with water molecules, enhancing the electrostatic attractions between charged groups trapped inside. Therefore, to strengthen the electrostatic attractions in the water medium, it is crucial to introduce multivalent interactions among multiply-charged groups through pH and ionic strength modulation.

Recent studies have demonstrated that electrostatic interactions play an essential role in triggering LLPS. For example, it was found that the occurrence of tau LLPS is predominantly driven by the electrostatic attraction interactions between the positively charged middle/C-terminal and negatively charged N-terminal regions [81]. Such roles indeed are more evident in the formation of complex coacervates based on polyelectrolytes [82, 83]. Additionally, electrostatic interactions are also crucial for the growth of self-assembly. On the one hand, electrostatic attraction interactions are responsible for the high-order self-assembly of primary assemblies. A prominent example illustrates a pH-modulated self-assembly, where the lateral association of nanofibers can be achieved through the electrostatic attraction interactions between surfaces with alternating positive and negative charges [84]. On the other hand, electrostatic repulsion interactions are also found to be an important driving force for the formation of high-order assemblies. For instance, a seminal study on a series of amphiphilic peptides, such as AAAAAAEEE, VVAAEEGGREDKETV, VVAAEEGGTKREEVD, and AAEEGGREDKETV, illustrates that the electrostatic repulsion interactions between charged fibers facilitate the formation and stabilization of a fibrillar network. Notably, even at much lower concentrations, long-range ordered fibrillar bundles can still form with extra charges induced by X-ray irradiation [85]. Moreover, hexagonal liquid crystal phases can also form through HOO-mediated growth of rigid helical nanofilaments driven by the electrostatic repulsion interactions [86].

3.4.2.2 Hydrogen Bonding Interaction

Hydrogen bonding interactions are universal in nature and especially prevalent in living systems [87]. They involve dipole–dipole attraction between a partially positive charged hydrogen atom and a substantially electronegative atom such as nitrogen or oxygen. Hydrogen bonds are highly directional and selective, making them essential for stabilizing peptide secondary structures and driving peptide self-assembly. Peptides have potent hydrogen bond formation ability, thanks to their amide groups, which facilitates the formation of various self-assemblies with long-range order. Just like electrostatic interactions, the strength of hydrogen bonding can be reduced in a polar environment. Therefore, to strengthen hydrogen bonding interactions for peptide self-assembly in water medium, hydrophobic shielding is necessary [88].

It is recognized that hydrogen bonding interactions usually play a pronounced role in the formation of peptide self-assemblies with long-range order [89]. Traditionally, research has focused on the hydrogen bonding interactions between peptides, but recent studies have shifted to the role of hydrogen bonds between peptides and solvents in the self-assembly process [90–92]. Recently, our group has discovered the predominant role of solvent-bridged hydrogen bonding in directing fiber formation of FF molecules. The hydrogen bonds between peptides are sensitive to the introduction of solvents with hydrogen-bond-forming ability such as ethanol and water. Such directional hydrogen bonding interactions enable molecular arrangement of peptides with long-range order, preferentially facilitating the formation of nanofibers or nanobelts (Figure 3.7a) [93]. Further study reveal that trace water is involved in the peptide self-assembly process, acting as a crucial factor in triggering the formation of stable peptide oligomers at the initial stage prior to nucleation (Figure 3.7b) [94].

3.4.2.3 π–π Stacking

Van der Waals interactions refer to interactions taking place between uncharged moieties, mainly including dipolar and London dispersion interactions. In general,

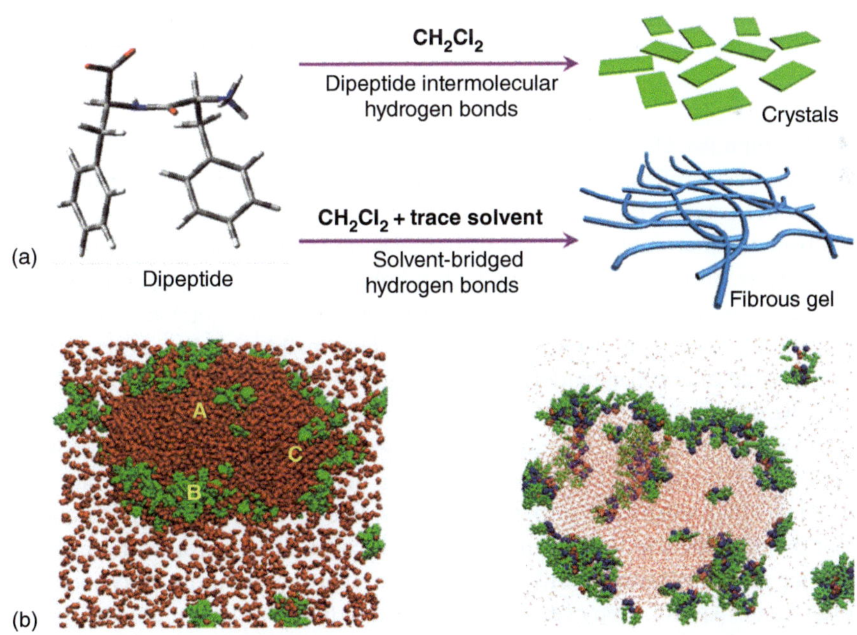

Figure 3.7 Solvent-bridged hydrogen bonding directing fiber formation of peptide. (a) Trace solvent with strong hydrogen-bond-formation ability facilitating fibers formation. (b) Water molecules involved in the peptide clusters formed at the initial stage prior to nucleation. Source: (a) Adapted with permission from Ref. [93]; © 2016, ACS. (b) Adapted with permission from Ref. [94]; © 2017, Wiley-VCH.

Van der Waals interactions contribute most to the stabilization of nonpolar components [69]. π–π Stacking interactions occur between large π-conjugated surfaces as a particular instance of Van der Waals interactions, although its exact nature is still being debated [95, 96]. It was discovered that the attraction between the π-surfaces with electron-rich and -poor motifs plays an essential role [97]. For peptide self-assembly systems, π–π stacking interactions widely exist in structures containing aromatic amino acids of tryptophan, tyrosine, and phenylalanine. π–π Stacking interactions are particularly persistent in water owing to the limited solubility of aromatic compounds. As such, π–π stacking interactions significantly contribute to the overall stability of peptide supramolecular polymers based on noncovalent interactions [98].

Numerous pieces of evidence have suggested that long-range ordered π–π stacking interactions can also facilitate the directional growth of peptide self-assemblies, especially short peptides modified with aromatic groups such as pyrene and 9-fluorenylmethyloxycarbonyl [99–101]. It should be noted that π–π stacking interactions could have an evident influence on the pathway of peptide self-assembly. A typical example in this case is illustrated by the self-assembly of FF conjugated with pyrene (Py-FF) and naphthalene (Nap-FF), respectively [102]. Intriguingly, different degrees of π–π stacking have a pronounced influence on the molecular arrangements and final morphology. One-dimensional nanofibers are observed for Py-FF, while bundled nanofibers are detected for Nap-FF. This can be attributed to the fact that an antiparallel arrangement has difficulty in shielding the hydrophobic naphthalene core in a polar solvent environment, which promotes the lateral growth of nanofibers.

3.4.2.4 Hydrophobic Effect

Hydrophobic effect refers to the rapid and disordered aggregation of hydrophobic motifs into the interior of aggregates. Hydrophobic collapse or π–π stacking may occur between the aromatic residues of peptides. It has been discovered that the arrangement of aromatic residues in hydrophobic aggregates is generally disordered, while it is ordered in π–π stacking. The rearrangement of hydrogen bond networks around the hydrophobic molecules is responsible for the occurrence of the hydrophobic effect. It is widely acknowledged that the intrinsic origin of the hydrophobic effect is entropy [103]. This phenomenon is attributed to entropy, as the overall number of hydrogen bonds between water molecules changes little with the presence of hydrophobic molecules. Given this condition, water molecules would adjust their organization patterns to that with reduced mobility to afford space for the hydrophobic molecules. As such, it is considered that static water clusters with ice-like structures should be around nonpolar groups at the early stage. In fact, the modern spectroscopy result has suggested that the water clusters around hydrophobic molecules are moderately slower dynamics [104].

Since the mobility reduction is entropically unfavorable, the entropy gain of the system is achieved through the release of hydrated water molecules into the bulk via the rapid aggregation of hydrophobic molecules. Previous thermodynamic investigations have proven the entropy nature of the hydrophobic effect for small hydrophobes. Therefore, the hydrophobic effect usually governs the initial stage of peptide self-assembly in aqueous solutions, which is consistent with the entropy-driven liquid droplets formation process dominated by hydrophobic interaction [27]. It is evident that enhanced hydrophobic interaction improves the stability of liquid droplets and inhibits their transition to nanofibers. A similar phenomenon was further confirmed by altering the amino-acid type, where amino acids with more hydrophobicity possess more stable liquid droplets. The important role of hydrophobic interaction has been supported by the coarse-grained molecular dynamics simulation results [105].

The hydrophobic effect can also direct the whole self-assembly process. A prominent example of this is the formation of nanotubes from a simple amino-acid amphiphile, N-R-lauryl-lysyl-aminolauryl-lysyl-amide (C12-β12), which illustrates the crucial role of the hydrophobic effect in self-assembly [106]. It was found that the initial formation of monolayer fibers is predominantly driven by the hydrophobic effect. Hydrophobic motifs of the monolayer exposed to the surrounding solvent facilitate the lateral association along the width direction, resulting in the generation of twisted ribbons, helical ribbons, and nanotubes. Despite the contribution of the hydrophobic effect, other types of intermolecular interactions, including electrostatic and hydrogen bonding interactions, are also responsible for self-assembly.

3.5 Design Rules for Materials Based on Peptide Self-assembly

Based on the principles described above, we propose several design rules for peptide self-assembling materials including the chemical structure design of peptides, modulation of intermolecular interactions, and regulation of self-assembly thermodynamics and kinetics.

3.5.1 Designing Amphiphilic Peptide Structures with Desired Functions

Peptide self-assembling materials are attracting attention due to their good biocompatibility, biodegradability, and tunable structures and mechanical properties. The first factor to consider when designing peptides is their good self-assembly ability. As such, amphiphilic characteristics are essential for peptide design. Solvophobic moieties of peptides tend to aggregate, while solvophilic sections tend to be solvent-exposed, resulting in the proceeding of peptide self-assembly. Chemical modification with alkyl chains has been developed as an effective strategy to obtain amphiphilic peptides. Peptides conjugated with alkyl chains have emerged as an elegant building block to fabricate versatile supramolecular self-assemblies such as nanofibers, nanobelts, and nanoribbons in aqueous solutions [107, 108]. For example, an amphiphilic peptide with an alternating hydrophobic–hydrophilic sequence can self-assemble into flat and wide nanobelts [109]. Further structural transformations can be achieved through modulation of external conditions such as pH and concentration.

Another factor to consider is designing and selecting peptide building blocks according to the desired functions. For instance, some peptides with biological and pharmacological activities are also good options for serving as building blocks [110, 111]. Cyclic peptides (CPs), which usually exist in bioactive natural products, have been reported to possess exceptional bioactivities, such as immune regulation and antimicrobial and antitumor activities [112, 113]. Moreover, the unexpected structural rigidity and superior hydrogen-bond-forming capability of the CPs enable them to have more stable metabolism and better bioavailability than their linear counterparts [114–116]. Another prominent type of peptides are those conjugated with photosensitizers, which can self-assemble into peptide-based nanodrugs for photothermal or photodynamic treatment. Our group has pioneered in the design and development of a series of functional peptides covalently and noncovalently modified with various photosensitizers, such as porphyrin and phthalocyanine [117–120].

3.5.2 Modulation of Intermolecular Interactions

The formation of various peptide self-assemblies, such as nanofibers, nanosheets, and nanotubes, is associated with the compromise and balance between multiple intermolecular interactions. The solute–solute and solvent–solute interactions play an essential role in shaping the energy landscape and determining the self-assembly pathway. The modulation of intermolecular interactions is crucial for the adjustment of peptide self-assembly behaviors. It is highly desired to achieve a balance between solvophobic and solvophilic interactions of peptides. An imbalance of such interactions would either result in the generation of disordered aggregates or keep the peptides in a solution state.

To achieve rational design, multiple types of intermolecular interactions should be introduced to the designed peptides, facilitating hierarchical self-assembly. For instance, strong hydrophobic interactions are a predominant driving force in the generation of spherical nanoparticles. Electrostatic interactions are also responsible for the colloidal stability. Prime examples of this case have been demonstrated by the design and development of well-dispersed nanoparticles self-assembled from diphenylalanine conjugated with hydrophobic porphyrins and chlorophyll derivatives, which are stabilized by the balance between hydrophobic and electrostatic interactions [118, 121]. In contrast to colloidal nanoparticles, directional interaction plays a crucial role in nanofiber formation. For example, the introduction of an amino-acid sequence with strong hydrogen-bond-forming ability, such as aspartic acid (D) and tyrosine (Y), facilitates the formation of long-range ordered nanofibers [121]. Moreover, hierarchical self-assembly is usually driven by multiple types of intermolecular interactions, such as hydrophobic effect, electrostatic, hydrogen bonding, and π–π stacking interactions, which can be modulated by peptide sequence or solvent types [122–124]. For example, multichambered microspheres are formed through co-assembly of hydrophobic dipeptide FF and tetraphenylporphinesulfonate (TPPS), while fiber bundles are generated through co-assembly of hydrophilic dipeptide KK and TPPS (Figure 3.8) [125, 126]. The former is the result of competition between long-range electrostatic repulsion and short-range π–π

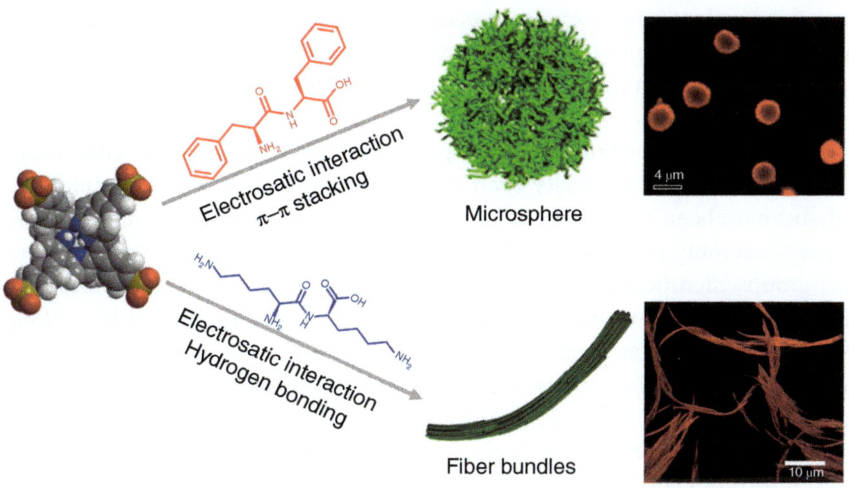

Figure 3.8 Controlled peptide self-assembly through intermolecular interactions modulation. Source: Adapted with permission from Refs. [125, 126]; © 2014, 2015, John Wiley & Sons.

stacking interactions. In contrast, the replacement of FF with KK results in short-range hydrogen bonding interactions dominating the formation of fiber bundles.

It should be noted that anisotropic growth of peptide self-assembly, associated with the directional arrangement of peptides, usually derives from the unbalanced competitive bonding between solvent molecules and the attractive sites of nanostructures [127–129]. Therefore, another effective strategy to modulate the intermolecular interactions governing peptide self-assembly can be achieved through the introduction of external fields, such as ultrasonic, electric, or magnetic fields, and even shear force [130–136]. The application of external forces could promote the oriented organization of peptides and primary assemblies, thus leading to the formation of hierarchical self-assemblies with anisotropy on a macroscopic scale. For example, anisotropic hydrogels have been obtained under external electric field and shear force [130, 134, 136]. Cavitation effects produced by ultrasonication irradiation can result in local high temperature and high pressure, which could cleave the weak inter- or intramolecular interactions while strengthening the hydrophobic effect. Such effects are beneficial for molecular rearrangement, hence leading to the formation of anisotropic peptide self-assemblies [132].

3.5.3 Regulating Self-assembly Thermodynamics and Kinetics

It is desirable to obtain various self-assemblies based on the same peptide building block. In real-life systems, one distinguishing characteristic of them is that they usually reside in kinetically trapped states, rather than the thermodynamic equilibrium, which extremely enriches their complexity and functionality [3, 137]. Therefore, self-assembly polymorphisms can be achieved through kinetically controlled, namely pathway-dependent, self-assembly. In general, self-assembly driven by weak noncovalent interactions results in a thermodynamically stable product. In contrast, strong noncovalent interactions could result in nonequilibrium assemblies with structural variation by kinetic trapping of the same molecular building block. Therefore, self-assembly pathway selection can be served as a powerful method to acquire kinetically trapped products. For example, peptides conjugated with groups possessing strong hydrophobicity such as phthalocyanine, perylene diimide, and alky chains have been reported to demonstrate pathway-dependent kinetically controlled self-assembly [138–140]. This is because the introduction of strong hydrophobic groups significantly increases the intermolecular interactions between peptides, which endows the formation of several kinetic products. Taking diphenylalanine-conjugated zinc (II) phthalocyanine (PF) as an example, rapid mixing of water and PF-concentrated solution in DMSO yielded nanoparticles. In contrast, increasing the initial DMSO/water volume ratio and extending the aging time to several days gave rise to helical nanofibrils. Such nanofibrils displayed near-infrared absorption with a superlarge redshift of 105 nm compared to that of nanoparticles (Figure 3.9) [138]. Notably, the formation of PF nanoparticle with random molecular orientation was an entropy-driven fast process, while the formation of

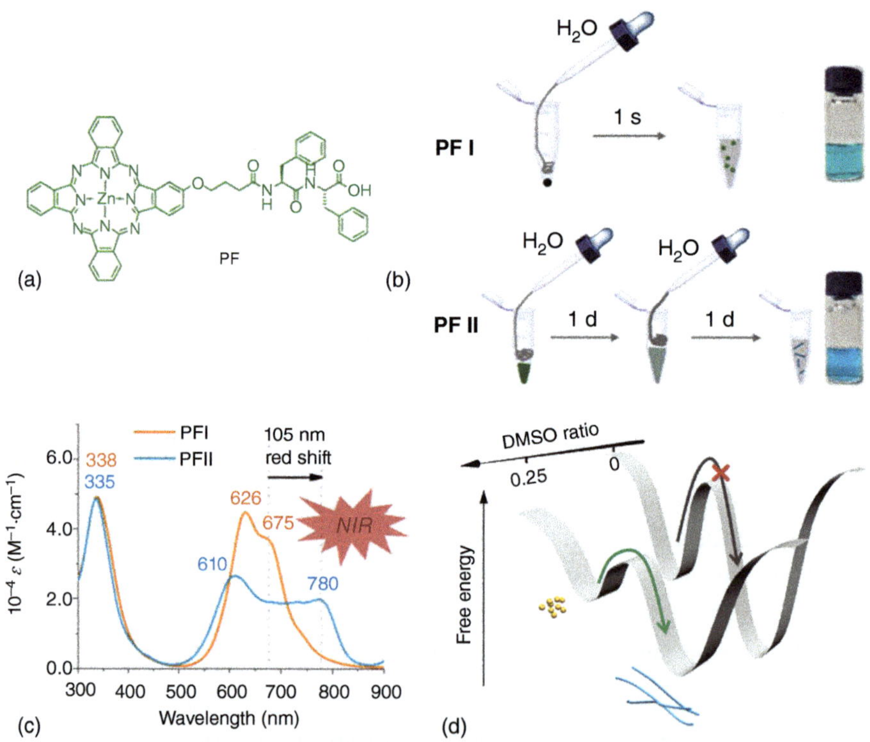

Figure 3.9 Kinetically controlled peptide self-assembly. (a) Chemical structure of peptide conjugate. (b) Schematic illustration for the self-assembly pathways. (c) UV–vis spectra of different self-assemblies. (d) Schematic illustration for the free-energy landscape of different self-assembly pathways. Source: Adapted with permission from Ref. [138]; © 2019, CCS.

nanofibrils with long-range ordered molecular arrangement was an enthalpy-driven process. Additionally, it is widely acknowledged that peptide hydrogel is usually in a kinetically controlled state. This implies that any approach, such as increasing the energy barrier of further phase transition, improving the kinetic stability of hydrogel is welcomed if stable hydrogel is desired.

3.6 Conclusions and Outlook

In conclusion, this chapter summarized the LLPS-mediated nucleation and HOO-mediated growth mechanism of peptide self-assembly, a newly developed nonclassical multistep self-assembly mechanism. We analyzed the multistep nucleation and growth process from the perspective of thermodynamics and kinetics and the role of intermolecular interactions, aiming to unravel the general principles governing peptide multistep self-assembly. Finally, we proposed design rules for supramolecular materials based on peptide self-assembly.

The multistep mechanisms of peptide self-assembly integrated in this chapter provide an effective strategy for the precise self-assembly and envisioning new avenues for the exploration of novel peptide-based biomaterials with sophisticated phase structures through facile solution-processable bottom-up fabrication. For example, primary self-assembly polymorphisms can be obtained by modulating the LLPS-formed liquid droplets. Lamination can be achieved through HOO-mediated growth, which has been an effective strategy to fabricate superstructures combining flexibility, strength, and toughness, such as Japanese Katana and Damascus steel. Peptide crystals with lamellar structures may have potential applications in advanced devices in optoelectronics and bio-optics.

Although experimental evidence has illustrated that peptide self-assembly follows LLPS-mediated nucleation and HOO-mediated growth, the nonclassical mechanism is still not fully understood. For instance, direct experimental determination of the three-dimensional structures of different intermediates, especially the amorphous intermediates, plays an essential role in understanding the structural evolution process, yet remains extremely challenging. The role of water molecules, especially water structures and dynamic interaction with peptides, during peptide self-assembly still remains elusive [141]. Mapping the free-energy landscape of peptide self-assembly is crucial to elaborate on the thermodynamic and kinetic basis of peptide multistep self-assembly, yet remains inherently challenging. Moreover, probing the dynamic changes of intermolecular interactions driving the LLPS-mediated self-assembly process remains challenging, mainly due to the complex coupling of multiple microscopic steps.

To address these challenges, direct and available experimental evidence is urgently needed. These call for integrated methods of *in situ* structure determination and intermolecular interaction probe, mainly including imaging and spectrum tracking. Therefore, *in situ* structure determination of different intermediates and final assemblies through high-resolution imaging is crucial toward the construction of reasonable models describing peptide self-assembly in a solution environment. Liquid cell *in situ* TEM, which has been widely used to investigate the formation process of nanocrystals, inorganic minerals, and soft matters [142–145], is expected to elucidate the molecular underpinning behind peptide self-assembly. Structural evolution of a series of intermediates can also be detected by cryo-TEM, especially cryo-electron tomography, which has the potential to unravel the internal structural characteristics of intermediates and final assemblies in their native environment [146–148]. Additionally, spectrum techniques with high spatiotemporal resolution that can probe the dynamic interwork of peptides and solvent molecules and multi-scale theoretical modeling are of equal importance. Despite many critical challenges, we firmly believe that the self-assembly principle and material design rules described in this chapter will pave the way for the future development of smart and multifunctional peptide materials.

Acknowledgments

We acknowledge the funding support from the National Science Fund for Distinguished Young Scholars of China (No. 22025207), National Natural Science Foundation of China (Nos. 22172172 and 22232006), Youth Innovation Promotion Association of CAS (Grant No. 2022049), China Scholarship Council (CSC, 202104910187), IPE Project for Frontier Basic Research (Grant No. QYJC-2022-011), and Natural Science Foundation of Hebei Province (No. B2020103036).

References

1 Zhang, S. (2003). Fabrication of novel biomaterials through molecular self-assembly. *Nat. Biotechnol.* 21: 1171–1178.
2 Aida, T., Meijer, E.W., and Stupp, S.I. (2012). Functional supramolecular polymers. *Science* 335: 813–817.
3 Mattia, E. and Otto, S. (2015). Supramolecular systems chemistry. *Nat. Nano* 10: 111–119.
4 Tao, K., Makam, P., Aizen, R., and Gazit, E. (2017). Self-assembling peptide semiconductors. *Science* 358: eaam9756.
5 Levin, A., Hakala, T.A., Schnaider, L. et al. (2020). Biomimetic peptide self-assembly for functional materials. *Nat. Rev. Chem.* 4: 615–634.
6 Knowles, T.P.J., Oppenheim, T.W., Buell, A.K. et al. (2010). Nanostructured films from hierarchical self-assembly of amyloidogenic proteins. *Nat. Nanotechnol.* 5: 204.
7 Karsenti, E. (2008). Self-organization in cell biology: a brief history. *Nat. Rev. Mol. Cell Biol.* 9: 255.
8 Yao, H.-B., Fang, H.-Y., Wang, X.-H., and Yu, S.-H. (2011). Hierarchical assembly of micro-/nano-building blocks: bio-inspired rigid structural functional materials. *Chem. Soc. Rev.* 40: 3764–3785.

9 Gardel, M.L., Shin, J.H., MacKintosh, F.C. et al. (2004). Elastic behavior of cross-linked and bundled actin networks. *Science* 304: 1301–1305.

10 Yin, P., Choi, H.M.T., Calvert, C.R., and Pierce, N.A. (2008). Programming biomolecular self-assembly pathways. *Nature* 451: 318–322.

11 Abbas, M., Lipiński, W.P., Wang, J., and Spruijt, E. (2021). Peptide-based coacervates as biomimetic protocells. *Chem. Soc. Rev.* 50: 3690–3705.

12 Sun, B., Tao, K., Jia, Y. et al. (2019). Photoactive properties of supramolecular assembled short peptides. *Chem. Soc. Rev.* 48: 4387–4400.

13 Wang, J., Liu, K., Xing, R., and Yan, X. (2016). Peptide self-assembly: thermodynamics and kinetics. *Chem. Soc. Rev.* 45: 5589–5604.

14 Sheehan, F., Sementa, D., Jain, A. et al. (2021). Peptide-based supramolecular systems chemistry. *Chem. Rev.* 121: 13869–13914.

15 Bernstein, S.L., Dupuis, N.F., Lazo, N.D. et al. (2009). Amyloid-β protein oligomerization and the importance of tetramers and dodecamers in the aetiology of Alzheimer's disease. *Nat. Chem.* 1: 326–331.

16 Adamcik, J., Jung, J.-M., Flakowski, J. et al. (2010). Understanding amyloid aggregation by statistical analysis of atomic force microscopy images. *Nat. Nanotechnol.* 5: 423–428.

17 Liang, C., Ni, R., Smith, J.E. et al. (2014). Kinetic intermediates in amyloid assembly. *J. Am. Chem. Soc.* 136: 15146–15149.

18 Jonkheijm, P., van der Schoot, P., Schenning, A.P.H.J., and Meijer, E.W. (2006). Probing the solvent-assisted nucleation pathway in chemical self-assembly. *Science* 313: 80–83.

19 Klug, A. (1999). The tobacco mosaic virus particle: structure and assembly. *Philos. Trans. R. Soc. Lond. Biol. Sci.* 354: 531–535.

20 Krishnan, R. and Lindquist, S.L. (2005). Structural insights into a yeast prion illuminate nucleation and strain diversity. *Nature* 435: 765–772.

21 Touve, M.A., Carlini, A.S., and Gianneschi, N.C. (2019). Self-assembling peptides imaged by correlated liquid cell transmission electron microscopy and MALDI-imaging mass spectrometry. *Nat. Commun.* 10: 4837.

22 Gnanasekaran, K., Korpanty, J., Berger, O. et al. (2021). Dipeptide nanostructure assembly and dynamics via in situ liquid-phase electron microscopy. *ACS Nano* 15: 16542–16551.

23 Levin, A., Mason, T.O., Adler-Abramovich, L. et al. (2014). Ostwald's rule of stages governs structural transitions and morphology of dipeptide supramolecular polymers. *Nat. Commun.* 5: 5219.

24 Liang, Y., Lynn, D.G., and Berland, K.M. (2010). Direct observation of nucleation and growth in amyloid self-assembly. *J. Am. Chem. Soc.* 132: 6306–6308.

25 Mañas-Torres, M.C., Gila-Vilchez, C., González-Vera, J.A. et al. (2021). In situ real-time monitoring of the mechanism of self-assembly of short peptide supramolecular polymers. *Mater. Chem. Front.* 5: 5452–5462.

26 Fichman, G., Guterman, T., Damron, J. et al. (2016). Spontaneous structural transition and crystal formation in minimal supramolecular polymer model. *Sci. Adv.* 2: e1500827.

27 Yuan, C., Levin, A., Chen, W. et al. (2019). Nucleation and growth of amino acid and peptide supramolecular polymers through liquid–liquid phase separation. *Angew. Chem. Int. Ed.* 58: 18116–18123.

28 Linse, S. (2021). Toward the equilibrium and kinetics of amyloid peptide self-assembly. *Curr. Opin. Struct. Biol.* 70: 87–98.

29 Erdemir, D., Lee, A.Y., and Myerson, A.S. (2009). Nucleation of crystals from solution: classical and two-step models. *Acc. Chem. Res.* 42: 621–629.

30 Knowles, T.P.J., Waudby, C.A., Devlin, G.L. et al. (2009). An analytical solution to the kinetics of breakable filament assembly. *Science* 326: 1533–1537.

31 Cohen, S.I.A., Vendruscolo, M., Welland, M.E. et al. (2011). Nucleated polymerization with secondary pathways. I. Time evolution of the principal moments. *J. Chem. Phys.* 135: 065105.

32 Hsieh, M.-C., Lynn, D.G., and Grover, M.A. (2017). Kinetic model for two-step nucleation of peptide assembly. *J. Phys. Chem. B* 121: 7401–7411.

33 Pellach, M., Atsmon-Raz, Y., Simonovsky, E. et al. (2015). Spontaneous structural transition in phospholipid-inspired aromatic Phosphopeptide nanostructures. *ACS Nano* 9: 4085–4095.

34 Chen, C., Tan, J., Hsieh, M.-C. et al. (2017). Design of multi-phase dynamic chemical networks. *Nat. Chem.* 9: 799–804.

35 Cohen-Gerassi, D., Arnon, Z.A., Guterman, T. et al. (2020). Phase transition and crystallization kinetics of a supramolecular system in a microfluidic platform. *Chem. Mater.* 32: 8342–8349.

36 Yuan, C., Li, S., Zou, Q. et al. (2017). Multiscale simulations for understanding the evolution and mechanism of hierarchical peptide self-assembly. *Phys. Chem. Chem. Phys.* 19: 23614–23631.

37 Narang, N. and Sato, T. (2022). Unique phase behaviour and self-assembly of a lysine derivative, Fmoc-homoarginine, in water–DMSO mixtures. *Soft Matter* 18: 7968–7974.

38 Rengifo, R.F., Sementilli, A., Kim, Y. et al. (2020). Liquid-like phases preorder peptides for supramolecular assembly. *ChemSystemsChem* 2: e2000007.

39 Narang, N. and Sato, T. (2021). Liquid-liquid phase separation and self-assembly of a lysine derivative Fmoc-L-lysine in water-DMSO mixtures. *Polym. J.* 53: 1413–1424.

40 Wu, A., Guo, Y., Li, X. et al. (2021). Co-assembled supramolecular gel of dipeptide and pyridine derivatives with controlled chirality. *Angew. Chem. Int. Ed.* 60: 2099–2103.

41 Chakraborty, P., Tang, Y., Yamamoto, T. et al. (2020). Unusual two-step assembly of a minimalistic dipeptide-based functional hypergelator. *Adv. Mater.* 32: 1906043.

42 Yuan, C., Xing, R., Cui, J. et al. (2023). Multistep desolvation as a fundamental principle governing peptide self-assembly through liquid–liquid phase separation. *CCS Chem.* https://doi.org/10.31635/ccschem.023.202302990.

43 Yuan, C.Q., Ji, W., Xing, R.R. et al. (2019). Hierarchically oriented organization in supramolecular peptide crystals. *Nat. Rev. Chem.* 3: 567–588.

44 Yan, X., Zhu, P., and Li, J. (2010). Self-assembly and application of diphenylalanine-based nanostructures. *Chem. Soc. Rev.* 39: 1877–1890.

45 Yan, X., Li, J., and Möhwald, H. (2011). Self-assembly of hexagonal peptide microtubes and their optical waveguiding. *Adv. Mater.* 23: 2796–2801.

46 Yan, X., Su, Y., Li, J. et al. (2011). Uniaxially oriented peptide crystals for active optical waveguiding. *Angew. Chem. Int. Ed.* 50: 11186–11191.

47 Adamcik, J., Sánchez-Ferrer, A., Ait-Bouziad, N. et al. (2016). Microtubule-binding R3 fragment from tau self-assembles into Giant multistranded amyloid ribbons. *Angew. Chem. Int. Ed.* 55: 618–622.

48 Wang, Y., Huang, R., Qi, W. et al. (2015). Capillary force-driven, hierarchical co-assembly of dandelion-like peptide microstructures. *Small* 11: 2893–2902.

49 Su, Y., Yan, X., Wang, A. et al. (2010). A peony-flower-like hierarchical mesocrystal formed by diphenylalanine. *J. Mater. Chem.* 20: 6734–6740.

50 Li, Y., Yan, L., Liu, K. et al. (2016). Solvothermally mediated self-assembly of ultralong peptide nanobelts capable of optical waveguiding. *Small* 12: 2575–2579.

51 Frederix, P.W.J.M., Scott, G.G., Abul-Haija, Y.M. et al. (2014). Exploring the sequence space for (tri-)peptide self-assembly to design and discover new hydrogels. *Nat. Chem.* 7: 30–37.

52 Lampel, A., McPhee, S.A., Park, H.-A. et al. (2017). Polymeric peptide pigments with sequence-encoded properties. *Science* 356: 1064–1068.

53 Chan, K.H., Xue, B., Robinson, R.C., and Hauser, C.A.E. (2017). Systematic moiety variations of ultrashort peptides produce profound effects on self-assembly, nanostructure formation, hydrogelation, and phase transition. *Sci. Rep.* 7: 12897.

54 Bortolini, C., Klausen, L.H., Hoffmann, S.V. et al. (2018). Rapid growth of acetylated Aβ(16–20) into macroscopic crystals. *ACS Nano* 12: 5408–5416.

55 Tang, C., Ulijn, R.V., and Saiani, A. (2011). Effect of glycine substitution on Fmoc–diphenylalanine self-assembly and gelation properties. *Langmuir* 27: 14438–14449.

56 Ejgenberg, M. and Mastai, Y. (2018). Hierarchical superstructures of l-glutathione. *Cryst. Growth Des.* 18: 5063–5068.

57 Cenker, Ç.Ç., Bomans, P.H.H., Friedrich, H. et al. (2012). Peptide nanotube formation: a crystal growth process. *Soft Matter* 8: 7463–7470.

58 Lu, K., Jacob, J., Thiyagarajan, P. et al. (2003). Exploiting amyloid fibril lamination for nanotube self-assembly. *J. Am. Chem. Soc.* 125: 6391–6393.

59 Schwahn, D., Ma, Y., and Cölfen, H. (2007). Mesocrystal to single crystal transformation of d,l-alanine evidenced by small angle neutron scattering. *J. Phys. Chem. C* 111: 3224–3227.

60 Medina, D.D. and Mastai, Y. (2008). Synthesis of dl-alanine mesocrystals with a hollow morphology. *Cryst. Growth Des.* 8: 3646–3651.

61 Nemtsov, I., Mastai, Y., and Ejgenberg, M. (2018). Formation of hierarchical structures of l-glutamic acid with an l-arginine additive. *Cryst. Growth Des.* 18: 4054–4059.

62 Ma, Y., Cölfen, H., and Antonietti, M. (2006). Morphosynthesis of alanine mesocrystals by pH control. *J. Phys. Chem. B* 110: 10822–10828.

63 Ejgenberg, M. and Mastai, Y. (2012). Biomimetic crystallization of l-cystine hierarchical structures. *Cryst. Growth Des.* 12: 4995–5001.

64 Rajbhandary, A., Raymond, D.M., and Nilsson, B.L. (2017). Self-assembly, hydrogelation, and nanotube formation by cation-modified phenylalanine derivatives. *Langmuir* 33: 5803–5813.

65 Liyanage, W. and Nilsson, B.L. (2016). Substituent effects on the self-assembly/coassembly and hydrogelation of phenylalanine derivatives. *Langmuir* 32: 787–799.

66 Rajbhandary, A., Brennessel, W.W., and Nilsson, B.L. (2018). Comparison of the self-assembly behavior of Fmoc-phenylalanine and corresponding Peptoid derivatives. *Cryst. Growth Des.* 18: 623–632.

67 Liyanage, W., Brennessel, W.W., and Nilsson, B.L. (2015). Spontaneous transition of self-assembled hydrogel fibrils into crystalline microtubes enables a rational strategy to stabilize the hydrogel state. *Langmuir* 31: 9933–9942.

68 Guterman, T., Levin, M., Kolusheva, S. et al. (2019). Real-time in-situ monitoring of a tunable pentapeptide gel–crystal transition. *Angew. Chem. Int. Ed.* 58: 15869–15875.

69 Krieg, E., Bastings, M.M.C., Besenius, P., and Rybtchinski, B. (2016). Supramolecular polymers in aqueous media. *Chem. Rev.* 116: 2414–2477.

70 Matern, J., Dorca, Y., Sánchez, L., and Fernández, G. (2019). Revising complex supramolecular polymerization under kinetic and thermodynamic control. *Angew. Chem. Int. Ed.* 58: 16730–16740.

71 Kashchiev, D. (2003). Thermodynamically consistent description of the work to form a nucleus of any size. *J. Chem. Phys.* 118: 1837–1851.

72 Tao, K., Levin, A., Jacoby, G. et al. (2016). Entropic phase transitions with stable twisted intermediates of bio-inspired self-assembly. *Chem. Eur. J.* 22: 15237–15241.

73 Manoharan, V.N. (2015). Colloidal matter: packing, geometry, and entropy. *Science* 349: 1253751.

74 Onsager, L. (1949). The effects of shape on the interaction of colloidal particles. *Ann. N. Y. Acad. Sci.* 51: 627–659.

75 Palermo, V. and Samorì, P. (2007). Molecular self-assembly across multiple length scales. *Angew. Chem. Int. Ed.* 46: 4428–4432.

76 Knowles, T.P., Fitzpatrick, A.W., Meehan, S. et al. (2007). Role of intermolecular forces in defining material properties of protein nanofibrils. *Science* 318: 1900–1903.

77 Mendes, A.C., Baran, E.T., Reis, R.L., and Azevedo, H.S. (2013). Self-assembly in nature: using the principles of nature to create complex nanobiomaterials. *Wiley Interdiscip. Rev. Nanomed. Nanobiotechnol.* 5: 582–612.

78 González-Rodríguez, D., van Dongen, J.L.J., Lutz, M. et al. (2009). G-quadruplex self-assembly regulated by coulombic interactions. *Nat. Chem.* 1: 151–155.

79 Faul, C.F.J. and Antonietti, M. (2003). Ionic self-assembly: facile synthesis of supramolecular materials. *Adv. Mater.* 15: 673–683.

80 Rehm, T.H. and Schmuck, C. (2010). Ion-pair induced self-assembly in aqueous solvents. *Chem. Soc. Rev.* 39: 3597–3611.

81 Boyko, S., Qi, X., Chen, T.-H. et al. (2019). Liquid-liquid phase separation of tau protein: the crucial role of electrostatic interactions. *J. Biol. Chem.* 294: 11054–11059.

82 Chen, S. and Wang, Z.-G. (2022). Driving force and pathway in polyelectrolyte complex coacervation. *Proc. Natl. Acad. Sci. U.S.A.* 119: e2209975119.

83 Zhao, F., Shen, G., Chen, C. et al. (2014). Nanoengineering of stimuli-responsive protein-based biomimetic protocells as versatile drug delivery tools. *Chem. Eur. J.* 20: 6880–6887.

84 Chen, Y., Gan, H.X., and Tong, Y.W. (2015). pH-controlled hierarchical self-assembly of peptide amphiphile. *Macromolecules* 48: 2647–2653.

85 Cui, H., Pashuck, E.T., Velichko, Y.S. et al. (2010). Spontaneous and X-ray–triggered crystallization at long range in self-assembling filament networks. *Science* 327: 555–559.

86 Wang, Y., Qi, W., Wang, J. et al. (2018). Columnar liquid crystals self-assembled by minimalistic peptides for chiral sensing and synthesis of ordered mesoporous silica. *Chem. Mater.* 30: 7902–7911.

87 Jeffrey, G.A. and Saenger, W. (2012). *Hydrogen Bonding in Biological Structures.* Springer Science & Business Media.

88 Brunsveld, L., Folmer, B.J.B., Meijer, E.W., and Sijbesma, R.P. (2001). Supramolecular polymers. *Chem. Rev.* 101: 4071–4098.

89 Mason, T.O., Chirgadze, D.Y., Levin, A. et al. (2014). Expanding the solvent chemical space for self-assembly of dipeptide nanostructures. *ACS Nano* 8: 1243–1253.

90 Thirumalai, D., Reddy, G., and Straub, J.E. (2012). Role of water in protein aggregation and amyloid polymorphism. *Acc. Chem. Res.* 45: 83–92.

91 Stephens, A.D. and Kaminski Schierle, G.S. (2019). The role of water in amyloid aggregation kinetics. *Curr. Opin. Struct. Biol.* 58: 115–123.

92 Camino, J.D., Gracia, P., and Cremades, N. (2021). The role of water in the primary nucleation of protein amyloid aggregation. *Biophys. Chem.* 269: 106520.

93 Wang, J., Liu, K., Yan, L. et al. (2016). Trace solvent as a predominant factor to tune dipeptide self-assembly. *ACS Nano* 10: 2138–2143.

94 Wang, J., Yuan, C., Han, Y. et al. (2017). Trace water as prominent factor to induce peptide self-assembly: dynamic evolution and governing interactions in ionic liquids. *Small* 13: 1702175.

95 Grimme, S. (2008). Do special noncovalent π–π stacking interactions really exist? *Angew. Chem. Int. Ed.* 47: 3430–3434.

96 Martinez, C.R. and Iverson, B.L. (2012). Rethinking the term "pi-stacking". *Chem. Sci.* 3: 2191–2201.

97 Cockroft, S.L., Hunter, C.A., Lawson, K.R. et al. (2005). Electrostatic control of aromatic stacking interactions. *J. Am. Chem. Soc.* 127: 8594–8595.

98 Li, J., Du, X., Hashim, S. et al. (2017). Aromatic–aromatic interactions enable α-helix to β-sheet transition of peptides to form supramolecular hydrogels. *J. Am. Chem. Soc.* 139: 71–74.

99 Gazit, E. (2002). A possible role for π-stacking in the self-assembly of amyloid fibrils. *FASEB J.* 16: 77–83.

100 Ma, M., Kuang, Y., Gao, Y. et al. (2010). Aromatic–aromatic interactions induce the self-assembly of pentapeptidic derivatives in water to form nanofibers and supramolecular hydrogels. *J. Am. Chem. Soc.* 132: 2719–2728.

101 Tao, K., Levin, A., Adler-Abramovich, L., and Gazit, E. (2016). Fmoc-modified amino acids and short peptides: simple bio-inspired building blocks for the fabrication of functional materials. *Chem. Soc. Rev.* 45: 3935–3953.

102 Ghosh, G., Kartha, K.K., and Fernández, G. (2021). Tuning the mechanistic pathways of peptide self-assembly by aromatic interactions. *Chem. Commun.* 57: 1603–1606.

103 Chandler, D. (2005). Interfaces and the driving force of hydrophobic assembly. *Nature* 437: 640–647.

104 Ball, P. (2008). Water as an active constituent in cell biology. *Chem. Rev.* 108: 74–108.

105 Hazra, M.K. and Levy, Y. (2021). Biophysics of phase separation of disordered proteins is governed by balance between short-and long-range interactions. *J. Phys. Chem. B* 125: 2202–2211.

106 Ziserman, L., Lee, H.-Y., Raghavan, S.R. et al. (2011). Unraveling the mechanism of nanotube formation by chiral self-assembly of amphiphiles. *J. Am. Chem. Soc.* 133: 2511–2517.

107 Cui, H., Cheetham, A.G., Pashuck, E.T., and Stupp, S.I. (2014). Amino acid sequence in constitutionally isomeric tetrapeptide amphiphiles dictates architecture of one-dimensional nanostructures. *J. Am. Chem. Soc.* 136: 12461–12468.

108 Hartgerink, J.D., Beniash, E., and Stupp, S.I. (2001). Self-assembly and mineralization of peptide-amphiphile nanofibers. *Science* 294: 1684–1688.

109 Cui, H., Muraoka, T., Cheetham, A.G., and Stupp, S.I. (2009). Self-assembly of Giant peptide nanobelts. *Nano Lett.* 9: 945–951.

110 Li, S., Zhang, W., Xing, R. et al. (2021). Supramolecular nanofibrils formed by Coassembly of clinically approved drugs for tumor photothermal immunotherapy. *Adv. Mater.* 33: 2100595.

111 Liu, Y., Chang, R., Xing, R., and Yan, X. (2023). Bioactive peptide nanodrugs based on supramolecular assembly for boosting immunogenic cell death-induced cancer immunotherapy. *Small Methods* 7: 202201708. https://doi.org/10.1002/smtd.202201708

112 Yang, M., Yuan, C., Shen, G. et al. (2019). Cyclic dipeptide nanoribbons formed by dye-mediated hydrophobic self-assembly for cancer chemotherapy. *J. Colloid Interface Sci.* 557: 458–464.

113 Zhao, K., Xing, R., and Yan, X. (2021). Cyclic dipeptides: biological activities and self-assembled materials. *Pept. Sci.* 113: e24202.

114 Manchineella, S. and Govindaraju, T. (2016). Molecular self-assembly of cyclic dipeptide derivatives and their applications. *ChemPlusChem* 82: 88–106.

115 Hu, K., Jiang, Y., Xiong, W. et al. (2018). Tuning peptide self-assembly by an in-tether chiral center. *Sci. Adv.* 4: aar5907.

116 Fears, K.P., Kolel-Veetil, M.K., Barlow, D.E. et al. (2018). High-performance nanomaterials formed by rigid yet extensible cyclic β-peptide polymers. *Nat. Commun.* 9: 4090.

117 Liu, K., Xing, R., Zou, Q. et al. (2016). Simple peptide-tuned self-assembly of photosensitizers towards anticancer photodynamic therapy. *Angew. Chem. Int. Ed.* 55: 3036–3039.

118 Zou, Q., Abbas, M., Zhao, L. et al. (2017). Biological photothermal nanodots based on self-assembly of peptide–porphyrin conjugates for antitumor therapy. *J. Am. Chem. Soc.* 139: 1921–1927.

119 Li, S., Zhao, L., Chang, R. et al. (2019). Spatiotemporally coupled photoactivity of phthalocyanine–peptide conjugate self-assemblies for adaptive tumor theranostics. *Chem. Eur. J.* 25: 13429–13435.

120 Sun, B., Chang, R., Cao, S. et al. (2020). Acid-activatable transmorphic peptide-based nanomaterials for photodynamic therapy. *Angew. Chem. Int. Ed.* 59: 20582–20588.

121 Chang, R., Zou, Q., Zhao, L. et al. (2022). Amino-acid-encoded supramolecular photothermal nanomedicine for enhanced cancer therapy. *Adv. Mater.* 34: 2200139.

122 Ma, X., Zhao, Y., He, C. et al. (2021). Ordered packing of β-sheet nanofibrils into nanotubes: multi-hierarchical assembly of designed short peptides. *Nano Lett.* 21: 10199–10207.

123 Zhao, Y., Yang, W., Wang, D. et al. (2018). Controlling the diameters of nanotubes self-assembled from designed peptide bolaphiles. *Small* 14: 1703216.

124 Banwell, E.F., Abelardo, E.S., Adams, D.J. et al. (2009). Rational design and application of responsive α-helical peptide hydrogels. *Nat. Mater.* 8: 596–600.

125 Zou, Q., Zhang, L., Yan, X. et al. (2014). Multifunctional porous microspheres based on peptide–porphyrin hierarchical co-assembly. *Angew. Chem. Int. Ed.* 53: 2366–2370.

126 Liu, K., Xing, R., Chen, C. et al. (2015). Peptide-induced hierarchical long-range order and photocatalytic activity of porphyrin assemblies. *Angew. Chem. Int. Ed.* 54: 500–505.

127 Dudukovic, N.A., Hudson, B.C., Paravastu, A.K., and Zukoski, C.F. (2018). Self-assembly pathways and polymorphism in peptide-based nanostructures. *Nanoscale* 10: 1508–1516.

128 Freeman, R., Han, M., Álvarez, Z. et al. (2018). Reversible self-assembly of superstructured networks. *Science* 362: 808.

129 Ramos Sasselli, I., Halling, P.J., Ulijn, R.V., and Tuttle, T. (2016). Supramolecular fibers in gels can be at thermodynamic equilibrium: a simple packing model reveals preferential fibril formation versus crystallization. *ACS Nano* 10: 2661–2668.

130 Lu, Q., Bai, S., Ding, Z. et al. (2016). Hydrogel assembly with hierarchical alignment by balancing electrostatic forces. *Adv. Mater. Interfaces* 3: 1500687.

131 Hill, R.J.A., Sedman, V.L., Allen, S. et al. (2007). Alignment of aromatic peptide tubes in strong magnetic fields. *Adv. Mater.* 19: 4474–4479.

132 Song, J., Xing, R., Jiao, T. et al. (2018). Crystalline dipeptide nanobelts based on solid–solid phase transformation self-assembly and their polarization imaging of cells. *ACS Appl. Mater. Interfaces* 10: 2368–2376.

133 Pappas, C.G., Frederix, P.W.J.M., Mutasa, T. et al. (2015). Alignment of nanostructured tripeptide gels by directional ultrasonication. *Chem. Commun.* 51: 8465–8468.

134 Zhang, S., Greenfield, M.A., Mata, A. et al. (2010). A self-assembly pathway to aligned monodomain gels. *Nat. Mater.* 9: 594–601.

135 Zhan, J., Cai, Y., Ji, S. et al. (2017). Supramolecular silk from a peptide hydrogel. *Mater. Chem. Front.* 1: 911–915.

136 Wall, B.D., Diegelmann, S.R., Zhang, S. et al. (2011). Aligned macroscopic domains of optoelectronic nanostructures prepared via shear-flow assembly of peptide hydrogels. *Adv. Mater.* 23: 5009–5014.

137 Yan, X. and Möhwald, H. (2017). Organized peptidic nanostructures as functional materials. *Biomacromolecules* 18: 3469–3470.

138 Zhao, L., Li, S., Liu, Y. et al. (2019). Kinetically controlled self-assembly of phthalocyanine–peptide conjugate nanofibrils enabling superlarge redshifted absorption. *CCS Chem.* 1: 173–180.

139 Tidhar, Y., Weissman, H., Wolf, S.G. et al. (2011). Pathway-dependent self-assembly of perylene diimide/peptide conjugates in aqueous medium. *Chem. Eur. J.* 17: 6068–6075.

140 Tantakitti, F., Boekhoven, J., Wang, X. et al. (2016). Energy landscapes and functions of supramolecular systems. *Nat. Mater.* 15: 469–476.

141 Ribeiro, S.S., Samanta, N., Ebbinghaus, S., and Marcos, J.C. (2019). The synergic effect of water and biomolecules in intracellular phase separation. *Nat. Rev. Chem.* 3: 552–561.

142 Nielsen, M.H., Aloni, S., and De Yoreo, J.J. (2014). In situ TEM imaging of CaCO3 nucleation reveals coexistence of direct and indirect pathways. *Science* 345: 1158–1162.

143 Ianiro, A., Wu, H., van Rijt, M.M.J. et al. (2019). Liquid–liquid phase separation during amphiphilic self-assembly. *Nat. Chem.* 11: 320–328.

144 Loh, N.D., Sen, S., Bosman, M. et al. (2017). Multistep nucleation of nanocrystals in aqueous solution. *Nat. Chem.* 9: 77–82.

145 Le Ferrand, H., Duchamp, M., Gabryelczyk, B. et al. (2019). Time-resolved observations of liquid–liquid phase separation at the nanoscale using in situ liquid transmission electron microscopy. *J. Am. Chem. Soc.* 141: 7202–7210.

146 Doerr, A. (2017). Cryo-electron tomography. *Nat. Methods* 14: 34–34.

147 Turk, M. and Baumeister, W. (2020). The promise and the challenges of cryo-electron tomography. *FEBS Lett.* 594: 3243–3261.

148 Bhalla, S., Melnekoff, D.T., Aleman, A. et al. (2021). Patient similarity network of newly diagnosed multiple myeloma identifies patient subgroups with distinct genetic features and clinical implications. *Sci. Adv.* 7: eabg9551.

4

Molecular Simulations and Computational Chemistry of Peptide Self-assembly

Peng Zhou[1], Hai Xu[2], and Xuehai Yan[1,3,4]

[1] *State Key Laboratory of Biochemical Engineering, Institute of Process Engineering, Chinese Academy of Sciences, Beijing, P. R. China*
[2] *State Key Laboratory of Heavy Oil Processing and Department of Biological and Energy Chemical Engineering, China University of Petroleum (East China), Qingdao, P. R. China*
[3] *School of Chemical Engineering, University of Chinese Academy of Sciences, Beijing, P. R. China*
[4] *Center for Mesoscience, Institute of Process Engineering, Chinese Academy of Sciences, Beijing, P. R. China*

4.1 Introduction

The concept of computational chemistry is typically used to describe the application of computational technology in chemistry, including quantum mechanics at the molecular scale and the dynamics of molecular aggregates at the mesoscopic scale. Molecular simulation refers to the general process of describing complex chemical systems based on atomic or molecular models, aiming to explain the microscopic mechanism at the atomic scale and predict the material properties on the macroscopic scale. An important application field of molecular simulation is the development of novel materials, which can predict and screen their physical and chemical properties before actual experiments.

The self-assembly process of peptides is a continuous multistage phase transition process. The complete process includes the transition from the solution phase to the sol phase, gel phase, and crystal state at the macroscopic scale [1] and the formation of a variety of nanostructures such as nanospheres [2], cylindrical nanofibers and twisted or spiral nanobelts [3], flat nanobelts [4], and nanotubes at the microscopic scale [5].

Peptides with different amino-acid sequences can assemble into nanostructures with different shapes. The same peptide can present various self-assembled structures under different environmental conditions (concentration, temperature, pH, and solvent), or at different time points under the same conditions, which is called "polymorphism" in self-assembly [6, 7]. It requires precise regulation of the structure and performance of peptide self-assembled materials as well as the development of molecular self-assembly theory.

Peptide Self-Assembly and Engineering: Fundamentals, Structures, and Applications, First Edition.
Edited by Xuehai Yan.

Several theoretical models have been proposed to describe the self-assembly process of peptides. Taking self-assembled nanofibers as an example, it is generally believed that the self-assembly process is similar to the classical nucleation process of crystals: (i) peptides form ordered oligomers through hydrogen bonds (H-bond) between molecular backbones; (ii) these oligomers further grow to form protofibrils; (iii) Under high concentration and long-time incubation, these protofibrils form a mesoscopic hierarchical assembled structure through electrostatic, hydrophobic, and π–π noncovalent interactions [1]. Recently, researchers have found liquid–liquid phase separation (LLPS) in the peptide self-assembly system [8], indicating that the peptide self-assembly process is not necessarily a classical nucleation process. However, to explore the early aggregation and nucleation process of self-assembly, it is challenging to study the details at the molecular scale using experimental routines alone [9]. Therefore, as an important supplement to experimental research, molecular simulation has been widely applied to study the peptide self-assembly system, drawing on the thermodynamic phase diagram of peptide self-assembly [10] and the multiscale method of phase separation process [11].

4.2 Molecular Models and Computational Methods

4.2.1 Theory

4.2.1.1 Quantum Mechanics

The quantum mechanics (QM) method is based on solving the Schrodinger equation of the wave function to calculate the exact molecular structure of an assembly. Since solving the Schrodinger equation of all possible nuclear arrangements in a self-assembly system (usually more than thousands of atoms) requires a huge amount of computation, it is still difficult in application. In the study of peptide self-assembly, the most widely used method is the density functional theory (DFT) method [12] in which the wave function is replaced with the electron density on calculating the energy of the system, thus reducing the complexity and difficulty. Through the DFT method, researchers can calculate the strength of the interaction between peptide molecules, such as the hydrogen bonding (H-bonding) interactions or dispersion interactions. On this basis, the molecular arrangement with the lowest energy and intermolecular binding energy can be obtained. Due to the low computational efficiency, the number of peptide molecules in a QM simulation system is usually less than 10.

4.2.1.2 Molecular mechanics

Because most self-assembly processes do not involve the electron transfer process, molecules can be described by the molecular mechanics (MM) model based on the classical mechanics theory [13]. In the MM model, the molecular structure and energy are calculated based on empirical and semiempirical parameters, which are defined in the force field (FF). In the FF, atoms are restricted by the stretching or compression of chemical bonds (Eq. 4.1),

$$V_b(r_{ij}) = \frac{1}{4} k_{ij}^b (r_{ij}^2 - \mathrm{b}_{ij}^2)^2 \tag{4.1}$$

where r_{ij} is the distance between atom i and atom j, b_{ij} is the length of the covalent bond, and k^b is the harmonic force constant of the covalent bond, and the distortion of bond angles (Eq. 4.2),

$$V_a(\theta_{ijk}) = \frac{1}{2} k^{\theta}_{ijk} (\theta_{ijk} - \theta^0_{ijk})^2 \tag{4.2}$$

where θ_{ijk} is the angles between a triplet of atoms i, atom j, and atom k, and k^{θ} is the harmonic force constant of the covalent bond-angle vibration.

The conformation of a real molecule is the balance of multiple interactions. The parameters in the FF are generally verified by quantum mechanical calculations or experimental measurements. Therefore, the MM model-based FF has been widely used to study the conformation and energy of molecules. Compared with the QM method, extending FF to the study of self-assembly behavior in peptide self-assembly can greatly expand the scale of the research system.

4.2.1.3 Molecular dynamics

Molecular dynamics (MD) is a molecular simulation method for studying the physical motion of atoms and molecules. In MD simulation, by calculating the interaction force between particles (atoms or molecules) in the system, the acceleration of each particle is obtained according to Newton's equation of motion (Eq. 4.3). The position and velocity of the next moment are calculated based on the current position, velocity, and acceleration of particles. The particle motion trajectory is calculated by cycling the above steps, thus realizing the dynamic evolution of the system [14, 15]:

$$m_i \frac{\partial^2 r_i}{\partial t^2} = F_i \tag{4.3}$$

The force between molecules and atoms is defined in the form of the MM FF through the potential energy between particles. As shown in Eq. 4.4, potential energy generally includes bond energy (bond expansion term, bond angle bending term, and dihedral angle distortion term) and nonbond energy (van der Waals force and Coulomb force):

$$\begin{aligned} V(\vec{r}) = & \sum\nolimits_{bonds} K_b (b - b_0)^2 + \sum\nolimits_{angles} K_{\theta} (\theta - \theta_0)^2 + \\ & \sum\nolimits_{dihedrals} \frac{V_n}{2} \left[1 + \cos(n\Phi - \delta)\right] + \sum\nolimits_{El_{ij}} \frac{q_i q_j}{\varepsilon_r r_{ij}} + \sum\nolimits_{vdw_{ij}} \frac{C^{12}_{ij}}{r^{12}_{ij}} - \frac{C^6_{ij}}{r^6_{ij}} \end{aligned} \tag{4.4}$$

The force can be obtained by calculating the partial derivative of the potential energy function ($V[r]$) (Eq. 4.5),

$$F_i = -\frac{\partial V}{\partial r_i} \tag{4.5}$$

4.2.2 Molecular Model of Force Field in Molecular Simulation

4.2.2.1 All-atom Force Field

The all-atom (AA) model is the most common molecular model of force field in which every atom of the molecule is represented explicitly (Figure 4.1a). AA model force fields, such as CHARMM, AMBER, OPLS, and Gromos, are often used in biological system simulation, and CFF, CVFF, PCFF, and COMPASS are commonly used in material field simulation. Generally, the partial charge of atoms in the AA force field is fixed during the MD simulation, which could not describe the instantaneous polarization produced by the environmental medium or induced by the atom with a high net charge. To overcome this limitation, the reactive force field has been developed to study the evolution of the protonation state of amino acids and peptides in solution [16]. In most cases that do not involve chemical reactions, the AA model has been widely used because of its high computational efficiency and its accuracy in describing molecular and atomic models.

The advantage of the AA force field is its high accuracy in describing the structure and properties of biomolecular systems such as peptides. However, the computational cost of the AA force field is high, and the simulation efficiency is low. Therefore, when computing resources are limited, researchers may have to reduce the size or shorten the evolution time of the simulation system, which could inevitably weaken the effectiveness of the simulation.

4.2.2.2 United-atom Force Field

In the early versions of the classical molecular force fields such as OPLS, Amber, and CHARMM, the united-atom (UA) model was used. In such models, as shown in Figure 4.1b, the hydrogen atoms that are directly connected to carbon atoms by covalent bonds are hidden, appearing in the force field indirectly. In the UA model, only the hydrogen atoms that are covalently connected with N, O, S, and other atoms, and can form H-bonds, are directly shown. In other words, groups such as CH4, CH3, CH2, CH, and C are regarded as a united atom. As a result, the computational efficiency of MD simulation can be improved by almost an order of

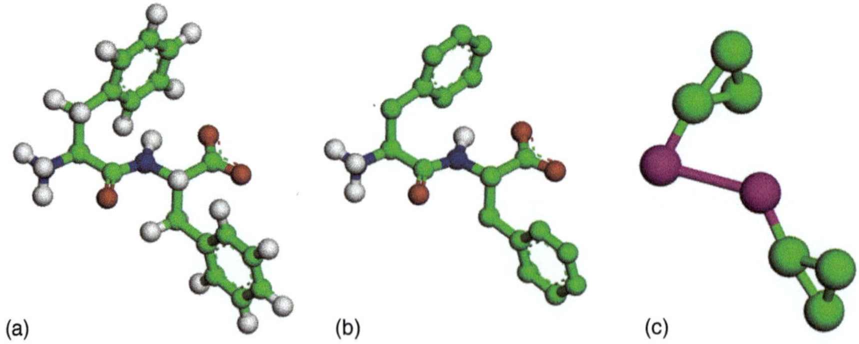

Figure 4.1 Molecular models of FF dipeptide. (a) All-atom model, (b) united-atom model, and (c) coarse-grained model.

magnitude. Therefore, with the same computing resources, researchers can apply the UA force field to simulate a larger system with a longer evolution time, obtaining simulation results that might be difficult to calculate using the AA force field. Common UA force fields include the GROMOS force field, COMPASS force field, and TraPPE force field [17].

4.2.2.3 Coarse-grained Force Field

To further improve the computational efficiency of molecular simulation and study mesoscopic phenomena on a larger scale, the coarse-grained (CG) model has been developed. As shown in Figure 4.1c, in the Martini CG force field, the phenylalanine residue of FF dipeptide is represented by four beads: one bead for the main chain and three beads for the side chain ring. Due to the simplification of the molecular model, the CG model can simulate large-scale system phase transition behavior and structural change phenomena. However, its accuracy is limited when simulating some problems involving atomic scale. Therefore, when using the CG method to simulate the structural evolution of peptide self-assemblies, it is sometimes necessary to introduce more restrictive variables to modify the CG model. Bellesia et al. successfully reproduced the early aggregation behavior of peptides and the transition from twisted nanoribbon to helical nanoribbon by introducing chiral degree and asymmetric limiting parameters when simulating peptide self-assembly using the CG model [18, 19].

4.2.3 Acceleration Method of Molecular Simulation

4.2.3.1 Replica-exchanged Molecular Dynamics

One problem that often occurs in conventional MD simulations is that the target molecule tends to be trapped in a local potential well because it cannot cross a high potential barrier, resulting in insufficient global sampling. Therefore, researchers have developed some unconventional MD methods to improve the sampling efficiency. Replica-exchanged molecular dynamics (REMD) is a powerful method to enhance the sampling of molecular simulations [20, 21]. The REMD method produces relatively independent molecular dynamics runs for copies at different temperatures and exchanges the configurations of adjacent copies at adjacent temperatures with a certain probability (Eq. 4.6) according to the Metropolis standard. In this way, the conformers of the low-temperature metastable state can jump out of the local potential energy well quickly, and the search ability for the lowest global potential energy point is enhanced [22]. Since the copies at different temperatures are calculated in parallel, REMD is also called parallel tempering [20]. Compared with conventional MD, REMD has high sampling efficiency and is suitable for studying systems with polymorphism and calculating the statistical probability of each state. Together with its advantages, however, REMD requires a significant amount of computation:

$$P_{ij} = \min\left[1, e(1/k_\beta T_i - 1/k_\beta T_j)(V_i - V_j)\right] \tag{4.6}$$

The REMD method can obtain the molecular conformation with the lowest free energy and the molecular binding mode with the highest probability by traversing

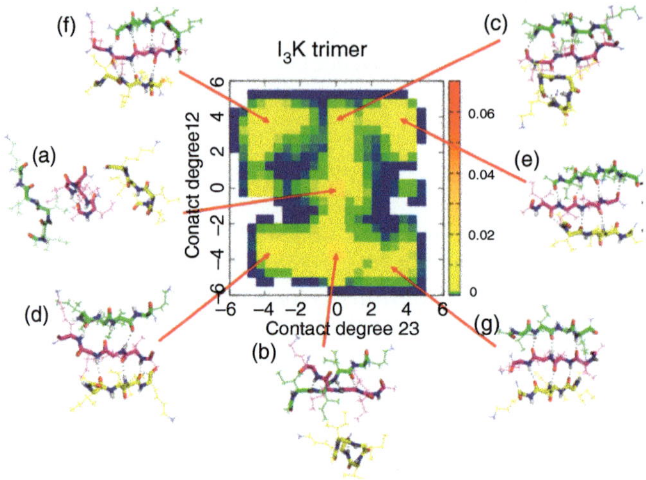

Figure 4.2 The REMD results of the packing modes of peptide trimer. *Source:* Reproduced with permission from Ref. [24]/American Chemical Society.

the conformational space. This method has been successfully applied in the study of the hairpin structure of long-chain peptides as well as peptide self-assembly systems [21, 23].

Gnanakaran et al. applied the REMD method to study the structure of Aβ(16–22) peptide dimer at 38 different temperatures and obtained the free energy surface of the packing modes of the dimer [25]. The simulation results showed that the peptide monomer showed a random coil structure similar to that of proline; Aβ(16–22) can form at least six stable dimers at low temperature, not limited to parallel or antiparallel structures; the structure of the dimer is sensitive to temperature; and the influence of temperature on the dimer is that it can alter the strength of different driving forces during the self-assembling process, thus determining the competitive results of various driving forces. REMD has also been used to study the molecular arrangement tendency, stability, and toxicity of another amyloid sequence, GNNQQNY [26, 27]. The distribution of trimers was calculated with different arrangements on the free energy surface through REMD, thus determining the parallel/antiparallel orientation propensity in the assembly β-sheet formation (Figure 4.2) [24, 26, 28].

4.2.3.2 Discrete molecular dynamics

To enhance the computing efficiency and simplify traditional molecular dynamics (MD), discrete molecular dynamics (DMD) has been developed [29]. Unlike traditional MD, where continuous potentials are used, DMD replaces them with discrete step function potentials. Instead of solving Newton's equations of motion for molecules, the DMD engine solves the ballistic equations of motion. As a result, fewer calculations are run per time step in DMD, allowing for simulation of peptide self-assembly with longer time and length scales. Using DMD simulations, the complete self-assembling process has been reproduced, and the phase diagram is fully explored [10].

4.2.4 Nonequilibrium Molecular Dynamics

Obtaining certain thermodynamic parameters directly from equilibrium simulations of the self-assembly process of peptides can be difficult due to limitations in time scale. To overcome this challenge, nonequilibrium kinetic simulation methods have been developed to calculate these parameters. One such method involves establishing a reasonable reaction coordinate and calculating the potential of mean force (PMF) of the system along the reaction coordinate. This can strengthen the sampling of high potential states [30]. The umbrella sampling method is a popular method to calculate PMF based on reaction coordinates [31]. In umbrella sampling, the system is divided into several "windows" along the reaction coordinate, and additional constraint potentials are added in each window to sample the reaction coordinate near the window, thereby obtaining the biased probability distribution of the system in each window. The weighted histogram analysis method can then be used to reconstruct the free energy surface of the true probability distribution [32]. By doing this way, the free energy difference between different states of the system can be obtained (Figure 4.3).

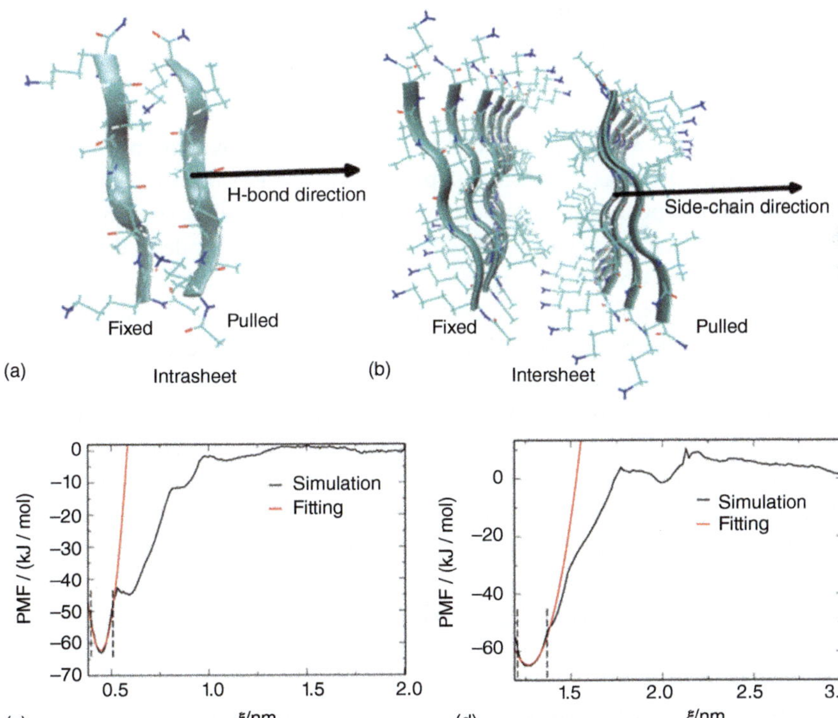

Figure 4.3 The free energy difference calculation of peptide self-assembly using umbrella sampling. Representative snapshots for the intra- (A) and inter-sheet (B) alignments studied by the steered MD simulations. Total PMFs and harmonic fittings of the potential wells for the intra- (C) and inter-sheet (D) alignments. *Source:* Reproduced with permission from Ref. [33]/Chinese Physical Society and IOP Publishing Ltd.

Wolf et al. applied the umbrella sampling method to calculate the free energy difference of peptide self-assembly in combination with REMD to simulate the dissolution and separation process of a peptide molecular monomer from a self-assembly and to obtain the dissociation/assembly equilibrium constant [34]. Deng et al. further qualified the inter-strand and inter-sheet driving force by means of steered molecular dynamics simulation combined with umbrella sampling, proving that the mechanical anisotropy comes from the fact that the intra-sheet alignment is stabilized by enthalpy, but the inter-sheet alignment is stabilized by entropy [33].

4.3 Analysis Object of Molecular Simulation in Peptide Self-assembly System

4.3.1 Analysis of Molecular Interactions

The main focus of molecular simulation is to study the microscopic interactions between molecules in self-assembly systems and the resulting orderness of assembled structures. In self-assembling peptide systems, the most important intermolecular interaction is the intermolecular H-bond on the amide backbone, which serves as the primary basis for structural order, i.e. secondary structure of the main chain in the system.

4.3.1.1 Types of Noncovalent Interaction in Peptide Self-assembly System

Peptides have the ability to assemble into supramolecular nanostructures due to the presence of multiple types of microscopic forces in the peptide self-assembly system. These microscopic forces can be categorized as noncovalent or covalent, depending on the objects of interaction.

Noncovalent microscopic forces, including H-bond, electrostatic interactions (such as ionic or salt bonds), van der Waals forces, hydrophobic effect, and aromatic stacking (π–π interaction) [2], facilitate molecular recognition and play an important role in forming ordered nanostructures. In contrast, covalent forces involve cross-linking reactions in some self-assembly systems, which can change not only the self-assembly ability of the peptide molecule itself but also the performance of the entire self-assembly system. For most peptide molecules, only intermolecular forces are present in their self-assembly systems. Nevertheless, when the peptide itself has a large size or relatively long main chain, its intramolecular force would play an important role in determining the self-assembly ability.

Hydrogen bonding (H-bonding)

H-bonding refers to the electrostatic attraction between polar groups [35, 36], which typically occurs between the hydrogen atom (H) of the electron acceptor (X atom) and the highly electronegative electron donor (Y atom) such as nitrogen (N), oxygen (O), or fluorine (F) atoms. In the peptide self-assembly system, H-bonding mainly occurs between the hydrogen atom on the amino group (N—H) and the oxygen atom of the carbonyl group (C=O) on the peptide backbone.

The element building blocks in peptide assemblies are generally rich in α-helix or β-sheet secondary structures that are stabilized by H-bonding. In addition to the H-bonds on the backbone, some amino-acid residues (such as Gln and Ser) can form H-bonds between the side chains. The energy of H-bonding between C=O and N—H is typically around 4–12 kJ mol^{-1} [2]. Unlike covalent bonds, there is no direct and explicit expression of H-bonding in many force fields. Instead, H-bonding is analyzed by measuring the distance and angle between donors and acceptors [35]. The X–H···Y angle is usually linear (180°). The closer the angle to 180° and the shorter the H···Y distance, the stronger is the H-bonding. In MD simulation, H-bonding is defined by the X–H•••Y angle being larger than 150°, and the H•••Y distance being less than 0.35 nm [37].

Electrostatic Interaction

As a long-range force, the energy of electrostatic interaction is about 0.5–4 kJ mol^{-1}. It is relatively flexible and widely exists in self-assembly systems, especially between amino-acid residues with electrostatic charge, without limitation of saturation and directionality. Electrostatic interaction can act either as an attraction to stabilize a self-assembled building block or as a repulsive force to maintain the dispersion of the system. The main factors affecting the electrostatic interaction are pH and ionic strength [2]. Adjusting pH and ionic strength can change and control the self-assembly behavior and self-assembled nanostructure of peptides [38, 39]. In MD simulation, the energy term of electrostatic interaction is calculated by measuring the Coulomb interaction between two atoms,

$$V_C(r_{ij}) = f\frac{q_i q_j}{\varepsilon_r r_{ij}} \tag{4.7}$$

van der Waals Interaction

van der Waals interaction (vdW interaction) is a short-range force that includes both attraction and repulsion [40]. It widely exists in the interactions among atoms and molecules. Although the strength of the vdW interaction is smaller than that of other types of microscopic interaction (i.e. H-bond and electrostatic interaction), it should not be ignored in peptide self-assembly system because it exists on both polar charged residues and hydrophobic residues. In MD simulation, the vdW interaction is usually calculated by measuring the Lennard-Jones potential between two atoms:

$$V_{LJ}(r_{ij}) = \frac{C_{ij}^{12}}{r_{ij}^{12}} - \frac{C_{ij}^{6}}{r_{ij}^{6}} \tag{4.8}$$

Hydrophobic Effect

The essence of the hydrophobic effect is caused by an increase in entropy. The concept was first proposed by Kauzmann when studying protein denaturation. Its core content is that two nonpolar surfaces in water (a polar solvent) tend to combine with each other to reduce the nonpolar surface area exposed to water (a polar solvent) and drain water (a polar solvent) to maximize entropy [41]. Hydrophobic interaction occurs between nonpolar surfaces and has no limitation of the direction of the interaction. It is a weak interaction between molecules (between small molecules and between and within large molecules). The energy of the hydrophobic effect is about 4–8 kJ mol^{-1} [2]. In the self-assembled peptide system, the hydrophobicity is mainly reflected in the contact of hydrophobic amino-acid residues. For example, half of the amino acids in the ionic complementary peptide sequence are hydrophobic. The hydrophobic interaction between the side chains of hydrophobic amino-acid residues greatly improves the tendency of ordered aggregation of ionic complementary peptide molecules, making such molecules have a relatively high self-assembly ability [42]. Surfactant-like peptides and lipopeptides show surfactant-like aggregation properties through their hydrophobic interaction [24, 43, 44]. In MD simulation, it is generally difficult to directly measure the contribution of the hydrophobic effect on energy term. An alternative way to evaluate the hydrophobic effect is to measure the changes in solvent accessible surface area (SASA) of the peptide aggregates compared with the fully dispersed solution [3, 45].

π–π Stacking

π–π Stacking is a noncovalent weak attraction between aromatic rings, and the energy is 8–12 kJ mol^{-1} [46]. π–π Stacking widely exists in the process of peptide self-assembly, protein folding, and molecular recognition in nucleic acid molecules (DNA and RNA). Although extensive experimental and theoretical studies have been carried out on studying π–π stacking, there is still no unified explanation for its mechanism. One explanation is the electrostatic model: the negative quadrupole of the aromatic ring with reduced electron-absorbing groups makes the aromatic ring easy to form parallel and sandwich structures. The negative quadrupole is added to the donor group, thus increasing the interaction intensity of the "T" conformation [47]. Another explanation is the interaction orientation model, that is one type of hydrophobic effect with specific configuration arrangement in space [48]. This explanation emphasizes the role of stacking direction, which includes face-to-face and side-to-side stacking, while face-to-face stacking also includes complete face-to-face stacking and partial face-to-face stacking. The side-to-side stacking can be regarded as a weak H-bond formed between the slightly electron-deficient hydrogen atom on one aromatic ring and the slightly electron-rich π electron cloud on the other aromatic ring [48]. In general, π–π stacking can be seen as the joint result of electrostatic force, dispersion force, and force orientation. π–π Stacking depends on the spatial geometric conformation, so it has different performance in different self-assembly systems. In MD simulation, one feasible way to analyze the π–π stacking is to calculate the radial distribution function (RDF) of the distance between aromatic ring centroids [49].

4.3.1.2 Inter- and Intramolecular Interactions

The microscopic noncovalent interactions in the peptide self-assembly system discussed above are mainly intermolecular interactions. For peptides with long chains, intramolecular interactions cannot be ignored and sometimes play an important role in specific self-assembly systems. For example, the coiled-coil self-assembled motif is mainly composed of α-helix secondary structure, which requires the formation of stable intramolecular H-bonds [50–52]. Even in the peptide self-assembly system of β-sheet secondary structure, intramolecular interaction also plays an important role. In a β-hairpin peptide system, the hairpin structure is stabilized by intramolecular H-bonding between two β-strands of one peptide [53–56]. On the other hand, for the linear peptide, take L_3K as an example, the intramolecular H-bonding stabilizes the coiled conformation of monomers, resulting in a low propensity to form β-sheet alignment [24]. Therefore, in MD simulation, it is important to distinguish between inter- and intramolecular interactions, which is generally difficult to do experimentally.

4.3.2 Molecular Conformation and Secondary Structure Analysis

As mentioned in the section above, driven by noncovalent interactions, peptide molecules can form well-defined secondary structures, which is a unique feature of peptide molecules that differentiates them from organic polymers. The formation of such secondary structures can stabilize the elemental ordered building block formed by the early aggregation of peptides, thus making it possible to further develop into hierarchical assemblies. The secondary structures of peptides in the self-assembly system include stable secondary structures such as α-helix and β-sheet, metastable secondary organization such as α-sheet, and unstable secondary structures such as random coil. An important advantage of molecular simulation is that it can analyze the conformation and secondary structure of each molecule, allowing for statistical analysis and comparison with experimental results.

4.3.2.1 α-Helix Secondary Structure

α-Helix is a common secondary structure in proteins and peptides. When forming an α-helix, the main chain adopts a right-handed helix which is stabilized by intramolecular H-bonding formed between the C=O of the "n" residue and the N—H of the "$n + 4$" residue [57, 58]. This secondary structure is also called the classic Pauling–Corey–Branson alpha helix or 3.6_{13}-helix. Each helix is composed of 3.6 amino-acid residues, with a total of 13 atoms participating in the intramolecular H-bonding. Multiple α-helices (usually 2–3 helices) can intertwine into a nanofiber, which is called "coiled-coil" motif [59–62], as shown in Figure 4.4.

In MD simulation, α-helix can be analyzed by calculating the intramolecular H-bonding and the dihedral angles (Phi/Psi angles). Additionally, since the helical wheel is centered around the origin and rotates around the Z-axis, other helix properties such as the average helix radius, average helical angle per residue, helical rise per residue, and the number of helical residues can be calculated to distinguish among the 3_{10}-helix, α-helix, and π-helix [64, 65].

4.3.2.2 β-Sheet Secondary Structure

β-Sheet, also called β-pleated sheet, is created by the way that β-strands are arranged transversely to form intermolecular H-bonds and form a lamellar structure of

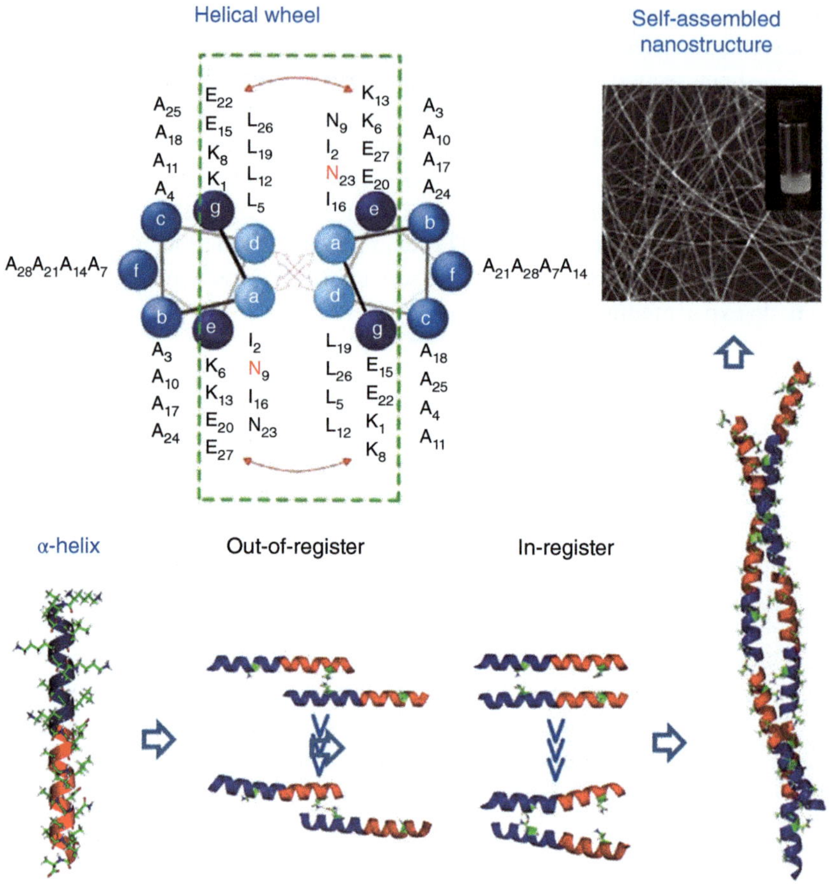

Figure 4.4 Nanofibril structures of coiled-coil motif based on α-helix secondary structure. *Source:* Reproduced with permission from Ref. [63]/John Wiley & Sons.

natural twisted pleated sheet [66–68]. In a β-strand, the main chain is close to the extended state, and the side chains of amino-acid residues are also stretched to both sides of the β-sheet plane. In MD simulation, similar to α-helix analysis, the determination of β-sheet is based on both H-bonding and the dihedral angles (Phi/Psi angles). Additionally, since the β-strands are parallelly packed, some other properties of β-sheet such as orientation (parallel or antiparallel) and residue shift between β-strands can be calculated. Moreover, as shown in Figure 4.5, inter-β-sheet packing modes based on specific inter-β-strand arrangements, which can affect both the self-assembling ability and assembled nanostructure, need to be carefully analyzed.

4.3.2.3 α-Sheet Secondary Structure

α-Sheet (also known as α-pleated sheet or polar pleated sheet) is an atypical secondary structure of a protein, first proposed by Linus Pauling and Robert Corey in 1951 [66]. In α-sheet, the arrangement between α-strands is similar to that of β-sheet, except for the direction of the H-bonds between α-strands. In α-sheet, the

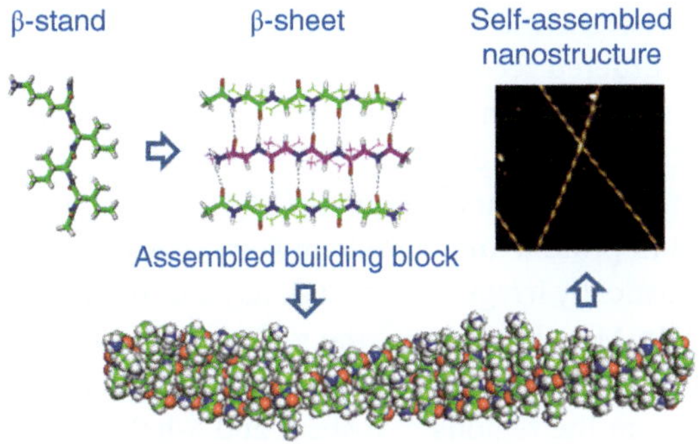

Figure 4.5 Nanofibril structures based on β-sheet secondary structure. *Source:* Reproduced with permission from Ref. [69]/Elsevier Inc.

H-bonds between adjacent α-strands are identical in direction, meaning that for each α-strand, the carbonyl group (C=O) on the main chain is located on one side of the molecule, while the amino group on the main chain is located on the other side of the molecule. As a result, the adjacent amino-acid residues in the α-sheet are alternately in α_R and α_L conformation. α-Sheet is rarely found in natural proteins and is considered to be an unstable intermediate that can rapidly form and transform into the β-sheet secondary structure [70]. In experiments, distinguishing between β-sheet and α-sheet is quite difficult due to their structural similarity in intermolecular packing, resulting in similar signals in X-ray diffraction [71]. In MD simulations, the α-sheet secondary structure can be recognized by analyzing the dihedral angles (Phi/Psi angles) of each strand, as shown in Figure 4.6.

4.3.2.4 Random Coil Secondary Structure

The definition of the random coil is derived from the conformational analysis of polymers. When there is no external force to stabilize the polymer chain, it is considered that the polymer chain can randomly traverse all possible conformations. Using this concept to describe the conformation of some peptide chains in protein

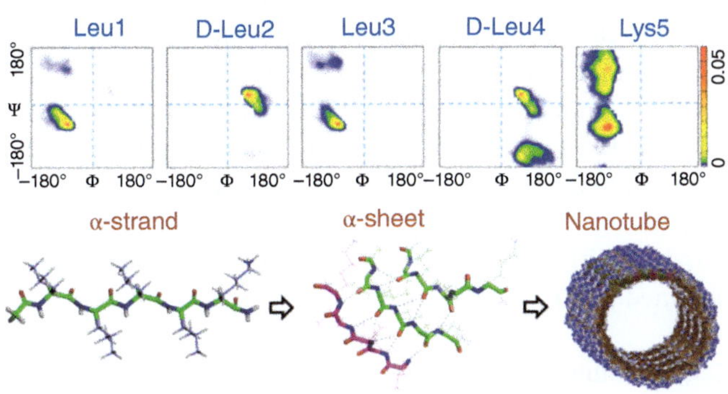

Figure 4.6 α-Sheet structure. Phi/Psi dihedral angle and conformation of α-strand and α-sheet. *Source:* Reproduced with permission from Ref. [71]/ American Chemical Society.

molecules, the conformation of peptide segments is irregular and the aggregation is relatively loose, unlike the stable and ordered secondary structures such as α-helix and β-sheet. The random coil is usually highly correlated with some zero-dimensional self-assembled nanostructures. For example, the peptides that form nanospheres are often rich in random coil secondary structures [24, 72]. However, it should be noted that the random coil in protein and peptide molecules is not truly or completely irregular; it is only "statistically irregular" because its conformational energy is high and unstable [73–75]. In MD simulations, the main routine to analyze the random coil is to calculate the distribution of dihedral angles (Phi/Psi angles), which is generally different from the regions of β-sheet and α-helix in the Ramachandran plot, with a wider distribution range.

4.4 Molecular Simulation of Self-assembling Process

4.4.1 Self-assembling Models of Peptide Systems

For different peptide systems and various self-assembled nanostructures, researchers have developed several theoretical models to explain the observed changes in the peptide assembly systems. These theoretical models could be regarded as the basic hypothesis for simulating and explaining the self-assembly phenomenon.

4.4.1.1 Hierarchical Self-assembly Model

The hierarchical self-assembly model describes the peptide self-assembly based on β-sheet secondary structure, with the peptide being regarded as a linear rod-like molecule. In this model, the rod-like molecules can form H-bonds with each other and form protofibrils, which further develop into a hierarchical self-assembled nanostructure (nanoribbon or nanofiber) with a larger size by stacking with each other under high concentration and long-term cultivation conditions [1].

4.4.1.2 Transition Model from Nanoribbon to Nanotube

In the MD simulation studies of peptide self-assembly systems, twisted nanoribbons, helical nanoribbons, and nanotubes are often observed. The mechanism driving these structural changes in self-assembly has always been attractive. Using the Monte Carlo simulation method (MC), Selinger et al. established a CG model of the chiral band, which shows a transition of the chiral band from twisted band to the helical band by adjusting the force in different directions in the chiral band [76]. Similarly, using MD simulation, Bellesia et al. established a single-layer belt nanostructure based on β-sheet packing. Through tuning the unsymmetrical parameters of the β-strand, the structural transition from the twisted belt to the helical belt with different shapes can be obtained [18, 19].

4.4.1.3 Surfactant-like Self-assembly Model

In the study of the self-assembly system of surfactant-like peptides, some phenomena similar to the aggregation of surfactants were observed. Therefore, to improve the understanding of the self-assembly of surfactant-like peptides, researchers try to employ the molecular stacking parameters in the surfactant self-assembly theory to quantify the peptide molecules and explain their self-assembly behavior [77, 78].

Xu et al. studied the self-assembly behavior of a series of surfactant-like peptides A_mK ($m = 3, 6, 9$). It was found that with the increase in the hydrophobic residue number, the self-assembled nanostructure changed from membrane, long nanofiber to short nanorod. This relationship between self-assembled nanostructure and molecular length can be reasonably explained using molecular stacking parameters [79].

4.4.1.4 Self-assembly Model Mediated by Phase Separation

In contrast to the classical nucleation–elongation mechanism, the LLPS is suggested to play a critical step prior to the nucleation of supramolecular nanofibrils through peptide self-assembly [81]. Through the local density fluctuation, LLPS takes place, and the droplets with different sizes and numbers can be obtained (as shown in Figure 4.7) [80]. Through the liquid–solid-phase transition (LSPT) from droplet to fibril, three types of prefibrils are formed: urchin-like fibrils, beaded spindle-like fibrils, and radial fibril clusters, which lead to different noncovalent cross-linking fibrillar network evolution during the transition from sol to gel. These fibrillar networks have different nanofibril diameters and cross-linking density, which makes the mechanical hierarchy of hydrogels tunable [80].

4.4.2 Simulation of Phase Separation

The self-assembly process, from the initial homogeneous solution to highly ordered self-assembled nanostructures, is a transition from the liquid phase to the solid phase. Therefore, the formation of solute-rich liquid droplets through LLPS appears to be a necessary step during the self-assembling process. Until recently, the LLPS

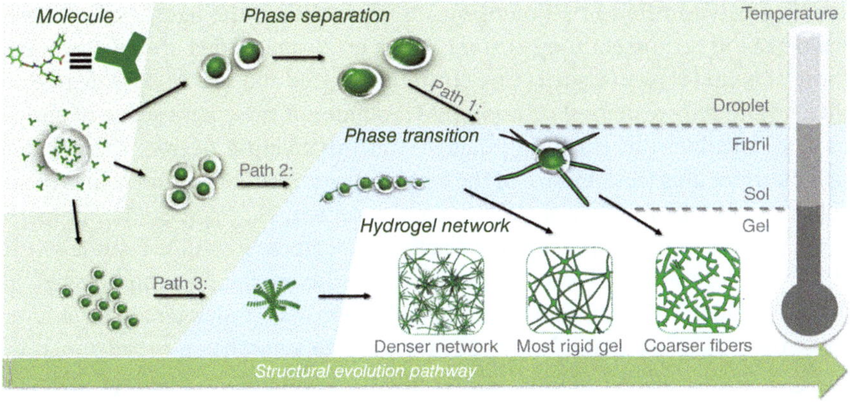

Figure 4.7 Schematic illustration of the three pathways of peptide self-assembly mediated by phase separation at different thermal histories and the resultant hydrogel network. *Source:* Reproduced with permission from Ref. [80]/Elsevier Inc.

phenomenon of peptide self-assembly systems had not been fully understood [81]. DLS and TEM can observe droplets with sizes greater than 100 nm, but it is difficult to detect droplets that form at the nanometer scale and that solidify and nucleate quickly.

Another approach to look for a liquid-like state in a self-assembly system is to simulate the properties of the liquid phase directly. Such simulation methods have shown success in probing the phase behaviors of intrinsically disordered proteins (IDPs) or proteins with intrinsically disordered regions (IDRs), including field-theoretic simulations, Gibbs ensemble simulations, and slab simulations [82].

4.4.2.1 Liquid–Liquid Phase Interface

One of the difficulties of simulating LLPS in self-assembling systems is how to judge the phenomenon on the microscale. In experiments, the liquid–liquid interfacial tension often makes droplets formed by phase separation that adopt a spherical shape. Therefore, it is necessary to simulate the interface of the liquid-like intermediate, which is the boundary between the concentrated solute phase and the dilute solute phase in the solution to study the relationship between the internal structure and properties of droplets systematically and carefully at the molecular scale. The microstructure of liquid–liquid two-phase interface and the aggregation state of molecules at the phase interface would have an important impact on the subsequent self-assembly nucleation process.

One simulation model that has been used to study the phase behavior is the slab geometry in which two coexisting phases are simulated in an elongated box with periodic boundary conditions [83, 84]. This strategy has been successfully used to study IDPs. The concentrations of a spherical droplet simulation are consistent with those from the slab geometry [85].

4.4.2.2 Aggregation and Fluidity in Liquid-like State

It has been proposed that the aggregation capability of a liquid-like intermediate is intermediate between that of a homogeneous solution and the liquid–solid-phase separated (LSPS) self-assembling system. Tang et al. quantified the aggregation capability of peptides by measuring the clustering degree and the collapse degree of an initially randomly dispersed solution [86]. To monitor the convergence of simulations and calculate the clustering degree and the collapse degree, the solvent accessible surface area (SASA), size of the largest cluster, number of molecules in all large clusters, and the number of clusters were recorded. The "aggregation propensity" (AP) parameter was defined as the fraction of decrease in SASA. Peptides with low clustering degree and AP values failed to aggregate, remaining a homogeneous solution, while high AP values indicated the formation of an ordered nanostructure. However, the liquid-like state, being an intermediate state, has a broad range of clustering degrees and AP values, which makes it difficult to distinguish from other states.

Although the aggregation capability is able to probe a peptide-rich dense phase, it cannot distinguish between LLPS and LSPS. To properly characterize the fluidity of the dense phase, liquid-like states can be distinguished from solid-like ones.

Liquid-like states generally show a rapid mass exchange between dense and dilute phases as well as the intermolecular interactions within the dense phase [86]. Two physical order parameters were used to characterize the microscopic fluidity and distinguish between liquid- and solid-like states: the fluctuation extent of clustering degree parameter (*FE*) and the exchange rate of interaction parameter (*ER*). *FE* is defined by the root-mean-square fluctuation (RMSF) of the dense phase size, while *ER* is defined as the fraction of intermolecular interactions lost per unit time within the dense phase. Based on clustering degree, collapse degree, *FE,* and *ER*, the liquid-like state could be well probed by molecular simulation, which is consistent with experimental observation.

4.4.3 Simulation of Structural Propensity and Evolution

The ultimate goal of molecular simulation is to completely reconstruct the self-assembling process and the nanostructure observed in experiments. For the peptide self-assembly system, the ideal scenario is when the initial fully dispersed molecular model is established, and the peptides can spontaneously form anisotropic self-assembled nanostructure through their own diffusion and aggregation. In this scenario, the simulation can replace the experiment and become a powerful prediction tool for self-assembly. However, due to computational limitations, the time scale and spatial scale of molecular simulation are much smaller than the experimental scale. Nevertheless, molecular simulation can still preliminarily determine the self-assembly propensity of peptides and find some structural characteristics of microstructures in simulation related to self-assembled macrostructures in experiments.

4.4.3.1 Aggregation Process

Benefiting from the high computational efficiency of the simplified model, CG-MD simulations could reach up to large systems and long times, allowing for the reproduction of the complete self-assembling process and the spontaneous formation of various nanostructures [87, 88]. As mentioned in Section 4.2.2, high AP values generally correlate with the formation of ordered nanostructures. By directly observing the morphology of molecular aggregation, the relationship between microscopic aggregation behavior and macroscopic assembly structure can be preliminarily established.

In a CG-MD simulation of FF, the process of spontaneously assembling into spherical vesicles and nanotubes was observed [87]. Furthermore, other ordered nanoarchitectures such as planar bilayers and various shapes of vesicle-like structures (toroid, ellipsoid, discoid, and pot-shaped vesicles) were also found in the CG-MD.

4.4.3.2 Molecular Alignment

In addition to simulating the spontaneous formation process of the self-assembled nanostructure, molecular simulations can study self-assembly by establishing a building block based on a specific secondary structure, based on the secondary structure measured by experiment [3]. The stability could be evaluated by monitoring the topological changes that occur in the molecular simulation, using the root mean square displacement (RMSD), radius of gyration (RG), and solvent-accessible surface area (SASA) analysis [89]. In these cases, the AA model with higher accuracy is usually adopted for the simulation.

Sometimes, it is hard to determine the intermolecular alignment between peptides by solely analyzing the secondary structure. Take β-sheet as an example, the residue shift between adjacent strands and the orientation of the β-sheet (parallel or antiparallel) are generally unknown. A simple and effective method is to build blocks with different packing modes and then calculate their energy and evaluate their stability. In this way, the intermolecular alignment with the lowest energy could be obtained. The self-assembly structure of these building blocks can be predicted by observing their structural propensity by measuring the twisting angle of building blocks (Figure 4.8).

When there is no experimental data on the secondary structure, some acceleration methods as mentioned in Section 2.3 are often needed to calculate the molecular conformation and the intermolecular alignment with the minimum energy. Taking the contact degree between strands as the coordinate [24, 69], the free energy landscape (FEL) could be calculated to analyze the propensity of parallel and antiparallel intermolecular alignments as well as the residue shifts between strands.

4.4.4 Simulation of Gelation

Self-assembled peptide hydrogels are promising biomaterials and have been widely used in many fields such as tissue engineering, drug delivery, and biosensing [90, 91]. The structural element of peptide hydrogels is the self-assembled nanofibers. These nanofibers generally entangle or cross-link with each other, resulting in three-dimensional (3D) networks that entrap large amounts of solvent [92–95]. The application of molecular simulation in peptide hydrogel systems is aimed to explore the relationship among molecular chemical structure, self-assembly process, and assembly morphology, to determine the ability of supramolecular gel. By improving the understanding of the hierarchical structural details of peptide hydrogels, the gelation propensity can be predicted. This section introduces the application of different molecular simulation methods in the research of nanofiber structure and nanofiber network formation, which can also be regarded as a part of the study on the simulation of the structural evolution of peptide systems.

4.4.4.1 Structural Propensity Toward Nanofiber

As mentioned in Section 4.3, based on the structural evolution and propensity simulation, we found that peptides could assemble to form 1D long supramolecular nanostructures with β-sheet secondary structure. Peptides with similar chemical structures could form different nanostructures. For example, KLVFFAE, KLV(Phg)(Phg)AE, and KLV(Cha)(Cha)AE could assemble into nanoribbons, nanofibers, and nanotubes, respectively [96]. When prealigned in multiple layers following the hierarchical self-assembly model mentioned in Section 4.1.1, the twisting angles follow an order of KLV(Phg)(Phg)AE > KLV(Cha)(Cha)AE > KLVFFAE, correlating to the

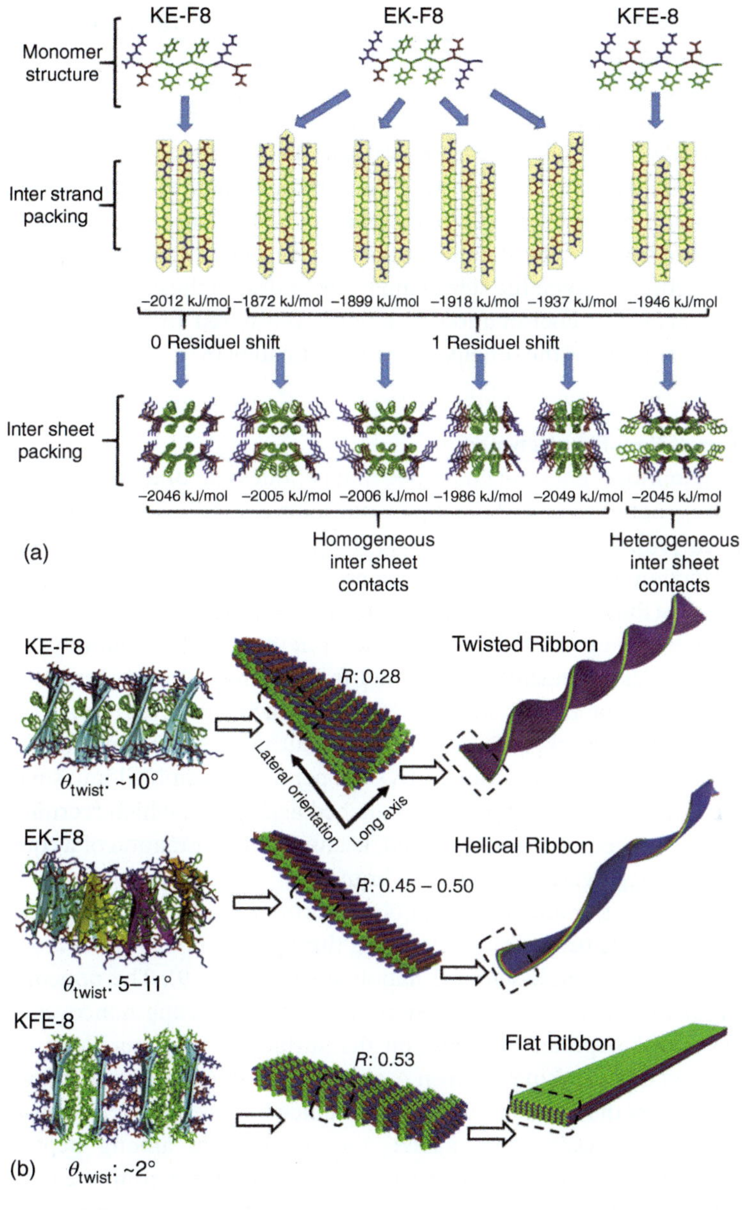

Figure 4.8 Molecular alignment and their structural propensity of building blocks. *Source:* Reproduced with permission from Ref. [3]/Elsevier Inc.

difference in the twisting nature of nanoribbon, nanotubes, and nanofibers. One important finding is that a high degree of twisting of the order could lead to the formation of the nanofiber architecture. The CG-MD simulation of another series of isomeric tripeptide sequences obtained similar results. Building blocks with different degrees of curved edge formed, correlating with the ribbon-like and cylindrical nanofibers in the experiment [97].

Another type of self-assembled nanofiber composed of long-chain peptide is the coiled-coil motif, which is based on α-helix secondary structure [50]. For such peptides, the task of molecular simulation is to evaluate the ability of single peptide molecules to form stable α-helix by analyzing the intramolecular H-bonds on the amide backbone. Additionally, it needs to simulate and calculate the interaction between α-helices and evaluate the stability of the coiled-coil motif, which requires analyzing the interaction between the side chains of peptides. In the coiled-coil system stabilized by metal ion coordination, it is also considered to use QM/MM method to study the binding of the α-helix with the metal ions [63].

4.4.4.2 Formation of Nanofiber Network

The formation of nanofibers does not guarantee that peptides can undergo gelation because it requires generating cross-linked polymer matrices that can absorb high amounts of solvent [98]. Therefore, to simulate the growth of the nanofibril network and the interaction between nanofibrils, using MD simulation is necessary. In the study of Fmoc–Lys(Fmoc)–Asp, the complete formation process of the nanofiber network was simulated using CG-MD [99]. Staring from the initially randomly dispersed state, the peptides aggregated quickly and formed ordered oligomers, thin nanofibers, and branched nanofibers step by step. When tracing the evolution of the growth of the nanofiber network, different evolution routines could be detected. In the CG-MD simulation of Fmoc–K(Fmoc)–RGD, the peptides assembled into small clusters at first. These clusters fused into big branched aggregates, which recruited the other small ones and became loosely packed, leading to the formation of a compact entangled fibrous aggregate with multiple branches.

Besides the spontaneous branched morphology formation, the growth of the nanofiber network can also result from entangling through the noncovalent cross-linking between preformed self-assembled nanofibers (Figure 4.9). The noncovalent cross-linking between nanofibers is generally due to the strong noncovalent interaction between the peptides molecule on the surface of the nanofibers. To study the noncovalent cross-linking between nanofibers, building the model of two preassembled nanofibers that are placed close in distance has been proven to be an effective approach [100]. Using this model, the relationship among peptide sequence, charge density on the nanofiber surface, and interfiber affinity can be studied quantitatively [101], which could provide the information to determine the assembly morphology as well as the viscosity of peptide hydrogel.

4.4.5 Simulation of Chirality Transfer

4.4.5.1 Chirality of Single Molecule

As biological molecules, peptides generally have complex chirality. Since amino acids have chirality, peptides composed of multiple amino-acid residues usually

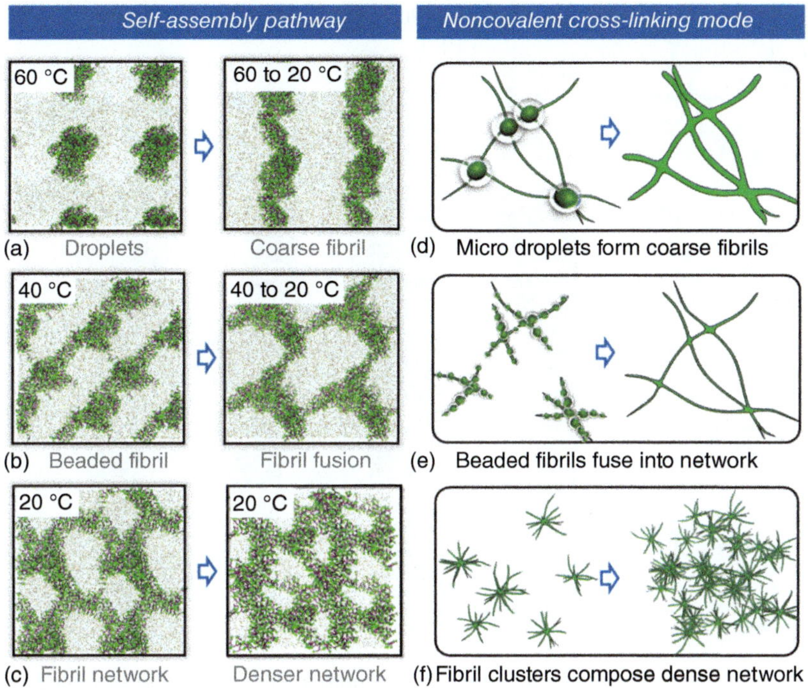

Figure 4.9 CGMD simulation of the noncovalent cross-linking modes through different self-assembly pathways. *Source:* Reproduced with permission from Ref. [80]/Elsevier Inc.

have multiple chirality centers, leading to the complex chirality of the nanostructures assembled by peptides.

Natural amino acids are generally L-type, which prefer to form a right-handed α-helix (α_R) conformation. As a result, a β-strand composed of L-type amino-acid residues represents right-handed twisting (as shown in Figure 4.10), which would induce a left-handed twist when they align together to form a β-sheet in MD simulation [103]. This is a complete chirality transfer from residue or peptide monomer to oligomer or building block. For some peptides, the strong interaction between side chains of peptides might hinder the twisting of the β-sheet, leading to a flat shape. In another case, although both experimental observations and theoretical studies have proved that right-handed twisting is energetically unfavorable for the β-sheet formed by L-peptides, there are still some nanostructures formed by natural peptides that show right-handed chirality [104–107], implying that the chirality transfer process might be more complex than expected.

4.4.5.2 Chirality of Building Blocks

When all L-type amino-acid residues are replaced by D-type ones, the resulting D-peptides usually form right-handed nanostructures based on the α_L conformation of residues [108, 109]. However, when only some L-type amino-acid residues are replaced by D-type ones, the chirality transfer between residue and peptide can be

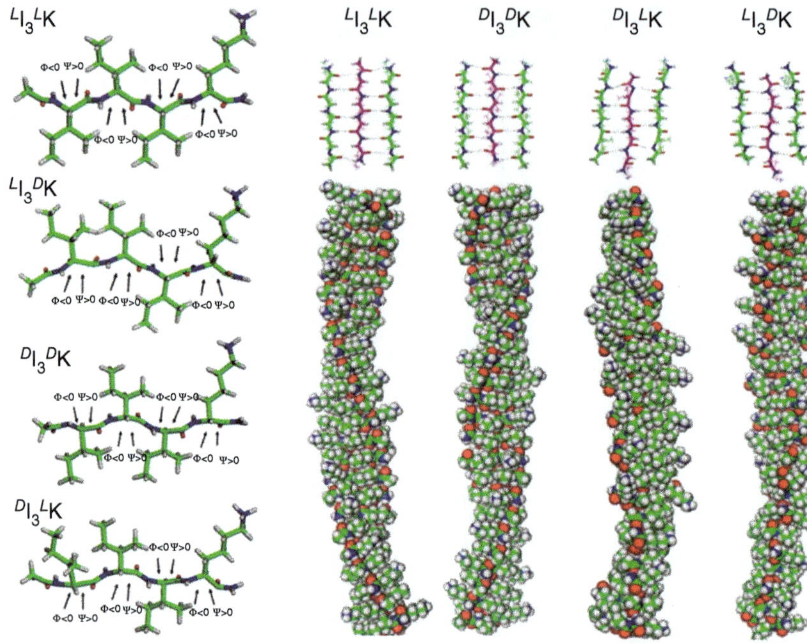

Figure 4.10 The structural relation full-atom MD calculation of peptide self-assembled structures. *Source:* Reproduced with permission from Ref. [102]/American Chemical Society.

altered, resulting in an unusual twisting propensity of a single molecule (as shown in Figure 4.10). Heterochiral peptides composed of both L- and D-type amino acids may exhibit stronger or weaker self-assembling ability [110, 111]. More importantly, the chirality transfer process becomes unpredictable. In the study of the I_3K peptide, $^{L}I_3{}^{L}K$ and $^{D}I_3{}^{L}K$ formed left-handed twisted ribbons, while $^{D}I_3{}^{D}K$ and $^{L}I_3{}^{D}K$ formed right-handed twisted ribbons (Figure 4.10) [102].

Benefiting from the high sampling of REMD simulation, the conformational propensity of the single molecule can be obtained as mentioned in Section 4.3.1, providing important information to correlate the chain geometry and molecular alignment. Aligned with the nontypical H-bonding orientation with unique main chain geometry and different residue shift, the building block of the $^{L}I_3{}^{D}K$ sheet shows a right-handed twisting similar to that of $^{D}I_3{}^{D}K$, which is opposite to that of $^{L}I_3{}^{L}K$ (Figure 4.10) [69, 102]. This result demonstrates the power of MD simulation in studying the chirality transfer process in a hierarchical self-assembly system.

4.5 Conclusion and Outlook

In this chapter, the application of molecular simulation in peptide supramolecular self-assembly systems has been summarized. Various molecular simulation methods probe microscopic insights into the self-assembled nanostructures at the microscopic scale, analyzing the intermolecular interactions that drive the self-assembly, thus becoming powerful supplements to experimental research.

To further improve the molecular simulation and prediction of peptide self-assembly, some challenges are still present ahead. First, it is necessary to improve the molecular model of force fields to balance the efficiency and accuracy of the simulation. Second, a peptide self-assembly database needs to be developed to collect information on the relationship among molecular structure, simulated morphology, and experimental characterization. Finally, Chemoinformatics tools need to be combined with molecular simulation to improve the accuracy of the prediction of properties of self-assembly.

Acknowledgments

This work was financially supported by the National Science Fund for Distinguished Young Scholars of China (No. 22025207), National Natural Science Foundation of China (Project Nos. 22372173, 51904322, 22172172, 22072154, and 22102187), Innovation Research Community Science Fund (No.21821005), the Youth Innovation Promotion Association of CAS (Grant No. 2022049), the Chinese Postdoctoral Science Foundation (Grant No. 2021M693205), and the Natural Science Foundation of Hebei Province (Nos. B2020103036 and B2020103025).

References

1 Aggeli, A., Nyrkova, I.A., Bell, M. et al. (2001). Hierarchical self-assembly of chiral rod-like molecules as a model for peptide β-sheet tapes, ribbons, fibrils, and fibers. *Proc. Natl. Acad. Sci. USA* 98: 11857–11862.

2 Mart, R., Osborne, R., Stevens, M., and Ulijn, R. (2006). Peptide-based stimuli-responsive biomaterials. *Soft Matter* 2: 822–835.

3 Zhou, P., Deng, L., Wang, Y. et al. (2016). Different nanostructures caused by competition of intra- and inter-β-sheet interactions in hierarchical self-assembly of short peptides. *J. Colloid Interface Sci.* 464: 219–228.

4 Tao, K., Wang, J., Zhou, P. et al. (2011). Self-assembly of short Aβ(16−22) peptides: effect of terminal capping and the role of electrostatic interaction. *Langmuir* 27: 2723–2730.

5 Zhao, Y., Wang, J., Deng, L. et al. (2013). Tuning the self-assembly of short peptides via sequence variations. *Langmuir* 29: 13457–13464.

6 Adamcik, J., Castelletto, V., Bolisetty, S. et al. (2011). Direct observation of time-resolved polymorphic states in the self-assembly of end-capped heptapeptides. *Angew. Chem. Int. Ed.* 50: 5495–5498.

7 Adamcik, J. and Mezzenga, R. (2018). Amyloid polymorphism in the protein folding and aggregation energy landscape. *Angew. Chem. Int. Ed.* 57: 8370–8382.

8 Yuan, C., Levin, A., Chen, W. et al. (2019). Nucleation and growth of amino acid and peptide supramolecular polymers through liquid–liquid phase separation. *Angew. Chem.* 131: 18284–18291.

9 Burke, K.A., Janke, A.M., Rhine, C.L., and Fawzi, N.L. (2015). Residue-by-residue view of in vitro FUS granules that bind the C-terminal domain of RNA polymerase II. *Mol. Cell* 60: 231–241.

10 Wang, Y., Bunce, S.J., Radford, S.E. et al. (2019). Thermodynamic phase diagram of amyloid-β (16–22) peptide. *Proc. Natl. Acad. Sci. USA* 116: 2091–2096.

11 Dignon, G.L., Zheng, W., Kim, Y.C. et al. (2018). Sequence determinants of protein phase behavior from a coarse-grained model. *PLoS Comp. Biol.* 14: e1005941.

12 Nagarajan, S., Rajadas, J., and Malar, E.J.P. (2010). Density functional theory analysis and spectral studies on amyloid peptide A[beta](28-35) and its mutants A30G and A30I. *J. Struct. Bio.* 170: 439–450.

13 Tuttle, T. (2015). Computational approaches to understanding the self-assembly of peptide-based nanostructures. *Israel J. Chem.* 55: 724–734.

14 Karplus, M. (2003). Molecular dynamics of biological macromolecules: a brief history and perspective. *Biopolymers* 68: 350–358.

15 Alder, B.J. and Wainwright, T.E. (1959). Studies in molecular dynamics. I. General method. *J. Chem. Phys.* 31: 459–466.

16 Monti, S., Corozzi, A., Fristrup, P. et al. (2013). Exploring the conformational and reactive dynamics of biomolecules in solution using an extended version of the glycine reactive force field. *Phys. Chem. Chem. Phys.* 15: 15062–15077.

17 Martin, M.G. (2006). Comparison of the AMBER, CHARMM, COMPASS, GROMOS, OPLS, TraPPE and UFF force fields for prediction of vapor–liquid coexistence curves and liquid densities. *Fluid Phase Equilib.* 248: 50–55.

18 Bellesia, G., Fedorov, M.V., and Timoshenko, E.G. (2008). Structural transitions in model beta-sheet tapes. *J. Chem. Phys.* 128: 195105–195107.

19 Bellesia, G. and Shea, J.-E. (2007). Self-assembly of beta-sheet forming peptides into chiral fibrillar aggregates. *J. Chem. Phys.* 126: 245104–245111.

20 Hansmann, U.H.E. (1997). Parallel tempering algorithm for conformational studies of biological molecules. *Chem. Phys. Lett.* 281: 140–150.

21 Sugita, Y. and Okamoto, Y. (1999). Replica-exchange molecular dynamics method for protein folding. *Chem. Phys. Lett.* 314: 141–151.

22 Park, S.-B., Yoon, J.-S., Jang, S.-M. et al. (2012). Computational study on oligomer formation of fibril-forming peptide of α-synuclein. *Bull. Korean Chem. Soc.* 33: 848–854.

23 Zhang, J., Qin, M., and Wang, W. (2006). Folding mechanism of β-hairpins studied by replica exchange molecular simulations. *Proteins: Struct. Funct. Bioinform.* 62: 672–685.

24 Zhou, P., Deng, L., Wang, Y. et al. (2016). Interplay between intrinsic conformational propensities and intermolecular interactions in the self-assembly of short surfactant-like peptides composed of leucine/isoleucine. *Langmuir* 32: 4662–4672.

25 Gnanakaran, S., Nussinov, R., and Garcia, A.E. (2006). Atomic-level description of amyloid β-dimer formation. *J. Am. Chem. Soc.* 128: 2158–2159.

26 Vitagliano, L., Esposito, L., Pedone, C., and De Simone, A. (2008). Stability of single sheet GNNQQNY aggregates analyzed by replica exchange molecular dynamics: antiparallel versus parallel association. *Biochem. Biophys. Re. Comm.* 377: 1036–1041.

27 De Simone, A., Esposito, L., Pedone, C., and Vitagliano, L. (2008). Insights into stability and toxicity of amyloid-like oligomers by replica exchange molecular dynamics analyses. *Biophys. J.* 95: 1965–1973.

28 Deng, L., Zhou, P., Zhao, Y. et al. (2014). Molecular origin of the self-assembled morphological difference caused by varying the order of charged residues in short peptides. *J. Phys. Chem. B* 118: 12501–12510.

29 Proctor, E.A., Ding, F., and Dokholyan, N.V. (2011). Discrete molecular dynamics. *WIREs Comput. Mol. Sci.* 1: 80–92.

30 Schmidt, R.K., Teo, B., and Brady, J.W. (1995). Use of umbrella sampling in the calculation of the potential of mean force for maltose in vacuum from molecular dynamics simulations. *J. Phys. Chem.* 99: 11339–11343.

31 Kästner, J. (2011). Umbrella sampling. *WIRes: Comput. Mol. Sci.* 1: 932–942.

32 Roux, B. (1995). The calculation of the potential of mean force using computer simulations. *Comput. Phys. Commun.* 91: 275–282.

33 Deng, L., Zhao, Y., Zhou, P. et al. (2017). Anisotropic formation mechanism and nanomechanics for the self-assembly process of cross-β peptides. *Chin. Phys. B* 26: 128701.

34 Wolf, M.G., Jongejan, J.A., Laman, J.D., and de Leeuw, S.W. (2008). Rapid free energy calculation of peptide self-assembly by REMD umbrella sampling. *J. Phys. Chem. B* 112: 13493–13498.

35 Arunan, E., Desiraju, G.R., Klein, R.A. et al. (2011). Definition of the hydrogen bond (IUPAC recommendations 2011). *Pure Appl. Chem.* 83: 1637–1641.

36 Sweetman, A.M., Jarvis, S.P., Sang, H. et al. (2014). Mapping the force field of a hydrogen-bonded assembly. *Nat. Commun.* 5: 3931.

37 Van Der Spoel, D., Lindahl, E., Hess, B. et al. (2005). GROMACS: fast, flexible, and free. *J. Comput. Chem.* 26: 1701–1718.

38 Hu, Y., Lin, R., Zhang, P. et al. (2016). Electrostatic-driven lamination and untwisting of β-sheet assemblies. *ACS Nano* 10: 880–888.

39 Caplan, M., Moore, P., Zhang, S. et al. (2000). Self-assembly of a [beta]-sheet protein governed by relief of electrostatic repulsion relative to van der Waals attraction. *Biomacromolecules* 1: 627–631.

40 Brevik, I., Marachevsky, V.N., and Milton, K.A. (1999). Identity of the van der Waals force and the Casimir effect and the irrelevance of these phenomena to sonoluminescence. *Phys. Rev. Lett.* 82: 3948–3951.

41 Chandler, D. (2005). Interfaces and the driving force of hydrophobic assembly. *Nature* 437: 640–647.

42 Gelain, F., Luo, Z., and Zhang, S. (2020). Self-assembling peptide EAK16 and RADA16 nanofiber scaffold hydrogel. *Chem. Rev.* 120: 13434–13460.

43 Xu, H., Wang, Y., Ge, X. et al. (2010). Twisted nanotubes formed from ultrashort amphiphilic peptide I3K and their templating for the fabrication of silica nanotubes. *Chem. Mater.* 22: 5165–5173.

44 Jia, D., Tao, K., Wang, J. et al. (2011). Dynamic adsorption and structure of interfacial bilayers adsorbed from lipopeptide surfactants at the hydrophilic silicon/water interface: effect of the headgroup length. *Langmuir* 27: 8798–8809.

45 Frederix, P.W.J.M., Scott, G.G., Abul-Haija, Y.M. et al. (2015). Exploring the sequence space for (tri-)peptide self-assembly to design and discover new hydrogels. *Nat. chem.* 7: 30–37.

46 Sinnokrot, M.O., Valeev, E.F., and Sherrill, C.D. (2002). Estimates of the ab initio limit for π–π interactions: the benzene dimer. *J. Am. Chem. Soc.* 124: 10887–10893.

47 Cockroft, S.L., Hunter, C.A., Lawson, K.R. et al. (2005). Electrostatic control of aromatic stacking interactions. *J. Am. Chem. Soc.* 127: 8594–8595.

48 Rashkin, M.J. and Waters, M.L. (2002). Unexpected substituent effects in offset π–π stacked interactions in water. *J. Am. Chem. Soc.* 124: 1860–1861.

49 Guo, K., Zhang, H., Sun, J., and Yuan, S. (2015). Molecular dynamics simulation on self-assembly of Fmoc-FF dipeptide. *Chem. J. Chinese U.* 36: 2171–2178.

50 Smith, A.M., Banwell, E.F., Edwards, W.R. et al. (2006). Engineering increased stability into self-assembled protein fibers. *Adv. Func. Mater.* 16: 1022–1030.

51 Vácha, R. and Frenkel, D. (2011). Relation between molecular shape and the morphology of self-assembling aggregates: a simulation study. *Biophys. J.* 101: 1432–1439.

52 Woolfson, D.N. and Alber, T. (1995). Predicting oligomerization states of coiled coils. *Protein Sci.* 4: 1596–1607.

53 Saladino, G., Pieraccini, S., Rendine, S. et al. (2011). Metadynamics study of a β-hairpin stability in mixed solvents. *J. Am. Chem. Soc.* 133: 2897–2903.

54 Zhou, R. and Berne, B.J. (2002). Can a continuum solvent model reproduce the free energy landscape of a β-hairpin folding in water? *Proc. Natl. Acad. Sci. USA* 99: 12777–12782.

55 Schneider, J.P., Pochan, D.J., Ozbas, B. et al. (2002). Responsive hydrogels from the intramolecular folding and self-assembly of a designed peptide. *J. Am. Chem. Soc.* 124: 15030–15037.

56 Kretsinger, J.K., Haines, L.A., Ozbas, B. et al. (2005). Cytocompatibility of self-assembled β-hairpin peptide hydrogel surfaces. *Biomaterials* 26: 5177–5186.

57 Pauling, L., Corey, R.B., and Branson, H.R. (1951). The structure of proteins: two hydrogen-bonded helical configurations of the polypeptide chain. *Proc. Natl. Acad. Sci. USA* 37: 205–211.

58 Kutchukian, P.S., Yang, J.S., Verdine, G.L., and Shakhnovich, E.I. (2009). All-atom model for stabilization of α-helical structure in peptides by hydrocarbon Staples. *J. Am. Chem. Soc.* 131: 4622–4627.

59 Oakley, M.G. and Hollenbeck, J.J. (2001). The design of antiparallel coiled coils. *Curr. Opin. Struct. Biol.* 11: 450–457.

60 Potekhin, S.A., Melnik, T.N., Popov, V. et al. (2001). De novo design of fibrils made of short α-helical coiled coil peptides. *Chem. Biol.* 8: 1025–1032.

61 Dong, H., Paramonov, S.E., and Hartgerink, J.D. (2008). Self-assembly of α-helical coiled coil nanofibers. *J. Am. Chem. Soc.* 130: 13691–13695.

62 Apostolovic, B., Danial, M., and Klok, H.-A. (2010). Coiled coils: attractive protein folding motifs for the fabrication of self-assembled, responsive and bioactive materials. *Chem. Soc. Rev.* 39: 3541–3575.

63 Li, J., Zhao, Y., Zhou, P. et al. (2020). Ordered nanofibers fabricated from hierarchical self-assembling processes of designed α-helical peptides. *Small* 16: 2003945.

64 Riek, R.P. and Graham, R.M. (2011). The elusive π-helix. *J. Struct. Bio.* 173: 153–160.

65 Tonlolo, C. and Benedetti, E. (1991). The polypeptide 310-helix. *Trends Biochem. Sci* 16: 350–353.

66 Pauling, L. and Corey, R.B. (1951). The pleated sheet, a new layer configuration of polypeptide chains. *Proc. Natl. Acad. Sci. USA* 37: 251–256.

67 Chothia, C. (1973). Conformation of twisted β-pleated sheets in proteins. *J. Mol. Biol.* 75: 295–302.

68 Zhou, P., Yuan, C., and Yan, X. (2022). Computational approaches for understanding and predicting the self-assembled peptide hydrogels. *Curr. Opin. Colloid Interface Sci.* 62: 101645.

69 Zhou, P., Wang, J., Wang, M. et al. (2019). Amino acid conformations control the morphological and chiral features of the self-assembled peptide nanostructures: young investigators perspective. *J. Colloid Interface Sci.* 548: 244–254.

70 Hopping, G., Kellock, J., Barnwal, R.P. et al. (2014). Designed α-sheet peptides inhibit amyloid formation by targeting toxic oligomers. *eLife* 3: e01681.

71 Zhou, P., Hu, X., Li, J. et al. (2022). Peptide self-assemblies from unusual α-sheet conformations based on alternation of d/l amino acids. *J. Am. Chem. Soc.* 144: 21544–21554.

72 Han, S., Cao, S., Wang, Y. et al. (2011). Self-assembly of short peptide amphiphiles: the cooperative effect of hydrophobic interaction and hydrogen bonding. *Chem. Eur. J.* 17: 13095–13102.

73 Smith, L.J., Fiebig, K.M., Schwalbe, H., and Dobson, C.M. (1996). The concept of a random coil: residual structure in peptides and denatured proteins. *Fold Des.* 1: R95–R106.

74 Baldwin, R.L. and Zimm, B.H. (2000). Are denatured proteins ever random coils? *Proc. Natl. Acad. Sci. USA* 97: 12391–12392.

75 Jha, A.K., Colubri, A., Freed, K.F., and Sosnick, T.R. (2005). Statistical coil model of the unfolded state: resolving the reconciliation problem. *Proc. Natl. Acad. Sci. USA* 102: 13099–13104.

76 Selinger, R.L.B., Selinger, J.V., Malanoski, A.P., and Schnur, J.M. (2004). Shape selection in chiral self-assembly. *Phys. Rev. Lett.* 93: 158103.

77 Yan, Y., Xiong, W., Li, X. et al. (2007). Molecular packing parameter in Bolaamphiphile solutions: adjustment of aggregate morphology by modifying the solution conditions. *J. Phys. Chem. B* 111: 2225–2230.

78 Nagarajan, R. (2002). Molecular packing parameter and surfactant self-assembly: the neglected role of the surfactant tail. *Langmuir* 18: 31–38.

79 Wang, J., Han, S., Meng, G. et al. (2009). Dynamic self-assembly of surfactant-like peptides A6K and A9K. *Soft Matter* 5: 3870–3878.

80 Zhou, P., Xing, R., Li, Q. et al. (2023). Steering phase-separated droplets to control fibrillar network evolution of supramolecular peptide hydrogels. *Matter* 6: 1945–1963.

81 Yuan, C., Levin, A., Chen, W. et al. (2019). Nucleation and growth of amino acid and peptide supramolecular polymers through liquid–liquid phase separation. *Angew. Chem. Int. Ed.* 58: 18297–18297.

82 Fawzi, N.L., Parekh, S.H., and Mittal, J. (2021). Biophysical studies of phase separation integrating experimental and computational methods. *Curr. Opin. Struct. Biol.* 70: 78–86.

83 Choi, J.-M., Dar, F., and Pappu, R.V. (2019). LASSI: a lattice model for simulating phase transitions of multivalent proteins. *PLoS Comp. Biol.* 15: e1007028.

84 Das, S., Eisen, A., Lin, Y.-H., and Chan, H.S. (2018). A lattice model of charge-pattern-dependent polyampholyte phase separation. *J. Phys. Chem. B* 122: 5418–5431.

85 Dignon, G.L., Zheng, W., and Mittal, J. (2019). Simulation methods for liquid–liquid phase separation of disordered proteins. *Curr. Opin. Chem. Eng.* 23: 92–98.

86 Tang, Y., Bera, S., Yao, Y. et al. (2021). Prediction and characterization of liquid-liquid phase separation of minimalistic peptides. *Cell Reports Phys. Sci.* 2: 100579.

87 Guo, C., Luo, Y., Zhou, R., and Wei, G. (2012). Probing the self-assembly mechanism of diphenylalanine-based peptide nanovesicles and nanotubes. *ACS Nano* 6: 3907–3918.

88 Batra, R., Loeffler, T.D., Chan, H. et al. (2022). Machine learning overcomes human bias in the discovery of self-assembling peptides. *Nat. chem.* 14: 1427–1435.

89 Noble Jesus, C., Evans, R., Forth, J. et al. (2021). Amphiphilic histidine-based oligopeptides exhibit pH-reversible fibril formation. *ACS Macro Lett.* 10: 984–989.

90 Zhao, X., Pan, F., Xu, H. et al. (2010). Molecular self-assembly and applications of designer peptide amphiphiles. *Chem. Soc. Rev.* 39: 3480–3498.

91 Matson, J.B., Zha, R.H., and Stupp, S.I. (2011). Peptide self-assembly for crafting functional biological materials. *Curr. Opin. Solid State Mater. Sci.* 15: 225–235.

92 Gelain, F., Luo, Z., and Zhang, S. (2021). Self-assembling peptide EAK16 and RADA16 nanofiber scaffold hydrogel. *Chem. Rev.* 121: 5093–5093.

93 Van Lommel, R., De Borggraeve, W.M., De Proft, F., and Alonso, M. (2021). Computational tools to rationalize and predict the self-assembly behavior of supramolecular gels. *Gels* 7: 87.

94 Li, J., Xing, R., Bai, S., and Yan, X. (2019). Recent advances of self-assembling peptide-based hydrogels for biomedical applications. *Soft Matter* 15: 1704–1715.

95 Wang, M., Zhang, Q., Jian, H. et al. (2020). Role of thermolysin in catalytic-controlled self-assembly of Fmoc-dipeptides. *CCS Chem.* 2: 317–328.

96 Wang, J., Tao, K., Zhou, P. et al. (2016). Tuning self-assembled morphology of the Aβ(16–22) peptide by substitution of phenylalanine residues. *Colloids Surf. B. Biointerfaces* 147: 116–123.

97 Chen, L., Feng, J., Yang, D. et al. (2019). Sequence isomerism-dependent self-assembly of glycopeptide mimetics with switchable antibiofilm properties. *Chem. Sci.* 10: 8171–8178.

98 Casalini, T. and Perale, G. (2019). From microscale to macroscale: nine orders of magnitude for a comprehensive modeling of hydrogels for controlled drug delivery. *Gels* 5: 28.

99 Chakraborty, P., Tang, Y., Yamamoto, T. et al. (2020). Unusual two-step assembly of a minimalistic dipeptide-based functional hypergelator. *Adv. Mater.* 32: 1906043.

100 Sather, N.A., Sai, H., Sasselli, I.R. et al. (2021). 3D printing of supramolecular polymer hydrogels with hierarchical structure. *Small* 17: 2005743.

101 Sasselli, I.R., Syrgiannis, Z., Sather, N.A. et al. (2022). Modeling interactions within and between peptide amphiphile supramolecular filaments. *J. Phys. Chem. B* 126: 650–659.

102 Wang, M., Zhou, P., Wang, J. et al. (2017). Left or right: how does amino acid chirality affect the handedness of nanostructures self-assembled from short amphiphilic peptides? *J. Am. Chem. Soc.* 139: 4185–4194.

103 Shamovsky, I.L., Ross, G.M., and Riopelle, R.J. (2000). Theoretical studies on the origin of β-sheet twisting. *J. Phys. Chem. B* 104: 11296–11307.

104 Rubin, N., Perugia, E., Goldschmidt, M. et al. (2008). Chirality of amyloid suprastructures. *J. Am. Chem. Soc.* 130: 4602–4603.

105 Ivan, U., Jozef, A., and Raffaele, M. (2013). Polymorphism complexity and handedness inversion in serum albumin amyloid fibrils. *Acs Nano* 7: 10465–10474.

106 Lara, C., Reynolds, N.P., Berryman, J.T. et al. (2014). ILQINS hexapeptide, identified in lysozyme left-handed helical ribbons and nanotubes, forms right-handed helical ribbons and crystals. *J. Am. Chem. Soc.* 136: 4732–4739.

107 Volpatti, L.R., Michele, V., Dobson, C.M., and Knowles, T.P.J. (2013). A clear view of polymorphism, twist, and chirality in amyloid fibril formation. *ACS Nano* 7: 10443–10448.

108 Tomoyuki, K., Miho, M., and Nobuyuki, H. (2005). Structural control of self-assembled nanofibers by artificial beta-sheet peptides composed of D- or L-isomer. *J. Am. Chem. Soc.* 127: 17596–17597.

109 Wadai, H., Yamaguchi, K.i., Takahashi, S. et al. (2005). Stereospecific amyloid-like fibril formation by a peptide fragment of β2-microglobulin. *Biochemistry* 44: 157–164.

110 Luo, Z., Wang, S., and Zhang, S. (2011). Fabrication of self-assembling d-form peptide nanofiber scaffold d-EAK16 for rapid hemostasis. *Biomaterials* 32: 2013–2020.

111 Marchesan, S., Easton, C.D., Styan, K.E. et al. (2014). Chirality effects at each amino acid position on tripeptide self-assembly into hydrogel biomaterials. *Nanoscale* 6: 5172–5180.

5

Stimuli-responsive Structural Transformations of Peptide Supramolecular Gels

Simona Bianco and Dave J. Adams

School of Chemistry, University of Glasgow, Glasgow, UK

5.1 Introduction

Gels constitute a type of soft material made of three-dimensional networks that entrap a liquid, such as water or organic solvents [1]. When gels are formed from water, they are referred to as hydrogels [2], while in the case of organic solvents, they are termed organogels [3]. These materials can retain large amounts of liquid, while still maintaining their three-dimensional structure [4]. Because of this, gels exhibit viscoelastic behavior: elasticity is given by their solid network, while the solution trapped in the pores gives rise to the viscous behavior [5]. Gels have been of particular interest for a broad range of applications, such as tissue engineering [6], medicine [7], drug delivery [8], cell culture [9, 10], and environmental remediation [11].

A variety of molecules can be used to synthesize gels, among which are naturally occurring and synthetic polymers [12]. Based on the forces holding the networks together, gels can be classified as "chemical" gels or "physical" gels [13]. In chemical gels, the polymer chains are linked together through covalent cross-linking (Figure 5.1a). Conversely, physical gels, or supramolecular gels, are held by noncovalent forces, such as hydrogen bonding, π–π stacking, and hydrophobic interactions between the monomers making up the chains (Figure 5.1b) [14]. Thus, compared to permanently cross-linked chemical gels, physical gels show lower mechanical stability and tend to break at lower strains. However, the presence of these noncovalent interactions means that the networks in physical gels can be reversed by application of external energy [15]. Supramolecular gels formed by the self-assembly of small molecules, namely low molecular weight gels (LMWGs), have gained significant attention in recent years [16].

Peptide Self-Assembly and Engineering: Fundamentals, Structures, and Applications, First Edition. Edited by Xuehai Yan.

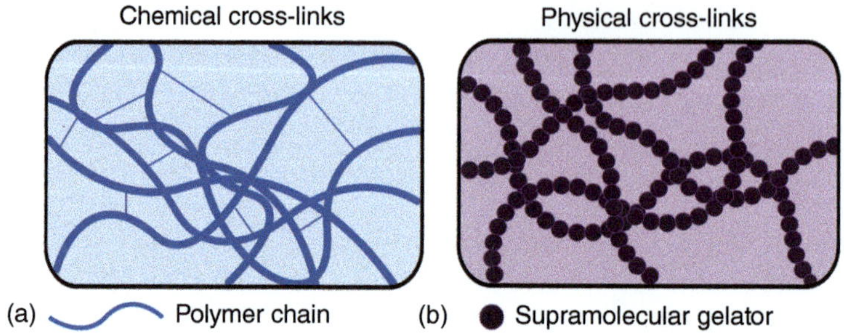

Figure 5.1 Schematic representing the underlying networks for (a) chemical gels formed by covalent cross-links and (b) physical gels formed by noncovalent interactions. Source: Simona Bianco.

To form a supramolecular gel, the gelator molecules are first dissolved or suspended in solution (Figure 5.2, left). Upon addition of a gelation trigger, self-assembly into long fibers in the nanometer scale via noncovalent interactions is promoted (Figure 5.2, center). In turn, the fibers will entangle to form a three-dimensional network at the micrometer scale, which entraps the solvent or water, resulting in a self-supporting material which is stable upon inversion (Figure 5.2, right) [16, 17]. The main driving forces leading self-assembly are typically strong hydrogen bonding, π–π stacking, donor–acceptor interactions, metal coordination, hydrophobic interactions, and van der Waals interactions [18].

A variety of triggers can be used to form supramolecular gels, such as temperature [19], pH changes [20], solvent changes [21], addition of salts [22], and enzymatic reactions [23]. Although different, all of these methods rely on changing the environment around the gelator, reducing its solubility and, thus, inducing self-assembly. It is worth noting that the choice of gelation trigger affects the self-assembly process, meaning that different gelation conditions lead to different final mechanical properties of the gels [24–26].

Due to the noncovalent nature of the interactions within supramolecular gels, the main advantage of such materials is that gel formation is reversible and highly

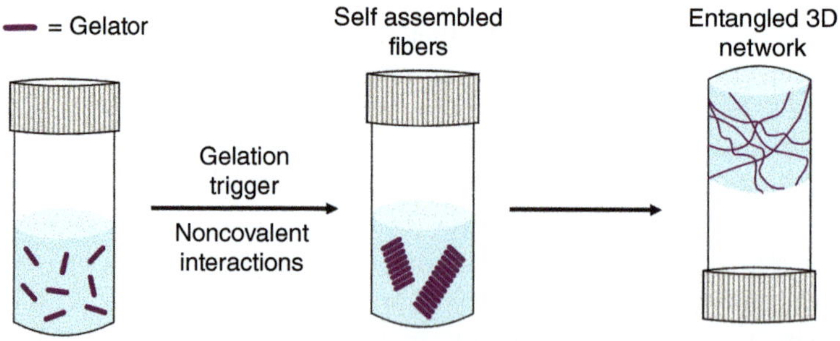

Figure 5.2 Cartoon representing the gelation mechanism for supramolecular gels. When a gelation trigger is applied, the gelator molecules undergo self-assembly, forming fibers (center). The fibers will entangle and grow, forming a three-dimensional network that traps the solvent. A self-supporting gel is formed as a result (right). Source: Simona Bianco.

sensitive to a variety of external stimuli, making them "smart" [15]. For instance, hydrogen-bonding motifs can be influenced by pH, often leading to a collapse of the gel upon addition of acid or base. Furthermore, reversible changes in aggregation, volume, or phase can be obtained through changes in heat, light, electrical fields, or mechanical stress [27–29]. As a result, these materials have been particularly interesting for applications in sensing, drug release, actuation, or self-healing. Among the various gelators that can lead to stimuli-responsive gels, gelators based on amino acids or peptides have attracted significant interest [30].

5.2 Designing Peptide-based Supramolecular Gelators

Peptide molecules provide an ideal candidate for the design of stimuli-responsive supramolecular gels, due to their ability to adopt different secondary structures such as α-helix, coiled-coils, and β-sheets. These can then further self-assemble hierarchically to form higher order aggregates, making them particularly suited for supramolecular gel design [30]. Furthermore, the different amino acids that form peptide sequences provide a wide range of physical properties [31]. For instance, incorporation of various noncovalent interactions can be possible, such as hydrogen bonding from polar amino acids, hydrophobic interactions from nonpolar amino acids, π-stacking from aromatic amino acids, electrostatic interactions from charged side chains, or even covalent bond formation through disulfide bonds (Figure 5.3) [32]. Hence, rational design of molecules containing these functions can allow control over molecular organization in response to applied stimuli, fine-tuning the material properties based on its design.

During self-assembly, peptides can also adopt different nanostructure shapes, such as micellar aggregates, fibrillar structures (fibers and nanotubes), and vesicles (Figure 5.4) [33]. Micellar structures typically form spontaneously above a critical micelle concentration and can be spherical, worm-like, or disk shaped [34].

Based on these considerations and peptide science, several small peptide gelators have been designed to contain sequences predisposed to from specific supramolecular structures. Short peptide sequences are advantageous due to ease of synthesis, cost, and scalability as well as the fact that they are easier to modify [35]. Common

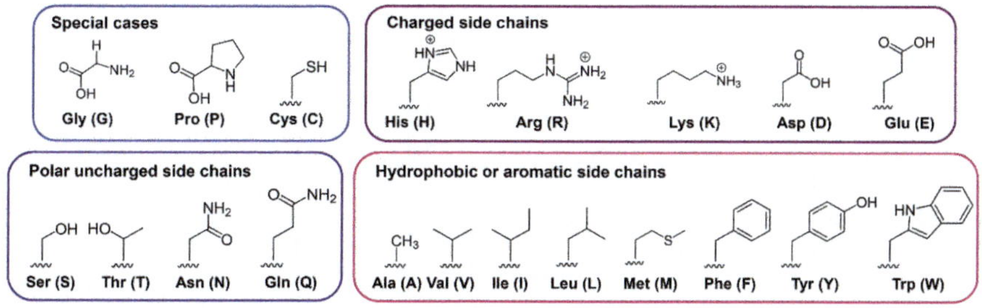

Figure 5.3 Representation of the side chains of the essential amino acids.
Source: Simona Bianco.

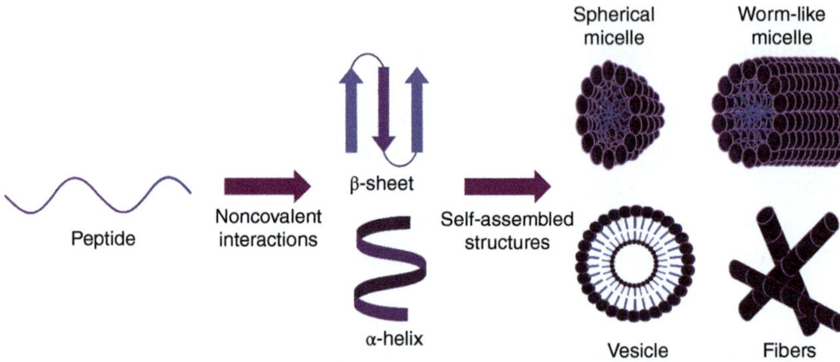

Figure 5.4 Different self-assembled structures that can be adopted by peptides through inter- or intramolecular noncovalent interactions. Source: Simona Bianco.

hydrogelators are peptide amphiphiles (PAs), where the peptide chains are typically functionalized with aliphatic or aromatic chains at the N-terminus (Figure 5.5) [36]. In such systems, self-assembly relies on the balance between hydrophilic and hydrophobic regions to ensure a balance between solubility and hydrophobicity.

The structure of a PA can typically be identified in four main regions: a hydrophobic region at the N-terminus, a linker, the peptide sequence, and the C-terminus, which can be either left unprotected or functionalized to impart bioactivity [37]. Firstly, the N-terminus is functionalized with a hydrophobic group, which typically drives self-assembly and gelation. For this purpose, long alkyl chains can be used. It is relevant to note that the chains must be of sufficient length, as shorter chains might not be hydrophobic enough to induce self-assembly and subsequent gelation [37, 38]. Another useful approach is to functionalize the N-terminus with aromatic groups. These allow an ulterior level of directionality, as self-assembly will further be directed by aromatic π–π stacking interactions. For self-assembling peptide-based hydrogels, the fluorenylmethoxycarbonyl (Fmoc) moiety is specifically common due to its role as a protecting group in peptide synthesis. The Fmoc group has been found to facilitate gelation for a variety of peptide-based systems [39–41]. Other synthetic aromatic groups have also been used to design peptide-based molecules capable of self-assembly such as naphthalene [42], anthracene [43], pyrene [44], or azobenzene [45].

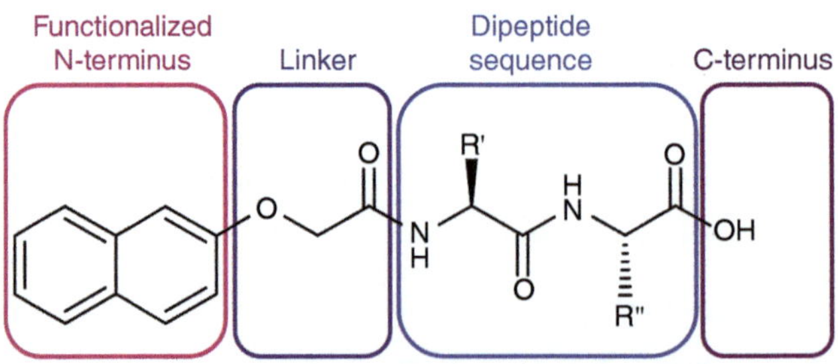

Figure 5.5 Representative structure of a peptide amphiphile with an aromatic group at the N-terminus. As a common example, a naphthalene dipeptide is shown. Source: Simona Bianco.

Groups such as naphthalenes can also be functionalized with different groups at various positions on the ring, affecting self-assembly and final gel properties [46].

The choice of linker between the peptide chain and the aromatic group has also been found to affect the self-assembly of such peptide-based gelators. For instance, Fleming et al. investigated the effect of the methoxycarbonyl linker for fluorenyl moieties on hydrogelation [47]. The methoxy linker of the Fmoc moiety was found to allow more extensive aromatic stacking interactions compared to fluorenyl analogues with shorter linkers. This was ascribed to fact that the shorter linkers restricted available molecular orientations in the molecule. Similarly, it was observed that the naphthoxy group allowed to form gels, while other naphthalene amphiphiles with different linkers failed to promote gelation [42]. This appeared to be a result of an increase in curvature of the molecules, suggesting that more linear geometries were necessary to allow optimal hydrogen bonding and π–π stacking. The methoxy group acts as an H-bond acceptor, which could further direct intermolecular interactions among the gelators.

As previously mentioned, the peptide sequence can have a significant effect on self-assembly, due to the wide range of possible molecular interactions that can be accessed. Therefore, they are an important aspect for the design of supramolecular gelators [35]. However, predicting whether a given molecule will form a hydrogel based on the amino-acid sequence is still challenging, and many gelators are often found by trial and error [48]. In part, this can be due to the fact that most gel systems are formed in metastable states, and thus, the self-assembly of the structures formed also depends on gelation pathway and environmental conditions [24].

Diphenylalanine is a popular choice due to Gazit's pioneering work [49], suggesting that the presence of aromatic side chains can facilitate gelation through π-stacking interactions. Aromatic residues in the peptide sequence have also been observed to increase the stiffness of the network formed as well as increase the rate of gelation [50]. Nevertheless, the order of the amino-acid sequences in the peptide backbone plays a substantial role in the process of self-assembly and, therefore, gelation. For instance, the Stupp group showed that the order of amino-acid sequence in a tetrapeptide amphiphile can significantly affect the type of 1D self-assembled structures by switching the position of two adjacent amino acids in the sequence [51]. Furthermore, simply switching the order of dipeptide sequences has been shown to lead to drastic changes and can, for example, turn gelator molecules into nongelators [52, 53].

5.3 Stimuli-responsive Changes in Peptide Supramolecular Gels

Generally, the self-assembled structures making up supramolecular gels tend not to change their properties with time once they are formed. However, due to the noncovalent interactions making up the network of the gels, these materials can show tunable and switchable behavior when exposed to external stimuli. This highly stimuli-responsive behavior means that the material properties can be modulated through external changes given by chemical, mechanical, optical, or electrical stimuli.

Upon application of a stimulus, changes in aggregation and morphology of the system are typically brought about as a response. Often, supramolecular gels collapse or undergo a gel–sol transition as a result of a reduction or increase in the solubility of the gelator molecules. However, it is possible for the molecules to reassemble into a different aggregate simultaneously, leading to materials that show interconversions between multiple physical states. Such "smart" materials that show changes such as swelling, shrinking, color, and extent of cross-linking are particularly interesting for applications in sensing, actuators, and drug delivery. Examples of supramolecular gels showing stimuli-responsive changes are summarized in Table 5.1.

Table 5.1 Representative examples of peptide-based supramolecular gel systems that show stimuli-responsive structural changes with different triggers.

Stimulus	Example gelator	Response	Ref.
Aging	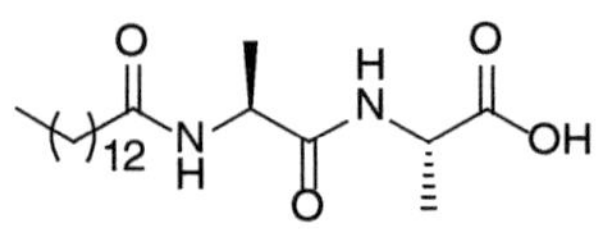	Turbidity change; increase in stiffness	[54]
Light	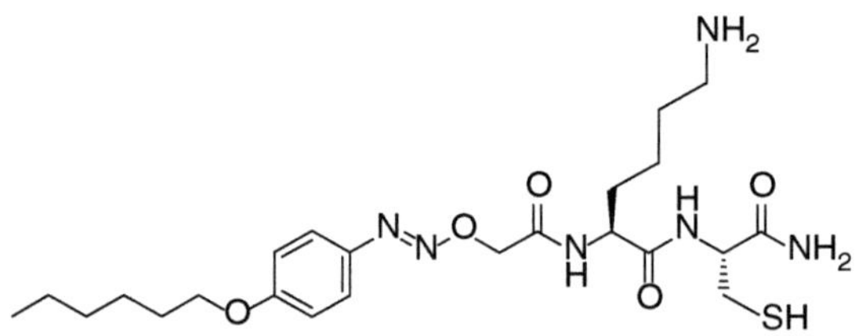	Volume decrease	[55]
Chemical	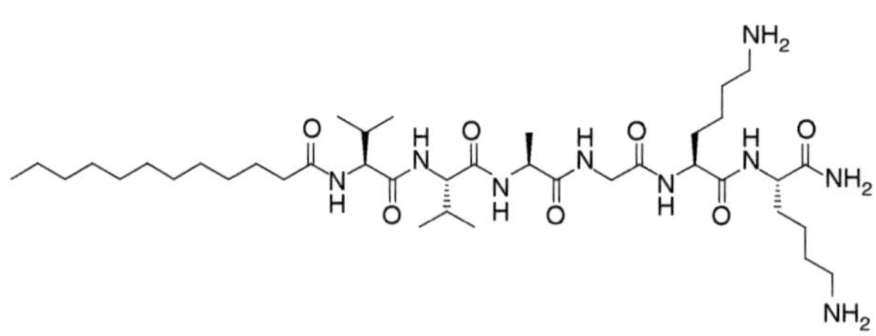	Color change; increase in stiffness.	[56]
Electrical	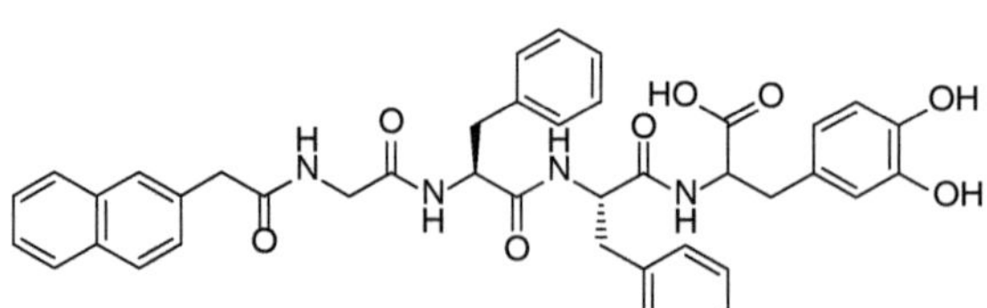	Volume decrease; increase in stiffness	[57]
Mechanical	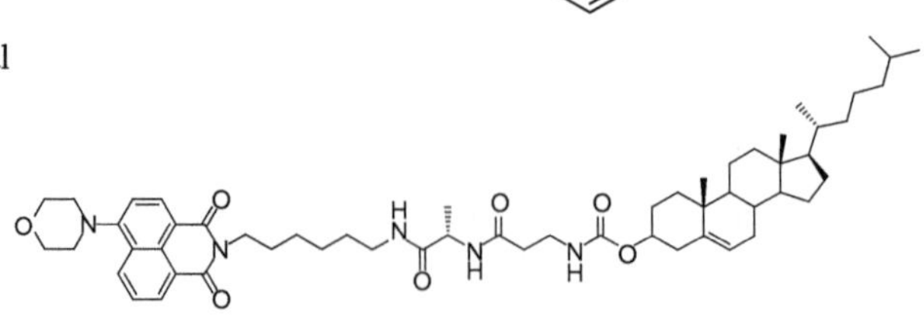	Morphological change; higher thermostability	[58]

It is relevant to note that the term "structural changes" is quite ambiguous when referring to these systems. The self-assembly process underpinning gelation is a hierarchical process, going from the intramolecular interactions between free gelator molecules at the angstrom length scale to the formation of 1D structures at the nanometer scale, which then form a three-dimensional network at the micrometer scale (Figure 5.6). It follows then that the term "structural change" can apply to several levels of changes. For instance, molecular changes can be brought about by a *trans* to *cis* transformation in a gelator molecule, while other changes might not affect the molecular scale but generate visual and bulk changes, for instance, in volume or color. Therefore, it is paramount to use appropriate techniques to characterize changes on each of these length scales [59]. On the molecular scale, infrared (IR) spectroscopy or nuclear magnetic resonance can be used to probe the interactions between molecular building blocks [60]. At the next length scale, the nanoscale 1D structures can be studied by microscopy, such as scanning electron microscopy (SEM) or atomic force microscopy (AFM) [61]. For these techniques, the gels need to be dried, which may lead to artifacts [62]. Because of this, small-angle X-ray or neutron scattering (SAXS or SANS) can provide more accurate complementary information on the 1D nanostructures in the wet gel state [63]. Finally, the macroscopic behavior from the gel network can be probed by rheology, giving information on the mechanical properties of the bulk gel [64].

Here, we discuss stimuli-responsive structural changes that occur across all different length scales accessed from hierarchical self-assembly. In the context of this chapter, we mainly focus on systems where such changes occur only as a transition between gel states without transiting through a solution phase. Hence, although being a common gelation trigger, temperature effects will not be discussed as most thermally responsive peptide-based supramolecular gel systems undergo gel–sol transitions upon heating [15]. Only one example of thermally induced structural change has been reported for a supramolecular polymer based on a glycosylated amino acid [65].

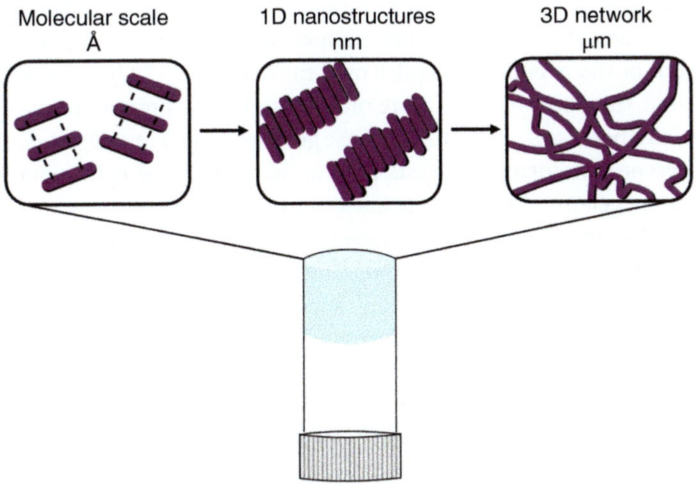

Figure 5.6 Cartoon representing the different length scales associated with self-assembly of small-molecule supramolecular gels.

5.3.1 Aging

Among the ways structural changes can occur within supramolecular gels, the process of aging plays a significant role. Although often overlooked, the effect of aging is crucial to determine the stability of these materials for their desired applications. Due to the dynamic nature of the self-assembly, the assembled network may rearrange over time or "self-correct" to a more homogeneous network. In particular, this can occur when the self-assembly process occurs under kinetic control and not in a thermodynamic minimum [66, 67]. As a result, the bulk properties of the gel can change over time, with changes in mechanical strength, visual appearance, and microstructure.

For instance, while studying a small library of small molecule hydrogelators, Shi et al. reported a phenylalanine-based compound that formed opaque gels at low pH [68]. The gel was found to turn into a clear hydrogel after aging for 10 days. It is hypothesized that the aging process aided the molecules to disperse and self-assemble over time. A similar unusual transition was also reported by the Adams group, where a 2-thiophene diphenylalanine LMWG was shown to go from a turbid gel to transparent over 3 days [69]. This change was accompanied by an increase in rheological properties compared to the turbid gel, suggesting a change in the bulk gel properties. SEM images before and after gel-to-gel transition show a morphological transition from large spherical structures to a fibrous network. The rearrangement of the fibers appeared to be triggered by an increased presence of carbonate, stressing the importance of conditions for specific applications.

Conversely, a time-dependent gradual transformation from transparent to turbid gel was observed for a peptide-based amphiphilic gelator molecule containing a long aliphatic chain at the N-terminus, **MAA** [54]. The gel was formed through a heat/cool process in a phosphate buffer solution. At a concentration of 4.5 mg/mL, the time-dependent transformation was observed across 2 days (Figure 5.7, left). However, the dynamic visual transition was found to be significantly dependent on gelator concentration, rate of cooling of the gelator solution, and time. SAXS and electron microscope images show that this transition occurred due to a compaction of the packing pattern of the gelator assemblies over time (Figure 5.7a,b). The data indicate that with time, the fibers become more ordered and show prominent twisting, leading to a fivefold increase in the stiffness of the gel with time. Time-dependent fluorescence with pyrene also further suggests that an increase in the hydrophobic environment around the gelator molecules occurs within the gel matrix.

Other than changes in turbidity, the hydrophobicity of gelator molecules can also lead to contraction of the gel over time, a phenomenon called "syneresis." Shrinking or swelling of materials can be important for many applications, such as purification of polluted water or controlled drug release. Shrinking in supramolecular hydrogels has been reported due to external stimuli, while "self-shrinking" hydrogels are seldom observed. Duraisamy et al. recently reported a self-shrinking hydrogel under physiological pH [70]. The LMWG consists of a fluorenylmethoxycarbonyl-functionalized β-L-phenylalanine (Fmoc-β-Phe) formed via a temperature switch. Here, the gel showed a reduction in size in 30 minutes, with expulsion of the buffer.

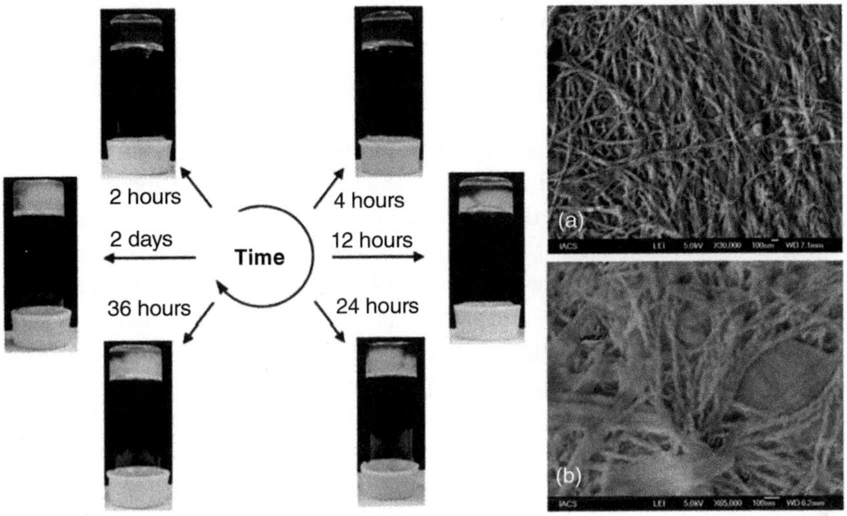

Figure 5.7 (a) Transparent to opaque gel-to-gel transformation of dipeptide **MAA** at a concentration of 4.5 mg/mL over time; (b) FE-SEM images showing twisted fibers for **MAA** dried gel freshly prepared (top) and after aging (bottom). Source: Reproduced with permission from Ref. [54]/Royal Society of Chemistry.

The morphological change in this system was proposed to be a minor rearrangement or adjustment from J-like aggregates to H-like aggregates with time. The shrinking was exploited to test the gel application for toxic dye removal in the presence of both anionic and cationic dyes. The Fmoc-β-Phe gel was shown to be able to trap cationic dyes efficiently, as >85% were trapped within the hydrogel.

Similarly, a triphenylalanine-based hydrogel capable of self-shrinking over time was reported for removal of organic dyes and lead (Pb^{2+}) ions from waste-water [71]. Here, shrinking occurred as a result of the rather high hydrophobic nature of the nanofibers obtained by self-assembly of the gelator. Over 7 days, the hydrophobic character of the gel increased, leading to fiber contraction and expulsion of water molecules. Interestingly, electron microscopy images showed that both the fresh and the shrunken gel contained intertwined nanofibers, indicating that syneresis in this system did not have a significant effect on the morphology of the gel. However, a negative effect on the final mechanical strength of the gel was observed after shrinking.

5.3.2 Light-triggered Response

Light can be a particularly suitable candidate for the design of stimuli-responsive materials as it can reversibly change the state of the hydrogel in a noninvasive manner. Furthermore, light allows to target specific areas in a spatially controlled manner, allowing formation of patterned gel surfaces [29]. Light-triggered dynamic changes in gels can be stimulated without the need to introduce any external chemical, making the system free from waste generation or contaminants. Such photoresponsive gel systems can be particularly useful for biomedical applications or optoelectronics [72]. Here, peptide-based supramolecular materials can be designed to incorporate photoresponsive moieties in their structure to form photoresponsive gelators or can interact noncovalently with photoresponsive molecules present within the gel matrix.

These photoresponsive groups can undergo reactions upon irradiation, such as bond cleavage, bond formation, or molecular switching. These bring about changes to the environment surrounding the gelator molecules. As a result, this causes a modulation of the supramolecular interactions at the molecular level. For this purpose, photoisomerizable molecules that switch between *trans*- and *cis*-forms are typically used, such as azobenzene and stilbene [18, 29]. Other photosensitive molecules that show light-induced ring opening or photodimerization such as spiropyrans [73], coumarin [74], and anthracene [43] have also been reported. In most of these cases, light-induced changes typically result in gel-to-sol transitions or gelation from a solution state. Cases in which light triggers cause an enhancement or change in the gel structure are instead less often reported.

5.3.2.1 Photodimerization

One powerful approach to induce structural changes using light is to make use of conjugated molecules that exhibit photodimerization. Molecules such as coumarins, alkenes, and anthracenes undergo photodimerization to highly stressed cyclobutene rings upon irradiation [75]. This covalent cross-linking between neighboring molecules can fix the orientation and conformation of different groups within the gel matrix. In general, due to the increase in molecule size, photodimerization can lead to a reduction in solubility and a disruption of the gel network. However, positive enhancement in the gel network can also occur due to a stabilization of the products through intermolecular forces.

For instance, a photoresponsive dipeptide-based LMWG showed an increase in rheological properties upon irradiation [76]. In this study, a coumarin moiety was attached to the N-terminus of a diphenylalanine peptide. Gelation of the LMWG was achieved through acidification using the hydrolysis of glucono-δ-lactone (GdL) as pH trigger and electrochemically using hydroquinone. Irradiation with a 365-nm LED caused photodimerization of the coumarin group between the neighboring fibers (Figure 5.8a), leading to an enhancement in the rheological properties of the gel. This change was hypothesized to be caused by stiffening within the fibers rather than fiber cross-linking, as the fiber diameter decreased from 42 to 39 nm. SEM images indicate that the network comprising the gel significantly changes after an hour of irradiation, showing less-defined structures (Figure 5.8d,e).

Similarly, stiffening of the individual nanofibers of hydrogels was observed for a 7-diethylamino coumarin-functionalized dilysine peptide [77]. Upon irradiation of the sample in just pure water with 365 nm light, an increase in the elastic storage modulus, G′, was observed from 20 to ~150 Pa. The data suggested formation of intrafiber cross-links due to the association between the individual molecules, resulting in stiffening of the fibers.

Additionally, ruthenium-complex-catalyzed photo-cross-linking has also been used to cross-link tyrosine-containing peptide supramolecular gels [78]. Here,

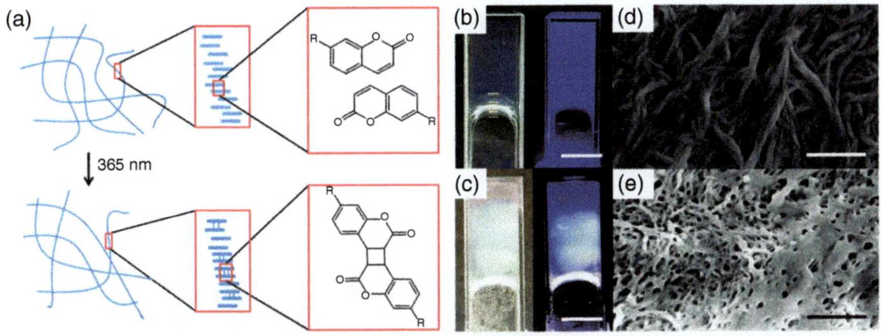

Figure 5.8 (a) Schematic showing the possible way in which dimerization can occur using UV light between coumarin dimers. Gels obtained using GdL before (b) and after (c) irradiation with 365 nm light. SEM images of GdL gel before d and after e irradiation. Scale bar represents 500 nm. Source: Adapted with permission from Ref. 76/Royal Society of Chemistry/CC BY 3.0.

photo-cross-linking seemed to stabilize the β-sheet structures of the self-assembled network by increasing entanglement of the peptide fibrils. This was further suggested by a decrease in pore size from ~100 to 10–50 nm.

5.3.2.2 Photoisomerization

Photoisomerization is a process by which molecules possessing *trans* or *cis* bonds can undergo *trans–cis* isomerization upon irradiation or vice versa. Such changes in structures are likely to cause a large effect on the packing between self-assembled molecules. For this purpose, azobenzene-containing low molecular weight gelators are typically used, where the azobenzene group can be directly incorporated in the short peptide sequences. Typically, due to dramatic changes in planarity of the molecule, the π–π stacking interactions among the azobenzene molecules are dramatically affected, ultimately leading to disruption or formation of gels [79, 80], whereas structural changes that do not affect the gel phase are less likely to occur.

Recently, the Das group reported instead a gel that exhibits a gel-to-gel structural change due to light-triggered *trans–cis* isomerization [55]. The peptide-based gel contains an azobenzene group and a cysteine residue. At basic pH, dimerization of the cysteine residue assists the self-assembly of the gelator as well as π–π stacking between the azobenzene groups. Under irradiation with 365 nm light, shrinkage of the initial gel was observed. HPLC shows that the shrunken gel is still constituted by the disulfide-linked dimer. However, based on circular dichroism (CD) and powder X-ray diffraction, the structural isomerization caused a significant change in the self-assembly state of the gel. In both cases, β-sheet-like arrangement can be observed, although the number of π–π interactions appeared to have increased in the shrunken gel. This change is further observed in the morphology of the gels, where the initial gel possessed thin fibers; ~1-μm-long rod-like structures are observed for the shrunken gel. It follows that presumably the packing of the molecules requires less water for stabilization, leading to shrinkage of the gel.

Another way to maintain the gel integrity upon irradiation is to introduce the photoresponsive group separately from the gelator molecule. For instance, a

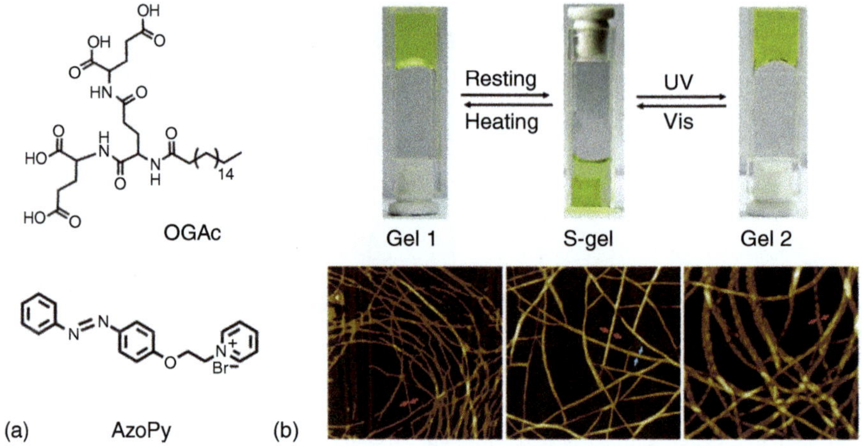

Figure 5.9 (a) Structures of AzoPy and the L-glutamic amphiphile (OGAc); (b) shrinking and photoinduced reswelling of the gel on resting and UV light irradiation (top) with representative 2 μm × 2 μm AFM images of the three gels. Source: Adapted with permission from Ref. 81/Royal Society of Chemistry.

multicomponent system was designed by Xie et al., comprising an azobenzene pyridinium salt (AzoPy) and an L-glutamic acid amphiphile (OGac) (Figure 5.9a) [81]. Initially, the two components form a gel by mixing in certain ratios by a heat/cool process. Shrinking of the gelator to a smaller gel (S-gel) occurs over time at room temperature at rest, expelling water (Figure 5.9b). Interestingly, a new swollen gel could be formed again by irradiating the S-gel with UV light. This process was found to be very reversible, and the shrunken gel could be reobtained by subjecting the gel to visible light (Figure 5.9b, top). The photoirradiated swollen gel was mainly composed of the *cis*-form azobenzene, indicating a *trans* to *cis* transformation from the S-gel upon irradiation. A change in morphology is further observed upon the change, as the swollen gel presented wider and twisted fibers compared to the long nanofibers in the initially formed gel (Figure 5.9b, bottom).

5.3.2.3 Photoreduction

Structural changes in peptide-based supramolecular gels can also be imparted where the light response is due to a change in the electronic configuration of a component in the gel. Specific gelators that show these properties either incorporate photoreducible metal ions or possess highly conjugated rings such as perylene (PBI) and naphthalene diimides (NDI). Upon irradiation, such molecules can form radical anions, altering the way the conjugated rings may stack, thereby affecting the electrostatic interactions in the systems [82, 83].

A rare response to irradiation as a result of photoreduction was reported by Draper et al. for an amino-acid-functionalized PBI gelator [84]. The molecule forms gels at low pH using a pH switch. Upon irradiation with 365 nm light, photoreduction leads to the formation of the radical anion, accompanied by a visible change in the color of the gel from red to dark purple. The change brought about an unusual enhancement in the rheological properties in both stiffness and mechanical strength. Small-angle neutron scattering (SANS) data suggests that the mechanical changes are due to a greater density of entanglement due to a reduced distance between cross-links.

5.3.3 Chemical-responsive Gels

Due to the possibility of appending functional groups to peptide-based precursors, control of the response of the material through chemical stimuli can further be obtained. By introducing functional groups that interact covalently or noncovalently with the environment through chemical reactions, the self-assembly process can be adapted in a predictable manner. Virtually, many different chemical reactions can be taken advantage of to rupture or make bonds, leading to gelation or dissolution. Furthermore, such stimuli may be reversible, and their response can be temporally programmable if multiple stimuli are included in the medium.

5.3.3.1 pH-triggered Changes

For peptide-based supramolecular hydrogels, pH is one of the most common ways to induce gelation if the gelator molecule contains ionizable pendant groups, such as amines or carboxylic acids. Depending on the pH of the medium surrounding the molecule, the protonation or deprotonation of such groups will control the degree of solubility of the molecules. In general, assembly or disassembly can be triggered in peptide hydrogels at pH above or below the apparent pK_a of the gelator molecule, depending on the type of functional group. For instance, deprotonation of a carboxylic acid moiety would lead to an increase in electrostatic repulsion and result in gel-to-sol transitions when the pH increases above the apparent pK_a [20, 85]. Conversely, if the pH of a gel based on an amine-functionalized gelator is lowered below the pK_a, formation of ammonium ions would also lead to transition to a solution state [86].

Taking advantage of the pH responsiveness of these molecules, several pH-triggered gel-to-sol-to-gel transitions have been reported in the literature [87–89]. Adaptable pH responses can be either tuned by direct addition of acid or base, or transient pH changes can further be induced by introducing triggers to controllably change the pH, such as enzymatic reactions or slow hydrolysis reactions [90, 91]. However, dynamic materials showing direct gel-to-gel transitions are limited, as the properties need to be varied in short pH ranges to avoid crossing the pK_a and dissolution of the assemblies.

One way to obtain such changes can be to incorporate multicomponent systems bearing groups with opposite pH responsiveness. Tena-Solsona et al. designed tetrapeptidic gelators containing ionizable groups with opposite charges (Figure 5.10a) [92]. At neutral pH, a co-assembled gel was formed (Figure 5.10b), which could be turned into single component gel by either lowering or increasing the pH. This would be obtained through either protonation of the amine or deprotonation of the carboxylic acid functionality, leading to the dissolution of one of the two components. The change in morphology can be studied by TEM, where fibrillar structures can be observed in all cases. After the addition of acid or base, the sample

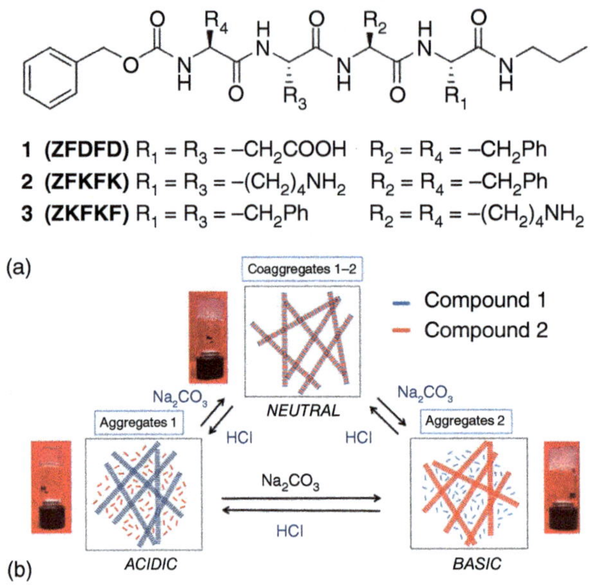

Figure 5.10 (a) Structures of the different pH-responsive gelators used in the study; (b) schematic representing the pH-responsive behavior of the mixtures of gelator **1** and gelator **2**. Source: Adapted with permission from Ref. [92]/Royal Society of Chemistry.

morphology resembled the one obtained for the pure compound that remained within the network.

A similar approach was used by Panja et al. to induce tune structural gel properties using oppositely charged functionalities [93]. Here, the two peptide-based amphiphiles are nongelling and bear opposite functionalities: a carboxylic acid moiety and an amine. The pH response of such system was also pathway dependent, based on the way deprotonation of the groups was obtained. Gelation occurs through a solvent switch at both pH 6.3 and 9.9 in DMSO/H_2O (20/80, vol/vol). Sequential addition of NaOH causes successive deprotonation of the terminal carboxylic acid at pH 6.3 and protonation of the ammonium ion at pH 9.9. Interestingly, a gel was obtained in both cases, where gel formation at lower pH was believed to be driven by electrostatic interactions between the NH_3^+ and COO−. A transition between these two gel states could be achieved by addition of 0.006 M NaOH on top of the lower pH gel. This slow diffusion method resulted in a gradual increase of both G' and G'', accompanied by a significant change in structures observed by SAXS and a decrease in volume by ~30%. Syneresis was ascribed to changes in the molecular packing, where the molecules were able to stack more effectively, causing an enhancement in the intermolecular interactions.

5.3.3.2 Dynamic Chemical Transformations

Changes in the molecular packing can be induced by affecting the noncovalent interactions in the systems. However, one issue for applications of such supramolecular materials is that they can often exhibit poor mechanical strength as they break at low strain. One way to overcome this is to drive *in situ* cross-linking of the gelator molecules driven by chemical reactions. When a gel is formed, they can be exposed to a chemical stimulus that instigates the covalent cross-linking, leading to changes in the material properties and improvement in the mechanical strength. Methods in which this can be obtained through light have been explored in Chapter 1.3.2, but they can further involve, for instance, disulfide, imine, and amide formation [94].

The Ikeda group recently reported a peptide-based supramolecular system that autonomously forms and shrinks to a second gel upon a single chemical input [56]. This was obtained through exploiting the hydrazone–oxime exchange in the presence of hydroxylamine. The precursor gelator was designed to contain a diphenylalanine dipeptide and a carboxybenzyl group at the N-terminus. To induce chemical responsiveness, the molecule was functionalized at the C-terminus with a hydrazone bond and a negatively charged hydrophilic sulfonate group. The electrostatic repulsion between sulfonate groups allows the sample to be dispersed in water. In the presence of hydroxylamine, removal of the charged sulfonate groups produced hydrazide-containing molecules and converted to a transparent gel at pH 5.5. Over 2 hours, the gel spontaneously shrunk, accompanied by morphological transformations of the nanostructures. Upon gel shrinking, the long fibers of the first gel appeared to increase in radius and form bundles. Interestingly, the appearance of the shrunk nanofiber network depended on the speed and the extent of the hydrazone–oxime reaction, further highlighting the pathway dependence of such chemoresponsive gels.

Another approach to impart chemical-stimulus changes in supramolecular systems is to take advantage of chemical cross-linking or coupling agents, such as glutaraldehyde or 1-ethyl-3-(3-dimethylaminopropyl)carbodiimide (EDC) [95]. Glutaraldehyde has been previously used to induce a dynamic imine covalent bond between amine and aldehyde groups by the Guler group [96]. The gelator was based on a PA molecule, Lauryl-VVAGKK-Am, that forms gels at pH >7. The cross-linking agent was added after gel formation, and UV–Vis spectroscopy confirmed the presence of the imine bond. Stiffer gels were observed by rheology, but interestingly the cross-linking stimulus did not appear to cause changes in the morphology and fiber structure of the gel. This is ascribed to the presence of the increased cross-linking within the fibers rather than between the fibers.

Marshall et al. used a similar approach to induce predictable changes in the mechanical properties of a peptide-based gel after formation by varying the position of the lysine residues within the gelator [97]. The response of the gelator to the glutaraldehyde was dependent on the position of the lysine residue, resulting in enhanced cross-linking where less steric hindering was present between the residues. As a result, different degrees of enhancement of the mechanical properties could be achieved with higher level of control. SANS data showed an increase in the radii upon cross-linking, suggesting that the increased stiffness of the gels could be related to the formation of imine bonds between adjacent fibers.

The chemical stimulus approach can be further expanded by introducing multicomponent systems with different responses to chemical triggers. In this way,

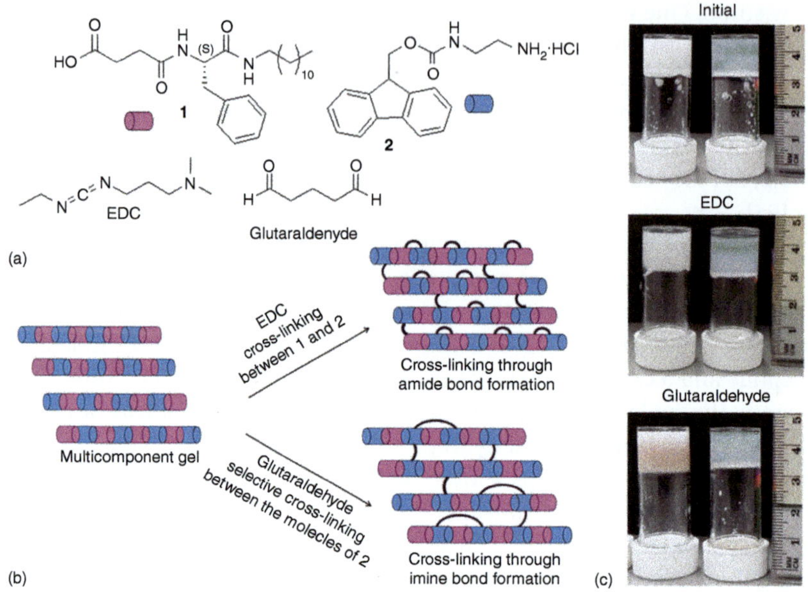

Figure 5.11 (a) Chemical structures of gelators used in the study and cross-linking agents EDC and glutaraldehyde; (b) schematic showing potential covalent cross-linking among the multicomponent gelators; c photographs highlighting the changes in gel volume after exposure to EDC and glutaraldehyde at pH 9.3 (left vial) and pH 10.8 (right vial). Source: Adapted with permission from Ref. [97]/Royal Society of Chemistry.

selective responses to a variety of reagents can be envisioned, allowing further control on the switching between structures. Our group recently explored this, by introducing a multicomponent system with gelators containing amino-acid residues bearing either a carboxylic acid or an amine group (Figure 5.11a) [98].

Here, cross-linking can be brought about using either EDC to link the two components or glutaraldehyde to selectively cross-link the molecules of the amine-functionalized gelator through formation of an imine bond. The gel forms through a solvent-switch method, and different gels could be obtained at both pH values of 9.3 and 10.8 by addition of different equivalents of NaOH. Cross-linking was induced postassembly, introducing either EDC or glutaraldehyde on top of the gels. The response of the system was found to be highly pH dependent. At pH 9.3, the mechanical properties of the gel did not appear to be affected by cross-linking. As the gel structure is maintained by electrostatic interactions at such pH, cross-linking is expected to destroy the salt bridge between the two gelators. This suggests that interfiber cross-linking allowed the system to retain the mechanical properties. However, dramatic changes in microstructure were observed, with more intertwined fibers observed for the cross-linked gels. This was further accompanied by a visual swelling of the gel (Figure 5.11c, left vial). At pH 10.8, the two cross-linking agents instead brought different responses, where the EDC caused an expansion in gel volume, and glutaraldehyde caused a contraction (Figure 5.11c, right vial). Therefore, such chemical-responsive systems allow control of the final structures of the gel as well as the macroscopic volume.

The approaches considered so far all rely on inducing structural changes postassembly from a starting gel state. However, another powerful approach for such chemical-triggered systems is to introduce reagents that cause preprogrammable modification of the building blocks over time. This method allows to design adaptable materials that change properties in an autonomous way with high degree of temporal control. This fuel-driven approach has emerged recently to design supramolecular peptide-based systems with tunable lifetimes by means of reversible chemical reaction cycles [99–104]. However, most of the work in this field is aimed at either inducing transient self-assembly through a sol-to-gel-to-sol cycle or to induce structural gel-to-gel changes by transiting through a disassembled state [88, 105]. Nevertheless, reversible network changes from an initial gel state to a new gel state can further be envisioned using this method. Examples of such systems have been recently published for polymer-based hydrogels [106, 107], although this remains largely unexplored for peptide-based materials.

One example of a amphiphile gelator containing an amino-acid residue undergoing a self-regulated gel-to-gel transition was reported by Panja et al. [108]. To induce the chemically fueled transition, EDC was used to promote reversible conversion to the anhydride form of the gelator, which would then slowly hydrolyze back to the original gelator. The anhydride form is further capable of forming a gel by co-assembly with the gelator molecule, resulting in a gel-to-gel transition. The kinetics of these unusual transitions could be controlled by the amount of fuel added, which also affected the underlying packing of the molecules at the end. This is further accompanied by a change in rheological properties and an improvement in mechanical strength. This behavior is ascribed to the presence of kinetically trapped co-assemblies at different concentrations of EDC, suggesting that the rate of hydrolysis would then govern the observed properties of the materials.

5.3.3.3 Redox Response

An interesting strategy to induce structural changes in gels is to use redox reactions. Generally, redox-responsive gelators are designed so that they bear redox-active units in their backbone, such as tetrathiafulvalenes [109], ferrocenes, or thiophenes, among others [110]. Redox-active metal ions can also be incorporated in supramolecular systems that change oxidation state upon electrical stimuli, causing a change in ion mobility and intermolecular interactions [111]. Redox changes can be achieved both chemically, by addition of reagents, and electrochemically, by exposing the gelators to externally applied electrical fields. In general, modification of the redox-active units of the gelator can lead to significant changes in solubility, triggering gel-to-sol transitions [112]. In some cases, changes in hydrophobicity of the fibers can also lead to volume changes, making these materials interesting for applications as actuators. Responsive hydrogel actuators have been designed based on covalently cross-linked hydrophilic homopolymer or copolymer networks [113, 114]. However, due to the reversible nature of the bonds in supramolecular gels, there is a potential to modulate such changes electrochemically in a reversible manner.

An example of this has been recently reported by Xue et al. using a tripeptide-based hydrogelator coupled to 3,4-dihydroxyphenylalanine (Dopa) at the C-terminus (Figure 5.12a, left) [57]. Dopa can be electrochemically oxidized to dopaquinone

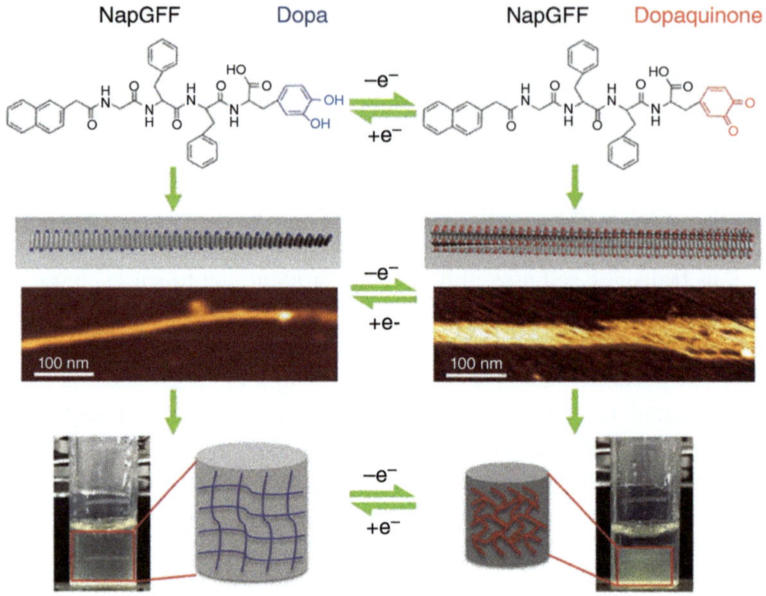

Figure 5.12 (a) Chemical structure of redox-responsive supramolecular tripeptide gelator containing a Dopa segment which can be oxidized to dopaquinone; changes in (b) morphology and (c) volume upon applied potential in the electroresponsive system. Source: Adapted with permission from Ref. [57], © 2016/John Wiley & Sons.

(Figure 5.12a, right), making this system redox-responsive. The Dopa-containing gelator formed gels via a solvent-switch method in DMSO and phosphate buffer saline. By applying a +5 V potential, oxidation of the gel was obtained, accompanied by a significant change both microscopically and macroscopically (Figure 5.12b,c). By AFM, an increase in height of the nanofibers formed by the gel was observed, indicating bundling of the fibers and increased interfiber cross-linking (Figure 5.12b, right). This further led to a dramatic decrease in the volume of the hydrogel by ~50% and increased stiffness. Based on these changes, a hydrogel actuator capable of unidirectional expansion and shrinkage by changing the redox states of the peptide was designed.

5.3.4 Mechanical Response

In some cases, self-assembly might be kinetically disfavored, and a stimulus might be necessary to overcome the high energy activation barrier and access different aggregation pathways. Mechanical stimuli such as ultrasound or sonication have recently emerged as an effective way of controlling supramolecular aggregation and disrupt noncovalent interactions to reshape interactions and aid gelation [15, 115]. For instance, Wang et al. observed that ultrasound treatment of a cyclohexane solution containing Fmoc-protected octylglycine (Fmoc-OG) resulted in gelation, due to a switch from intra- to intermolecular H-bonds [116]. Sonication has also shown potential to form interlocked gel states via intermolecular interactions by breaking intramolecular bonds or π-stacking [117].

Building on this behavior, sonication was further shown to be effective in inducing direct gel-to-gel transitions for peptide-based compounds containing L-alanine residues [58]. In particular, the gelator was found to undergo this instantaneous transition upon sonication in both acetonitrile and toluene due to changes in aggregation. In acetonitrile, SEM images show a transition from a core–shell structure surrounded by entangled fibers to just entangled nanofibers upon sonication. It is believed that under sonication, the entangled fibers around the core were stretched and cut, while the microball making up the core was incised into forming nanoballs. These further assembled into cross-linked fibers, resulting in the rapid gel to partial gel transformation. When the same process was performed in toluene, a transformation from 100 nm wide fibers to spindly fibers was observed, with a significant reduction in width to 20–50 nm. These final gels obtained through this method were also found to be more homogeneous and stable than gels obtained through sonication of high-temperature solutions of the molecules. It is assumed that the U-shaped dimers formed by the molecules aggregate in different initial conformations in the two solvents due to the solvophilic effect of toluene, leading to lamellar structures (Figure 5.13, center). Sonication appears not to cleave the intermolecular hydrogen bonding between the molecules but rather allows to make the π-stacking and hydrophobic interactions more homogeneous. This then results in a change in the morphologies to ones more favorable for gelation (Figure 5.13).

An ultrasound-induced transition between two gel states was also reported by the Yi group for a peptide-based organogel [118]. In the gelator structure, a lysine group was introduced into the linker, creating an umbrella molecule containing two cholesterol groups. In toluene, the molecule self-assembles into fully closed nanoring structures upon heating and cooling, forming a transparent gel. Upon treatment with ultrasound for 2 minutes, no gel was destroyed. By AFM and SEM, it appeared that the initial nanoring was destroyed while simultaneously smaller nanorings were formed from the original shells. This phenomenon was assumed to occur on a quicker timescale than observation, allowing retainment of the gel state across the change in structure.

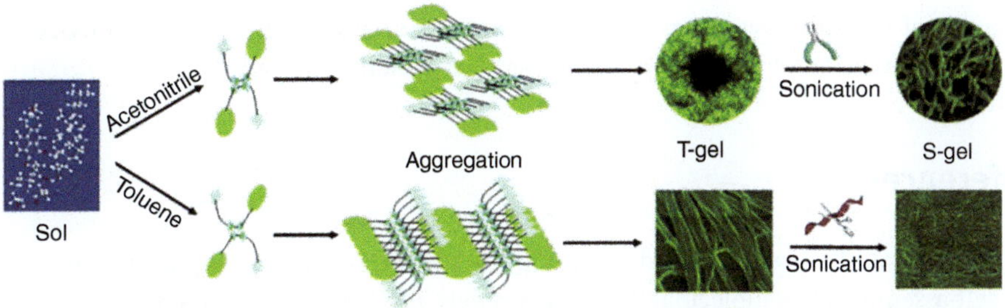

Figure 5.13 Schematic representation of the aggregation changes of the gelator in different solvents and upon sonication. The molecule adopts a different conformation in toluene, leading to lamellar structures in the first gel. Under sonication, both gels undergo a gel-to-gel transition to fibrous networks. Source: Adapted with permission from Ref. [58], © 2010/John Wiley & Sons.

5.4 Conclusion and Future Perspective

Supramolecular gels are advantageous materials due to the noncovalent interactions that hold their networks. This allows the materials to show responsive behavior and undergo transitions between different phases in a reversible manner. Peptide-based gelators have gained particular attention as they are easily functionalized and are biocompatible. Particular functional groups can be appended to their structures to induce responsive behavior for a wide range of applied stimuli. As a result, they are promising candidates for applications in sensing, actuation, drug delivery, catalysis, tissue engineering, and environmental remediation [119].

Dynamic structural transitions between different gel networks across the different gelation length scales can be achieved by application of external stimuli, such as light, pH, mechanical stress, or electricity. It follows that these gel-to-gel transitions are particularly promising to allow constructions of new gels from a single material, accessing gel states that might not be accessible directly in a controllable way [120]. Most stimuli-responsive dynamic gel systems rely on application of a stimulus after a gel has been formed. However, the field is currently moving toward constructing self-regulating gel-to-gel behavior to adapt material properties in a time-programmable manner exploiting dissipative self-assembly [89]. This allows to design materials exhibiting responses that can be controlled noninvasively and with temporal control.

Nevertheless, for most of these applications, better understanding of the way the self-assembled structures evolve upon the stimuli is paramount. For instance, morphology changes are often tracked before and after stimulus using SEM or AFM. Drying of the sample is necessary for these experiments, and this can further lead to changes in the self-assembled network. As such, techniques such as confocal laser scanning microscopy or small-angle scattering would provide a useful way to track changes in gel microstructures *in situ*, but they are often not performed [121, 122]. Moreover, testing the mechanical properties of these materials is important for specific applications. Time-dependent rheological measurements would provide useful information on the way the mechanical properties of the gels evolve, although they are rarely reported [123].

Further improved understanding and knowledge on how to directly control the behavior of the self-assembled structures through stimuli would allow to envision the next generation of bioinspired gel-based functional materials.

References

1 de Loos, M., Feringa, B.L., and van Esch, J.H. (2005). Design and application of self-assembled low molecular weight hydrogels. *Eur. J. Org. Chem.* 2005 (17): 3615–3631.

2 Estroff, L.A. and Hamilton, A.D. (2004). Water gelation by small organic molecules. *Chem. Rev.* 104 (3): 1201–1217.

3 Terech, P. and Weiss, R.G. (1997). Low molecular mass gelators of organic liquids and the properties of their gels. *Chem. Rev.* 97 (8): 3133–3159.

4 Kopeček, J. (2007). Hydrogel biomaterials: a smart future? *Biomaterials* 28 (34): 5185–5192.

5 Tirella, A., Mattei, G., and Ahluwalia, A. (2014). Strain rate viscoelastic analysis of soft and highly hydrated biomaterials. *J. Biomed. Mater. Res. A* 102 (10): 3352–3360.

6 Spicer, C.D. (2020). Hydrogel scaffolds for tissue engineering: the importance of polymer choice. *Polym Chem.* 11 (2): 184–219.

7 Saul, J.M. and Williams, D.F. (2011). Hydrogels in regenerative medicine. In: *Handbook of Polymer Applications in Medicine and Medical Devices*, 279–302. Oxford: Elsevier https://doi.org/10.1016/B978-0-323-22805-3.00012-8.

8 Hamidi, M., Azadi, A., and Rafiei, P. (2008). Hydrogel nanoparticles in drug delivery. *Adv. Drug Deliv. Rev.* 60 (15): 1638–1649.

9 Tibbitt, M.W. and Anseth, K.S. (2009). Hydrogels as extracellular matrix mimics for 3D cell culture. *Biotechnol. Bioeng.* 103 (4): 655–663.

10 Zanna, N., Focaroli, S., Merlettini, A. et al. (2017). Thixotropic peptide-based physical hydrogels applied to three-dimensional cell culture. *ACS Omega* 2 (5): 2374–2381.

11 Okesola, B.O. and Smith, D.K. (2016). Applying low-molecular weight supramolecular gelators in an environmental setting-self-assembled gels as smart materials for pollutant removal. *Chem. Soc. Rev.* 45 (15): 4226–4251.

12 Ahmed, E.M. (2015). Hydrogel: preparation, characterization, and applications: a review. *J. Adv. Res.* 6 (2): 105–121.

13 Hoffman, A.S. (2002). Hydrogels for biomedical applications. *Adv. Drug. Deliv. Rev.* 54 (1): 3–12.

14 Fichman, G. and Gazit, E. (2014). Self-assembly of short peptides to form hydrogels: design of building blocks, physical properties and technological applications. *Acta Biomater.* 10 (4): 1671–1682.

15 Jones, C.D. and Steed, J.W. (2016). Gels with sense: supramolecular materials that respond to heat, light and sound. *Chem. Soc. Rev.* 45 (23): 6546–6596.

16 Draper, E.R. and Adams, D.J. (2017). Low-molecular-weight gels: the state of the art. *Chem.* 3 (3): 390–410.

17 Chivers, P.R.A. and Smith, D.K. (2019). Shaping and structuring supramolecular gels. *Nat. Rev. Mater.* 4 (7): 463–478.

18 Zhang J, Hu Y, Li Y. Gel Chemistry: Interactions, Structures and Properties. Springer, 2018.

19 Cui, J., Liu, A., Guan, Y. et al. (2010). Tuning the helicity of self-assembled structure of a sugar-based organogelator by the proper choice of cooling rate. *Langmuir* 26 (5): 3615–3622.

20 Adams, D.J., Butler, M.F., Frith, W.J. et al. (2009). A new method for maintaining homogeneity during liquid-hydrogel transitions using low molecular weight hydrogelators. *Soft Matter* 5 (9): 1856–1862.

21 Mahler, A., Reches, M., Rechter, M. et al. (2006). Rigid, self-assembled hydrogel composed of a modified aromatic dipeptide. *Adv. Mater.* 18 (11): 1365–1370.

22 Cardoso, A.Z., Mears, L.L.E., Cattoz, B.N. et al. (2016). Linking micellar structures to hydrogelation for salt-triggered dipeptide gelators. *Soft Matter* 12 (15): 3612–3621.

23 Das, A.K., Collins, R., and Ulijn, R.V. (2008). Exploiting enzymatic (reversed) hydrolysis in directed self-assembly of peptide nanostructures. *Small* 4 (2): 279–287.

24 Raeburn, J., Cardoso, A.Z., and Adams, D.J. (2013). The importance of the self-assembly process to control mechanical properties of low molecular weight hydrogels. *Chem. Soc. Rev.* 42 (12): 5143–5156.

25 Colquhoun, C., Draper, E.R., Schweins, R. et al. (2017). Controlling the network type in self-assembled dipeptide hydrogels. *Soft Matter* 13 (9): 1914–1919.

26 Ding, B., Li, Y., Qin, M. et al. (2013). Two approaches for the engineering of homogeneous small-molecule hydrogels. *Soft Matter* 9 (18): 4672.

27 Vázquez-González, M. and Willner, I. (2020). Stimuli-responsive biomolecule-based hydrogels and their applications. *Angew. Chem. Int. Ed.* 59 (36): 15342–15377.

28 Li, Z., Zhou, Y., Li, T. et al. (2022). Stimuli-responsive hydrogels: fabrication and biomedical applications. *View* 3 (2): 20200112.

29 Pramanik, B. and Ahmed, S. (2022). Peptide-based low molecular weight photosensitive supramolecular gelators. *Gels* 8 (9): 1–26.

30 Dasgupta, A., Mondal, J.H., and Das, D. (2013). Peptide hydrogels. *RSC Adv.* 3 (24): 9117–9149.

31 Ulijn, R.V. and Smith, A.M. (2008). Designing peptide based nanomaterials. *Chem. Soc. Rev.* 37 (4): 664–675.

32 Mart, R.J., Osborne, R.D., Stevens, M.M., and Ulijn, R.V. (2006). Peptide-based stimuli-responsive biomaterials. *Soft Matter* 2 (10): 822–835.

33 Xian, S. and Webber, M.J. (2020). Temperature-responsive supramolecular hydrogels. *J. Mater. Chem. B* 8 (40): 9197–9211.

34 Tu, R.S. and Tirrell, M. (2004). Bottom-up design of biomimetic assemblies. *Adv. Drug Deliv. Rev.* 56 (11): 1537–1563.

35 Fleming, S. and Ulijn, R.V. (2014). Design of nanostructures based on aromatic peptide amphiphiles. *Chem. Soc. Rev.* 43 (23): 8150–8177.

36 Das, S. and Das, D. (2021). Rational design of peptide-based smart hydrogels for therapeutic applications. *Front. Chem.* 9 (November): 1–30.

37 Cui, H., Webber, M.J., and Stupp, S.I. (2010). Self-assembly of peptide amphiphiles: from molecules to nanostructures to biomaterials. *Biopolymers* 94 (1): 1–18.

38 Meijer, J.T., Roeters, M., Viola, V. et al. (2007). Stabilization of peptide fibrils by hydrophobic interaction. *Langmuir* 23 (4): 2058–2063.

39 Draper, E.R., Morris, K.L., Little, M.A. et al. (2015). Hydrogels formed from Fmoc amino acids. *CrstEngComm* 17 (42): 8047–8057.

40 Castelletto, V., Cheng, G., Greenland, B.W. et al. (2011). Tuning the self-assembly of the bioactive dipeptide l -carnosine by incorporation of a bulky aromatic substituent. *Langmuir* 27 (6): 2980–2988.

41 Jayawarna, V., Ali, M., Jowitt, T.A. et al. (2006). Nanostructured hydrogels for three-dimensional cell culture through self-assembly of fluorenylmethoxycarbonyl-dipeptides. *Adv. Mater.* 18 (5): 611–614.

42 Yang, Z., Liang, G., Ma, M. et al. (2007). Conjugates of naphthalene and dipeptides produce molecular hydrogelators with high efficiency of hydrogelation and superhelical nanofibers. *J. Mater. Chem.* 17 (9): 850–854.

43 Chivers, P.R.A., Dookie, R.S., Gough, J.E., and Webb, S.J. (2020). Photo-dissociation of self-assembled (anthracene-2-carbonyl)amino acid hydrogels. *Chem. Commun.* 56 (89): 13792–13795.

44 Zhang, V., Yang, Z., Yuan, F. et al. (2004). Molecular recognition remolds the self-assembly of hydrogelators and increases the elasticity of the hydrogel by 106-fold. *J. Am. Chem. Soc.* 126 (46): 15028–15029.

45 Ji, W., Tang, Y., Makam, P. et al. (2021). Expanding the structural diversity and functional scope of diphenylalanine-based peptide architectures by hierarchical coassembly. *J. Am. Chem. Soc.* 143 (42): 17633–17645.

46 Chen, L., McDonald TO, and Adams, D.J. (2013). Salt-induced hydrogels from functionalised-dipeptides. *RSC Adv.* 3 (23): 8714–8720.

47 Fleming, S., Sisir, D., Frederix, P.W.J.M. et al. (2013). Aromatic peptide amphiphiles: significance of the Fmoc moiety. *Chem. Commun.* 49 (90): 10587–10589.

48 Van Esch, J.H. (2009). We can design molecular gelators, but do we understand them? *Langmuir* 25 (15): 8392–8394.

49 Reches, M. and Gazit, E. (2003). Casting metal nanowires within discrete self-assembled peptide nanotubes. *Science* 300 (5619): 625–627.

50 Orbach, R., Mironi-Harpaz, I., Adler-Abramovich, L. et al. (2012). The rheological and structural properties of Fmoc-peptide-based hydrogels: the effect of aromatic molecular architecture on self-assembly and physical characteristics. *Langmuir* 28 (4): 2015–2022.

51 Cui, H., Cheetham, A.G., Pashuck, E.T., and Stupp, S.I. (2014). Amino acid sequence in constitutionally isomeric tetrapeptide amphiphiles dictates architecture of one-dimensional nanostructures. *J. Am. Chem. Soc.* 136 (35): 12461–12468.

52 Orbach, R., Adler-Abramovich, L., Zigerson, S. et al. (2009). Self-assembled Fmoc-peptides as a platform for the formation of nanostructures and hydrogels. *Biomacromolecules* 10 (9): 2646–2651.

53 Adams, D.J., Morris, K., Chen, L. et al. (2010). The delicate balance between gelation and crystallisation: structural and computational investigations. *Soft Matter* 6 (17): 4144–4156.

54 Baral, A., Basak, S., Basu, K. et al. (2015). Time-dependent gel to gel transformation of a peptide based supramolecular gelator. *Soft Matter* 11 (24): 4944–4951.

55 Das, B.K., Pramanik, B., Chowdhuri, S. et al. (2020). Light-triggered syneresis of a water insoluble peptide-hydrogel effectively removes small molecule waste contaminants. *Chem. Commun.* 56 (23): 3393–3396.

56 Sugiura, T., Kanada, T., Mori, D. et al. (2020). Chemical stimulus-responsive supramolecular hydrogel formation and shrinkage of a hydrazone-containing short peptide derivative. *Soft Matter* 16 (4): 899–906.

57 Xue, B., Qin, M., Wang, T. et al. (2016). Electrically controllable actuators based on supramolecular peptide hydrogels. *Adv. Funct. Mater.* 26 (48): 9053–9062.

58 Yu, X., Liu, Q., Wu, J. et al. (2010). Sonication-triggered instantaneous gel-to-gel transformation. *Chem. A Eur. J.* 16 (30): 9099–9106.

59 Nebot, V.J. and Smith, D.K. (2014). CHAPTER 2. Techniques for the characterisation of molecular gels. In: *Functional Molecular Gels*, 30–66. Royal Society of Chemistry Available from: http://ebook.rsc.org/?DOI=10.1039/9781849737371-00030.

60 Yu, G., Yan, X., Han, C., and Huang, F. (2013). Characterization of supramolecular gels. *Chem. Soc. Rev.* 42 (16): 6697–6722.

61 Denzer, B.R., Kulchar, R.J., Huang, R.B., and Patterson, J. (2021). *Advanced Methods for the Characterization of Supramolecular Hydrogels, Gels* 7 (4): 158.

62 Mears, L.L.E., Draper, E.R., Castilla, A.M. et al. (2017). Drying affects the fiber network in low molecular weight hydrogels. *Biomacromolecules* 18: 3531–3540.

63 Pauw, B.R. (2014). Everything SAXS: small-angle scattering pattern collection and correction. *J. Phys. Condens. Mat.* 26 (23): 239501.

64 Yan, C. and Pochan, D.J. (2010). Rheological properties of peptide-based hydrogels for biomedical and other applications. *Chem. Soc. Rev.* 39 (9): 3528. Available from: http://xlink.rsc.org/?DOI=b919449p.

65 Kiyonaka, S., Sugiyasu, K., Shinkai, S., and Hamachi, I. (2002). First thermally responsive supramolecular polymer based on glycosylated amino acid. *J. Am. Chem. Soc.* 124 (37): 10954–10955.

66 Grötsch, R.K. and Boekhoven, J. (2018). Unique properties of supramolecular biomaterials through nonequilibrium self-assembly. In: *Self-Assembling Biomaterials*, 235–250. Elsevier Available from: https://doi.org/10.1016/B978-0-08-102015-9.00012-5.

67 Tantakitti, F., Boekhoven, J., Wang, X. et al. (2016). Energy landscapes and functions of supramolecular systems. *Nat. Mater.* 15 (4): 469–476.

68 Shi, J., Gao, Y., Yang, Z., and Xu, B. (2011). Exceptionally small supramolecular hydrogelators based on aromatic-aromatic interactions. *Beilstein J. Org. Chem.* 7: 167–172.

69 Draper, E.R., McDonald, T.O., and Adams, D.J. (2015). A low molecular weight hydrogel with unusual gel aging. *Chem. Commun.* 51 (30): 6595–6597.

70 Duraisamy, D.K., Sureshbhai, P.D., Saveri, P. et al. (2022). A "self-shrinking" supramolecular hydrogel with a 3D shape memory performance from an unnatural amino acid derivative. *Chem. Commun.* 58 (96): 13377–13380.

71 Basak, S., Nandi, N., Paul, S. et al. (2017). A tripeptide-based self-shrinking hydrogel for waste-water treatment: removal of toxic organic dyes and lead (Pb2+) ions. *Chem. Commun.* 53 (43): 5910–5913.

72 Sheehan, F., Sementa, D., Jain, A. et al. (2021). Peptide-based supramolecular systems chemistry. *Chem. Rev.* 121 (22): 13869–13914.

73 Qiu, Z., Yu, H., Li, J. et al. (2009). Spiropyran-linked dipeptide forms supramolecular hydrogel with dual responses to light and to ligand-receptor interaction. *Chem. Commun.* 7345 (23): 3342–3344.

74 Liu, Q., Wang, H., Li, G. et al. (2019). A photocleavable low molecular weight hydrogel for light-triggered drug delivery. *Chinese Chem. Lett.* 30 (2): 485–488.

75 Ramamurthy, V. and Sivaguru, J. (2016). Supramolecular photochemistry as a potential synthetic tool: photocycloaddition. *Chem. Rev.* 116 (17): 9914–9993.

76 Draper, E.R., McDonald, T.O., and Adams, D.J. (2015). Photodimerisation of a coumarin-dipeptide gelator. *Chem. Commun.* 51 (64): 12827–12830.

77 Kim, S.H., Sun, Y., Kaplan, J.A. et al. (2015). Photo-crosslinking of a self-assembled coumarin-dipeptide hydrogel. *New J. Chem.* 39 (5): 3225–3228.

78 Ding, Y., Li, Y., Qin, M. et al. (2013). Photo-cross-linking approach to engineering small tyrosine-containing peptide hydrogels with enhanced mechanical stability. *Langmuir* 29 (43): 13299–13306.

79 Li, W., Park, I.S., Kang, S.K., and Lee, M. (2012). Smart hydrogels from laterally-grafted peptide assembly. *Chem. Commun.* 48 (70): 8796–8798.

80 Huang, Y., Qiu, Z., Xu, Y. et al. (2011). Supramolecular hydrogels based on short peptides linked with conformational switch. *Org. Biomol. Chem.* 9 (7): 2149–2155.

81 Xie, F., Qin, L., and Liu, M. (2016). A dual thermal and photo-switchable shrinking-swelling supramolecular peptide dendron gel. *Chem. Commun.* 52 (5): 930–933.

82 Roy, S., Kumar Maiti, D., Panigrahi, S. et al. (2012). A new hydrogel from an amino acid-based perylene bisimide and its semiconducting, photo-switching behaviour. *RSC Adv.* 2 (29): 11053–11060.

83 Das, A. and Ghosh, S. (2016). H-bonding directed programmed supramolecular assembly of naphthalene-diimide (NDI) derivatives. *Chem. Commun.* 52 (42): 6860–6872.

84 Draper, E.R., Schweins, R., Akhtar, R. et al. (2016). Reversible photoreduction as a trigger for photoresponsive gels. *Chem. Mater.* 28 (17): 6336–6341.

85 Hartgerink, J.D., Beniash, E., and Stupp, S.I. (2001). Self-assembly and mineralization of peptide-amphiphile nanofibers. *Science* 294 (5547): 1684–1688.

86 Van Bommel, K.J.C., Van Der Pol, C., Muizebelt, I. et al. (2004). Responsive cyclohexane-based low-molecular-weight hydrogelators with modular architecture. *Angew. Chem. Int. Ed.* 43 (13): 1663–1667.

87 Panja, S., Fuentes-Caparrós, A.M., Cross, E.R. et al. (2020). Annealing supramolecular gels by a reaction relay. *Chem. Mater.* 32 (12): 5264–5271.

88 Singh, N., Formon, G.J.M., De Piccoli, S., and Hermans, T.M. (2020). Devising synthetic reaction cycles for dissipative nonequilibrium self-assembly. *Adv. Mater.* 32 (20): 1906834.

89 Panja, S. and Adams, D.J. (2019). Gel to gel transitions by dynamic self-assembly. *Chem. Commun.* 55 (68): 10154–10157.

90 Heuser, T., Steppert, A.K., Molano Lopez, C. et al. (2015). Generic concept to program the time domain of self-assemblies with a self-regulation mechanism. *Nano Lett.* 15 (4): 2213–2219.

91 Panja, S. and Adams, D.J. (2021). Urea-urease reaction in controlling properties of supramolecular hydrogels: pros and cons. *Chem. A Eur. J.* 27 (35): 8928–8939.

92 Tena-Solsona, M., Alonso-De Castro, S., Miravet, J.F., and Escuder, B. (2014). Co-assembly of tetrapeptides into complex pH-responsive molecular hydrogel networks. *J. Mater. Chem. B* 2 (37): 6192–6197.

93 Panja, S., Seddon, A., and Adams, D.J. (2021). Controlling hydrogel properties by tuning non-covalent interactions in a charge complementary multicomponent system. *Chem. Sci.* 12 (33): 11197–11203.

94 Picchioni, F. and Muljana, H. (2018). Hydrogels based on dynamic covalent and non covalent bonds: a chemistry perspective. *Gels.* 4 (1): 1–11.

95 Schnitter, F., Bergmann, A.M., Winkeljann, B. et al. (2021). Synthesis and characterization of chemically fueled supramolecular materials driven by carbodiimide-based fuels. *Nat. Protoc.* 16 (8): 3901–3932.

96 Khalily, M.A., Goktas, M., and Guler, M.O. (2015). Tuning viscoelastic properties of supramolecular peptide gels via dynamic covalent crosslinking. *Org. Biomol. Chem.* 13 (7): 1983–1987.

97 Marshall, L.J., Matsarskaia, O., Schweins, R., and Adams, D.J. (2021). Enhancement of the mechanical properties of lysine-containing peptide-based supramolecular hydrogels by chemical cross-linking. *Soft Matter* 17 (37): 8459–8464.

98 Panja, S., Bianco, S., Dietrich, B., and Adams, D.J. (2022). Macroscopic volume phase transitions in supramolecular gels directed by covalent crosslinking. *Mater Chem Front.* 6 (18): 2703–2708.

99 Boekhoven, J., Hendriksen, W.E., Koper, G.J.M. et al. (2015). Transient assembly of active materials fueled by a chemical reaction. *Science.* 349 (6252): 1075–1079.

100 Tena-Solsona, M., Rieß, B., Grötsch, R.K. et al. (2017). Non-equilibrium dissipative supramolecular materials with a tunable lifetime. *Nat. Commun.* 8 (1): 15895.

101 Singh, N., Lainer, B., Formon, G.J.M. et al. (2020). Re-programming hydrogel properties using a fuel-driven reaction cycle. *J. Am. Chem. Soc.* 142 (9): 4083–4087.

102 Zhang, B., Jayalath, I.M., Ke, J. et al. (2019). Chemically fueled covalent crosslinking of polymer materials. *Chem. Commun.* 55 (14): 2086–2089.

103 Angulo-Pachón, C.A. and Miravet, J.F. (2016). Sucrose-fueled, energy dissipative, transient formation of molecular hydrogels mediated by yeast activity. *Chem. Commun.* 52 (31): 5398–5401.

104 Pappas, C.G., Sasselli, I.R., and Ulijn, R.V. (2015). Biocatalytic pathway selection in transient tripeptide nanostructures. *Angew. Chem. Int. Ed.* 54 (28): 8119–8123.

105 Tena-Solsona, M. and Boekhoven, J. (2019). Dissipative self-assembly of peptides. *Isr. J. Chem.* 59 (10): 898–905.

106 Xu H, Bai S, Gu G, Gao Y, Sun X, Guo X, et al. (2022). Bioinspired Self-Resettable Hydrogel Actuators Powered by a Chemical Fuel. *ACS Appl. Mater. Interfaces* 14 (38): 43825–43832.

107 Nakamoto, M., Kitano, S., and Matsusaki, M. (2022). Biomacromolecule-fueled transient volume phase transition of a hydrogel. *Angew. Chem. Int. Ed.* 61 (33).

108 Panja, S., Dietrich, B., and Adams, D.J. (2020). Chemically Fuelled self-regulating gel-to-gel transition. *ChemSystemsChem.* 2 (1): 2–6.

109 Nalluri, S.K.M., Shivarova, N., Kanibolotsky, A.L. et al. (2014). Conducting nanofibers and organogels derived from the self-assembly of tetrathiafulvalene-appended dipeptides. *Langmuir* 30 (41): 12429–12437.

110 Chu, C.-W. and Schalley, C.A. (2021). Recent advances on supramolecular gels: from stimuli-responsive gels to co-assembled and self-sorted systems. *Org Mater.* 03 (01): 025–040.

111 Sui, X., Feng, X., Hempenius, M.A., and Vancso, G.J. (2013). Redox active gels: synthesis, structures and applications. *J. Mater. Chem. B* 1 (12): 1658–1672.

112 Adhikari, B. and Kraatz, H.B. (2014). Redox-triggered changes in the self-assembly of a ferrocene-peptide conjugate. *Chem. Commun.* 50 (42): 5551–5553.

113 Shin, Y., Choi, M.Y., Choi, J. et al. (2021). Design of an electro-stimulated hydrogel actuator system with fast flexible folding deformation under a low electric field. *ACS Appl. Mater. Interfaces* 13 (13): 15633–15646.

114 Li, Y., Sun, Y., Xiao, Y. et al. (2016). Electric field actuation of tough electroactive hydrogels Cross-linked by functional triblock copolymer micelles. *ACS Appl. Mater. Interfaces* 8 (39): 26326–26331.

115 Bardelang, D., Camerel, F., Margeson, J.C. et al. (2008). Unusual sculpting of dipeptide particles by ultrasound induces gelation. *J. Am. Chem. Soc.* 130 (11): 3313–3315.

116 Wang, Y., Zhan, C., Fu, H. et al. (2008). Switch from intra- to intermolecular H-bonds by ultrasound: induced gelation and distinct nanoscale morphologies. *Langmuir* 24 (15): 7635–7638.

117 Naota, T. and Koori, H. (2005). Molecules that assemble by sound: an application to the instant gelation of stable organic fluids. *J. Am. Chem. Soc.* 127 (26): 9324–9325.

118 Yu, X., Cao, X., Chen, L. et al. (2012). Thixotropic and self-healing triggered reversible rheology switching in a peptide-based organogel with a cross-linked nano-ring pattern. *Soft Matter* 8 (12): 3329–3334.

119 Devika, V., Sreelekshmi, P.J., Rajeev, N. et al. (2022). Recent advances in peptides-based stimuli-responsive materials for biomedical and therapeutic applications: a review. *Mol. Pharm.* 19 (7): 1999–2021.

120 Panja, S. and Adams, D.J. (2021). Stimuli responsive dynamic transformations in supramolecular gels. *Chem. Soc. Rev.* 50 (8): 5165–5200.

121 Draper, E.R., Dietrich, B., McAulay, K. et al. (2020). Using small-angle scattering and contrast matching to understand molecular packing in low molecular weight gels. *Matter* 2 (3): 764–778.

122 Kubota, R., Nakamura, K., Torigoe, S., and Hamachi, I. (2020). The power of confocal laser scanning microscopy in supramolecular chemistry: in situ real-time imaging of stimuli-responsive multicomponent supramolecular hydrogels. *ChemistryOpen.* 9 (1): 67–79.

123 Bianco, S. Panja, S. Adams, D.J. (2022). Using rheology to understand transient and dynamic gels. *Gels* 8(2): 132.

6

Self-assembling Cyclic Peptide Nanotubes: Methods and Characterization

Ángel L. Fuentes de Arriba[1] *and Juan R. Granja*[2]

[1] *Organic Chemistry Department, University of Salamanca, Salamanca, Spain*
[2] *Organic Chemistry Department, Centro Singular de Investigación en Química Biolóxica e Materiais Moleculares (CIQUS), Rúa de Jenaro de la Fuente, Universidade de Santiago de Compostela, Santiago de Compostela, Spain*

6.1 Introduction

Self-assembling cyclic peptides nanotubes (SCPNs) are dynamic, tubular-shaped synthetic organic supramolecular structures formed by stacking cyclic peptides (CPs) through the formation of β-sheet like hydrogen-bonding interactions. Since the discovery of the first SCPNs by Ghadiri and coworkers in 1993 [1], the field of these materials has experienced a boost due to their interesting applications in medicine, material science, and nanotechnology [2].

This chapter focuses on the preparation of SCPNs, the different techniques used for their characterization, and their recent applications.

6.2 Preparation of Self-assembling Cyclic Peptide Nanotubes

6.2.1 Requirements

The formation of SCPNs by stacking of cyclic peptides requires to fulfill some requirements. Although different types of amino acids, both natural and noncanonical, have been used in the literature to build cyclic peptides, α-amino acids, β-amino acids, and γ-, δ-, and ε-cyclic amino acids (Figure 6.1), the relative chirality of the residues that constitute these CPs is fundamental for allowing their stacking.

The stacking of cyclic peptides to form nanotubes requires a planar ring-like conformation of the cyclic components, so that β-sheet hydrogen bonds can be established between the carbonyl and amide (NH) groups of the neighboring cyclic

Peptide Self-Assembly and Engineering: Fundamentals, Structures, and Applications, First Edition.
Edited by Xuehai Yan.

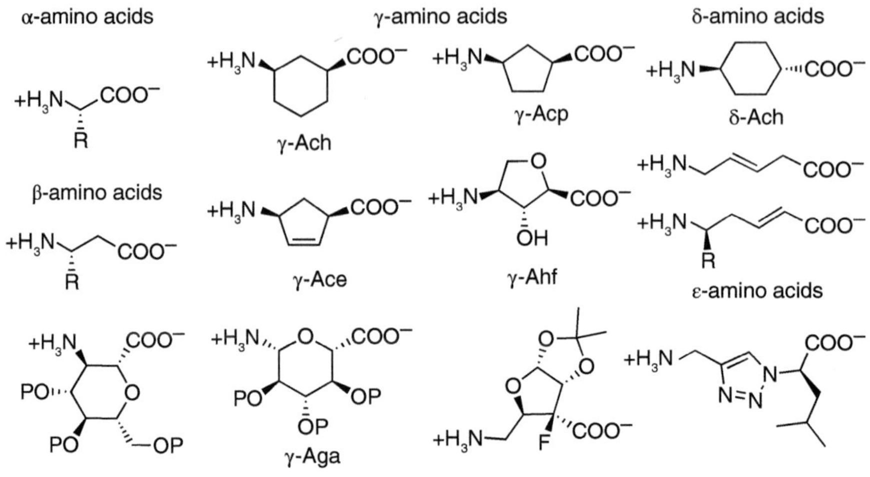

Figure 6.1 Amino acids reported in the literature used in the synthesis of SCPNs.

peptides. In this conformation, C=O and NHs groups must be placed perpendicularly with respect to the plane of the flat ring. This geometry leaves the side chains of the different amino acids pointing outward from the assembly in a pseudoequatorial arrangement and generates a hydrophilic pore inside the supramolecular structure.

To achieve this planar conformation, an even number of atoms are needed to avoid moebius-type geometries. In addition, in the original nanotubes, the CPs (α-[*D*,*L*]-CPs) must be composed of residues with alternating configuration [3].

Following these requirements, in 1993, Ghadiri's group prepared the first SCPNs using octapeptide cyclo-[(*L*-Gln-*D*-Ala-*L*-Glu-*D*-Ala-)$_2$] to provide nanotubes with an internal diameter of 7.5 Å and 4.73-Å separation between each cyclic peptide. Although pioneering work by Lorenzi had previously pursued the nanotube formation using nonionizable residues, the low solubility of these peptides prevented its assembly under appropriate conditions [3]. The clever approach of using glutamic acid that can be easily deprotonated under mild basic conditions to favor its solubility in aqueous media and the later use of the pH change to trigger nanotube formation was a key aspect in the success of Ghadiri's approach (Figure 6.2). An antiparallel-type arrangement of the cyclic peptides through β-sheet hydrogen bonds was proposed for these SCPNs. The supramolecular nanotubes were fully characterized by TEM, ED, FT-IR spectroscopy, and molecular modeling [1]. Since then, a myriad of CPs, made up of the combination of different amino acids and functionalized with diverse molecules, have been constructed (Figure 6.2).

The minimum number of α-amino acids of a CP reported in the literature that can build nanotubes is 6, while the maximum number is 12. Increasing the number of α-amino acids raises the possibility of having twisted structures, which complicates the stacking of cyclic peptides [4]. In particular, the stronger binding affinities between the cyclic peptides units have been reported for octapeptides [2e].

In 1997, Seebach reported the synthesis of CPs made up of β-amino acids following the same synthetic strategy employed by Ghadiri with α-amino acids [5]. In this case, the chirality of the residues is not determinant in the SCPN formation ability, and the

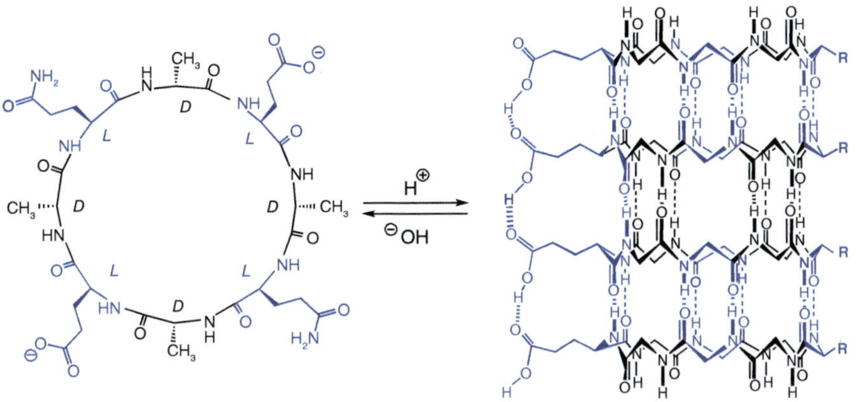

Figure 6.2 First SCPNs strategy reported by Ghadiri's group using the proton-triggering approach.

three stereoisomers (all-*S*, alternating chirality [*R*,*S*-CP], and the *R*,*R*,*S*,*S*-CP) were able to stack and form the corresponding nanotubes (Figure 6.3). Also, SCPNs were reported using CPs made by the combination of α- and β-amino acids [6]. However, β-amino acids have been less studied, as they have more problems to adopt the planar conformation due to the ability to adopt two different conformations (anti or gauche) with similar energetic costs. The most favorable, the anti-conformer, has all NHs pointing in the same direction and therefore stacking up to form parallel β-sheets, which in the case of the smaller CPs establishes nonlinear hydrogen bonds [7].

Cyclic γ-amino acids made up of five- or six-membered carbocycles are excellent building blocks for the preparation of SCPNs, especially when they are combined with α-amino acids. These structures, pioneered by Granja's group [8], introduce rigidity and flatness, thanks to the cycloalkane segments and its *cis*-configuration which allow the construction of SCPNs with larger pore diameter [9]. Also, these residues allow to modify the lumen of the nanotubes because one of the methylene moieties (Cβ) of each γ-residue is pointing into the cavity. In these structures, because of the alternation between α- and γ-residues, the NHs and carbonyl groups of each type of amino acids are pointing in opposite directions (Figure 6.4). Initially, it was proposed that due to the different distances between the hydrogen bond

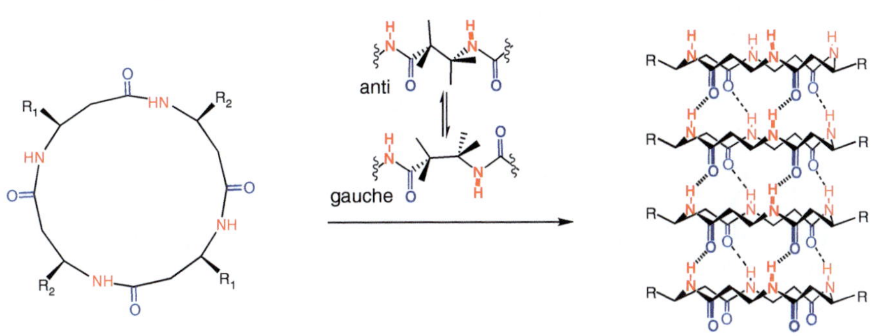

Figure 6.3 β-SCPNs model reflecting the larger conformational freedom of β-amino acids.

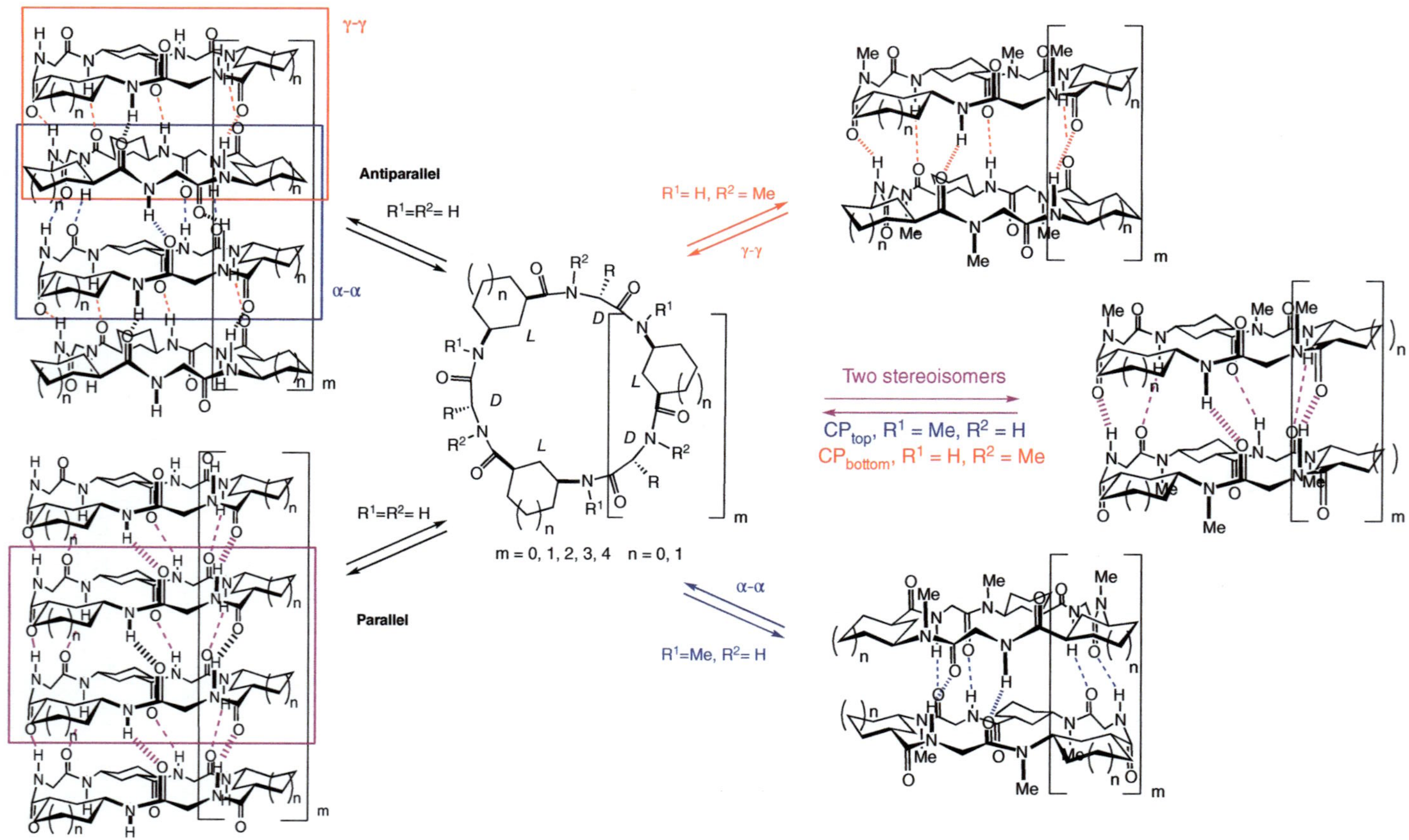

Figure 6.4 SCPN nanotubes made up of α,γ-CPs and the corresponding molecular models to study the different β-sheet-type interactions.

donors and acceptors of both residue types, the antiparallel stacking, hydrogen bonds between the same type of residues would be more favorable, but recently, it has been shown that the parallel arrangement, in which hydrogen bonds are formed between the two types of residues, is more stable [10].

Granja's group has also designed δ-amino acids able to form SCPNs when combined with α-amino acids. This kind of structure will be analyzed in the following section [11].

6.2.2 Design

One of the most attractive features of SCPNs is the ability to control the nature of the inner and outer surfaces of the nanotube as well as the internal diameter. This control can be achieved by selecting the type and number of amino acids used to build the cyclic peptide. As the CPs are stacked forming hydrogen bonds between the amide protons and carbonyl groups, the alternating amino-acid backbone configuration imposes conformational and steric constraints that leave the side chain of each amino acid pointing outward in a pseudoequatorial disposition. Although SCPNs are usually formed in water, a strong hydrogen bond competitor media, the cooperativity of forming several hydrogen bonds per pair of subunits and the presence of the amino-acid side chains pointing outward obstruct the access of water molecules and protect the SCPNs to be disassembled in this media. Therefore, by selecting the right amino acids that make up the CPs, it is possible to tune the surface nature of the nanotubes with a hydrophilic, hydrophobic, and positively or negatively charged character [2].

For example, Kimura has developed β-CPs made of sugar-derived amino acids (Figure 6.5) in which a pyranose ring behaves as lateral chain of the amino acid. In this way, the solubility is increased, and the stacking is favored due to the formation of a more planar structure. The resulting supramolecular hollow cylinders have shown good affinity for lectins [12].

Kimura has also combined these hydrophilic β-glucosamino acids with hydrophobic *trans*-2-amino cyclohexylcarboxylic acid (Figure 6.5, right, in red) to construct amphiphilic SCPNs that form triple bundles of the columnar assemblies [13].

The nature of the lateral chain of the amino acids also affects the intra- and intermolecular interactions established between CPs and nanotubes (hydrogen bonds, electrostatic, van der Waals interactions, and hydrophobic effect) that affect the

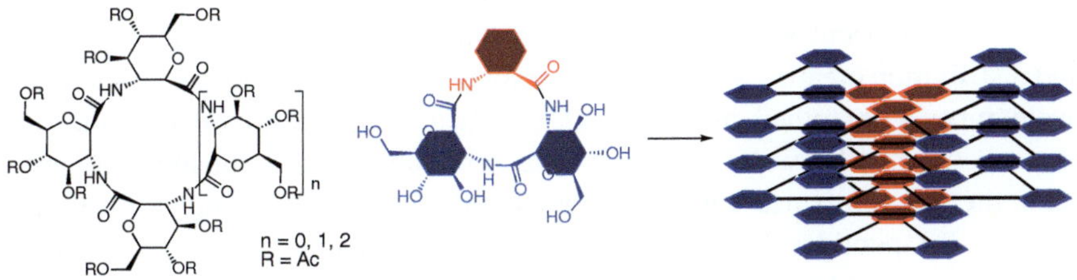

Figure 6.5 Saccharide-derived cyclic peptides developed by Kimura.

aggregation of the tubular ensembles. In this sense, the use of aromatic rings has been shown to help stabilizing not only the tubular structure but also the formation of bundles or hydrogels through the π,π-stacking of the aromatic moieties [14].

In addition, the use of covalent modifications on the amino-acid side chains increases the repertoire of tools that allow to modify the external properties of the SCPNs, for instance hybrid materials of peptide–polymer conjugates in which different types of polymers are attached to the side chain of glutamic acid or lysine [15]. These structures allow to improve the solubility of SCPNs, avoiding or diminishing their aggregation tendency and expanding their applications. Therefore, SCPN/polymer hybrids have found applications as smart antibacterial agents [16], artificial transmembrane ion channels [17], or drug delivery and controlled release systems [18]. Especially interesting is the possibility to behave as stimuli-responsive materials able to change the hydrophobicity of the SCPNs upon an external stimulus [19].

As already mentioned, the size of the internal diameter of the nanotube can be controlled by varying the number of amino acids that make up the cyclic peptide and also the type of amino acids. Consequently, cyclic peptides with an internal diameter from 1 to 30 Å have been described since their original discovery [20].

One of the smallest members of the SCPN-forming CPs was described by Granja's group, a cyclic tetrapeptide made up of two α-amino acids and two γ-amino acids with 1 Å of internal diameter [2e]. The SCPN with the wider diameter (13 Å) made up of α-amino acids is a dodecapeptide [2e]. Although larger α-(*D*,*L*)-CPs can be synthetized, their main drawback for SCPN preparation is their flexibility and torsional constrains that prevent these larger derivatives from adopting the flat conformation and, consequently, the resulting twisted structures are precluded of stacking to form nanotubes. However, it is possible to increase the size of the internal diameter of the SCPNs using noncanonical amino acids. Granja employed *cis*-3-aminocyclopentanecarboxylic acid to make CPs with a wider pore when combined with α-amino acids (Figure 6.6) [11]. These γ-amino acids possess an angle between the plane of the NH_2 and the plane of the COOH, which varies between 120° and 180°, and allow the construction of hexa-, octa-, deca-, dodeca-, and hexadecapeptides with different internal diameters up to 30 Å. Interestingly, recently, the authors have also reported the use of *trans*-4-aminocyclohexanecarboxylic acid (δ-Ach) for the preparation of a new class of SCPN with large diameter and with internal cavity of hydrophobic properties. For example, SCPNs made of dodecapeptides containing six δ-Ach residues were able to encapsulate C_{60} fullerenes, forming a new class of organic nanopeapods (Figure 6.6) [11]. In addition, the partially hydrophobic character of the nanotube lumen has a critical impact on the behavior of the nanoconfined ions and water molecules. Nanoconfined water in this α,δ-SCPN adopts a highly ordered distribution creating a clearly defined six-point star shape, formed by at least four water types. The studied ions (Na, K, Li, Cs, Ca, and Cl) accommodate inside the nanotube disturbing this water pattern as little as possible. In fact, the ions with less impact in the water organization remain for a longer time inside the nanotube [21].

Cyclic α,ε-*D*,*L*-peptides have also been constructed using a copper-catalyzed azide–alkyne cycloaddition of a 1,4-disubstituted-1,2,3-triazole ε-amino acid [22]. This CP chimera was also able to stack, forming a new class of peptide nanotubes.

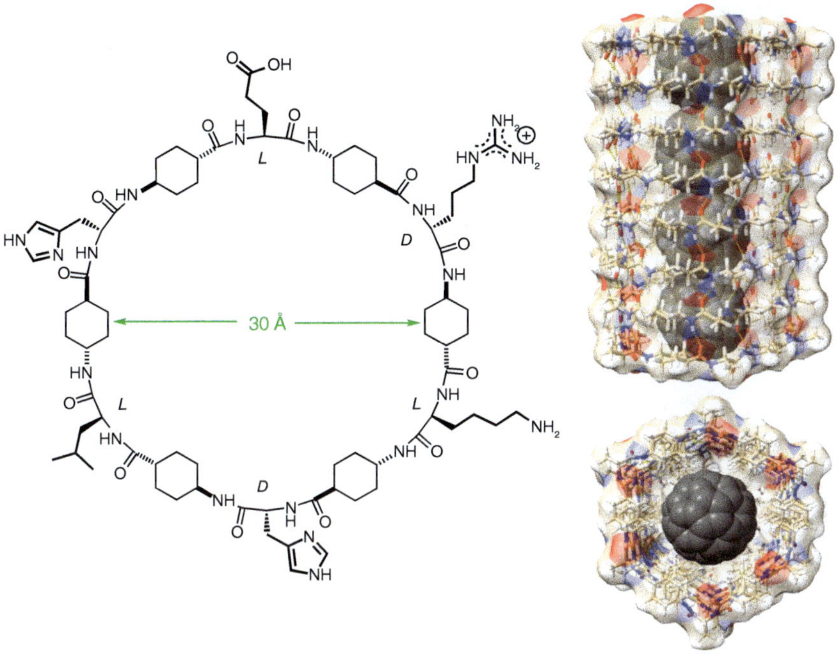

Figure 6.6 Cyclic peptide with 30 Å internal diameter synthesized by Granja's group and model of fullerene wire generated in the lumen of the resulting SCPN.

The special feature of this design is the orientation of the triazole rings that form aromatic walls along the pore of the SCPN. This represents one of the first examples described in the literature designed with the aim of tuning the inner properties of the nanotubes. The first example in this respect was the use of CPs constituted by only one residue of nonnatural chirality every fourth (Figure 6.7, top-left) [23]. In this type of sequence, the side chain of the second residue of this *L,L,L,D* repeat sequence points into the pore of the resulting tubular structure. The cyclic dodecapeptide with the sequence *cyclo*-[*L*-Trp-*L*-Dap-*L*-Leu-*D*-Ala-*L*-Trp-*L*-Ser-*L*-Val-*D*-Ala-*L*-Trp-*L*-Ser-*L*-Ile-Gly-] was proposed to adopt the planar conformation, leaving the side chains of Ser and Dap pointing toward the pore of the nanotube, facilitating the transmembrane ion transport. Alternatively, the internal cavity functionalization can be achieved using noncanonical amino acids. SCPNs made of α-amino acids possess a hydrophilic lumen with water molecules inside interacting with the antibonding orbital of peptide bonds. The employment of nonnatural amino acids allows to change the nature of the lumen of the nanotubes. For example, many of the γ-, δ-, and ε-amino acids shown in Figure 6.1 leave methylene groups pointing toward the inner part of the nanotube, at least increasing the hydrophobic character of the channel, and allowing the transport of hydrophobic molecules, as mentioned above.

These and other amino acids are also susceptible to chemical modifications that allow to tune the properties of the pore of the nanotubes (Figure 6.7). Granja has

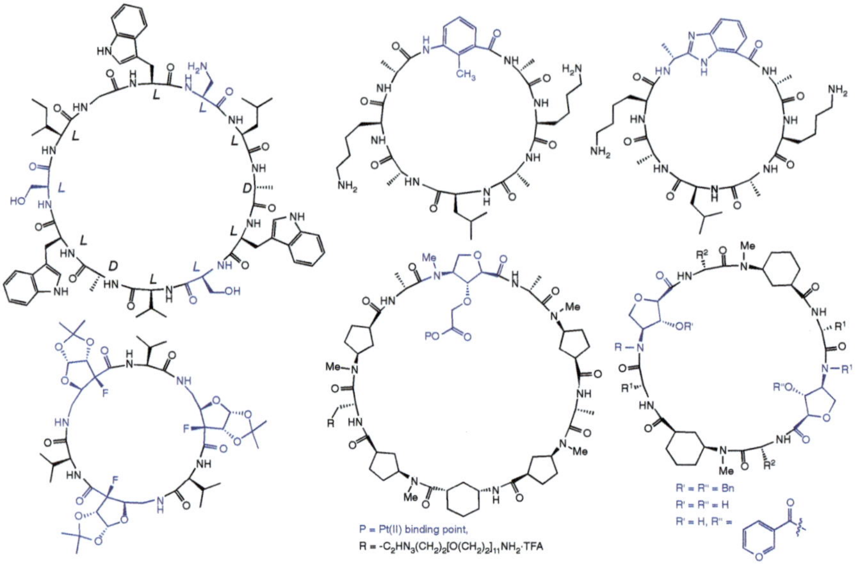

Figure 6.7 Examples of cyclic peptides with internal chemical modifications.

functionalized γ-amino acids with different functional groups, such as a carboxylic acid to bind platinum complexes (bottom-center) [24], or hydroxylic groups that require the presence of polar and protic solvents for self-assembly (bottom-right) [25]. The functionalization of the hydroxy group with picolinic acid provided a derivative whose pyridine moiety can complex metal ions as Ag(I). Dhavale group has also succeeded in using fluorinated sugar amino acids to incorporate fluorine atoms toward the internal part of nanotubes made of cyclic tetra- and hexapeptides (bottom-left) [26]. Xu has employed γ- or ε-amino acids containing aromatic moieties, such as tolyl or benzimidazole, to replace two α-amino acids from an α-*D,L*-CP, to modify the internal cavity of the resulting assemblies (top-center). In the case of benzimidazole derivative, it is possible to incorporate one ionizable nitrogen group into the internal cavity of the nanotube [27].

6.2.3 Synthesis

The manufacture of SCPNs requires first the synthesis of the CP units. Usually, the linear sequence is synthesized using solid-phase peptide synthesis (SPPS), which allows the fast coupling of the amino acids, and the purification step is carried out only at the end of the synthesis, after the final cyclization step [28]. After the linear sequence is synthesized, the cyclization can be performed on resin or in solution [29]. While cyclization on resin is faster, it usually requires several orthogonal protecting groups to deprotect only the functionalities susceptible for cyclization. Generally, it also needs an amino acid bearing an orthogonal reactive functional group, such as amino or carboxylic acid groups, normally at their side chain, to anchor it to the resin [30]. On the contrary, cyclization in solution permits the synthesis of potentially any cyclic peptide, although fewer orthogonal protecting groups are also needed. However, it requires dilution conditions that make the cyclization step slower. Additionally, solid-phase cyclization can also be achieved using special resins, such as oxime, anchoring that allow not only the growth of the linear peptide but also the cyclization as an interesting alternative to release the fully protected peptide from the resin [31]. Although four different cyclization couplings can be performed, head-to-tail, head-to-side chain, side chain-to-tail, and side chain-to-side chain, usually the first one is the more employed. The development of efficient coupling reagents improved SPPS procedure [32]; normally, the peptide elongation is based on the *in situ* preparation of an activated carboxylic acid derivative of the *N*-protected amino acid. Generally, these reagents are very stable; present a low toxicity; and coupling procedures take place at very mild conditions without racemization of the coupled amino acid.

6.3 Characterization of Self-assembling Cyclic Peptide Nanotubes

Different techniques have been described in the literature to characterize the self-assembled cyclic peptide nanotubes. While each method provides different information, usually many of them complement each other to have a more complete understanding of the nanotubular structure.

Commonly used strategy to facilitate the study of CP ensembles in solution is the use of simple synthetic mimetics that reproduce some of the characteristic properties of the nanotubes, such as dimers, generated by cyclic peptides that contain selectively methylated amide groups [2e]. Generally, these peptides have all the residues of the same chirality *N*-alkylated, so that all these groups point toward the same face and, consequently, can only interact, forming hydrogen bonds, through the opposite face. Therefore, these *N*-alkylated CPs have restricted the assembling process to discrete dimers. In general, this modification that inhibits the formation of nanotubes while favoring the formation of dimers, which is easier to study, allows the full characterization by techniques such as ^{1}H NMR. The structural similarity of dimers to the infinite nanotube allows, in many cases, to extrapolate the results obtained with these models to the supramolecular ensemble.

6.3.1 NMR

Nuclear magnetic resonance is a nondestructive technique that provides information of the atoms with spin quantum number different from zero [33]. ^{1}H NMR is one of the more powerful techniques for the characterization of organic compounds that allows the identification of all protons of a sample in solution. For example, it has been used to characterize the formation of hydrogen bonds between the CPs by comparing the chemical shift of amide protons of the corresponding monomer with those of SCPNs. ^{1}H NMR is especially suitable for studying dimeric models of CPs and to evaluate the importance of cross-strand side-chain interactions in nanotube stability [2e]. In addition to the changes in chemical shift of protons involved in hydrogen bonds, proton J coupling constant also provides additional information. The proposed disk-shaped conformation implies in most of the cases the antiperiplanar disposition of most of the relevant protons of the cyclic peptide. Therefore, typically, amide protons provide signals with coupling constant of 7–9 Hz.

The SCPN stability in water can be studied by measuring the half-lives of the backbone H-bonds using $^1H/D$ exchange. When CPs are dissolved in D_2O, backbone amide protons are exchanged with different rates than free monomers; therefore, the signal decay can be quantified to correlate with SCPN stability [34].

NOE (nuclear Overhauser effects) experiments are used to measure transfer of nuclear spin polarization via cross-relaxation within a molecule or within molecules in close proximity. These cross-peaks give information about protons that are closer than 4 Å in space [35]. This technique is very helpful to understand the relative arrangement of the CPs that make the nanotube, based on the cross-strand polarizations. For example, the formation of homo- and heterodimers was studied by this technique, finding cross-peaks between two different CPs that confirms the formation of the heteromeric aggregate [36]. Also, the use of metals that coordinate with pyridine or diphenylphosphino groups attached to the side chain of a CP restricts the equilibrium to only one rotamer [37], which was confirmed by the NMR experiments. Hence, after adding the coordinating metal, most of the NMR spectra are simplified, corroborating that interdimer equilibrium is delimited to one single dimer whose NOE cross-peak confirms that it corresponds to the chelating structure. More recently, nucleobase-regulated dimerization processes were studied by RMN [38]. ^1H NMR, DOSY, TOCSY, and ROESY experiments showed that the addition of an adenine dinucleotide to a solution containing an α,γ-cyclic hexapeptide functionalized at the side chain of one Ser residue with thymine allowed to control the equilibrium between nonequivalent dimers by T-A pairing. The complementary base-pairing process through hydrogen bonds restricts the equilibria between dimers. This process was shown to be reverted by the addition of Cs_2CO_3.

Also, ^1H NMR offers the possibility to measure association constants between cyclic peptides by titration, although this technique is usually employed with methylated cyclopeptides that can only dimerize; then, the dimerization constant is extrapolated to the nanotube formation. Association constants for original nanotube forming models up to $K_a = 1.4 \times 10^4 \, M^{-1}$ have been reported in CCl_4 [14a].

Additionally, diffusion-ordered spectroscopy (DOSY) is another NMR technique that can give information about the size of the SCPNs by measuring diffusion coefficients in solution through the differences in their molecular translational diffusion. In this respect, the formation of venturi-type nanotubes, those with a narrow section in the central part of their inner hole, constructed with a precise number of cyclic subunits, was studied by this technique [39].

6.3.2 X-ray Diffraction

Being able to obtain the X-ray diffraction structure of SCPNs is sometimes quite challenging, but when successful, it allows to unambiguously characterize the formation of SCPNs and analyze the different structural parameters of the nanotubes: hydrogen bonds, cyclic peptide conformation, pore and outer diameters, distance between CP units, nanotube length, etc. However, the geometry observed in the solid state does not necessarily have to match with the interactions expected in solution. The fast growth of the nanotube in only one dimension made the structural characterization of peptide nanotubes difficult, and other techniques had to be used for their characterization. In general, the crystals are usually needle shaped in which the nanotubes are oriented along the major axis of these microcrystals. In any case, some successes have been reported.

Seebach was the first one to obtain the X-ray diffraction structure of SCPNs showing the stacking of CPs through β-sheet hydrogen bonds in a parallel arrangement [5]. In fact, as mentioned above, the three stereoisomers gave rise to the formation of the endless hydrogen-bonding network characteristic of the supramolecular tubular structures. In any case, only the homochiral CPs form the structure in which all the carbonyl groups are pointing in one direction, while the amide protons are oriented in the opposite direction, providing a macrodipole. This was evaluated with channel-forming CPs through voltage gating and current rectification.

In 2003, Ghadiri's group reported the X-ray structure of an open-ended hollow tubular structure made up of cyclic peptides composed of alternating α- and ε-amino acids. The crystal structure showed that the cyclic peptides were held together by intermolecular hydrogen bonds between backbone amide groups, and the solvent molecules were filling the nanotubes pore.

In 2004, Dory's group was able to grow monocrystals of nanotubes made of lactams composed of 3 or 4 δ-amino acids (Figure 6.8) [40]. Rigidity was introduced using *E*-alkenes. Lactams **A** and **B** displayed a β-turn structure in the crystal state, and the CPs are stacked in a parallel fashion. While in lactam **A**, only two amide groups participated in the intermolecular hydrogen-bonding network, in tetramer **B**, all amide groups were involved. Lactam **C** adopted a rigid crown structure with all amide groups parallel to the nanotube axis and allowed the use of substituents at δ-position to prepare chiral nanotubes [41].

With regard to the original CPs made of α-amino acids, it was not until recently that Chalmers and Perrier have reported the crystal structures of continuous SCPNs with antiparallel and parallel stacking arrangements (Figure 6.9) [42]. An interesting aspect of this work is not only the confirmation of the tubular structure but also the possibility of adopting the parallel or the antiparallel stacking directed by small changes in the peptide sequences. In their work, the replacement of a Leu for Ala in

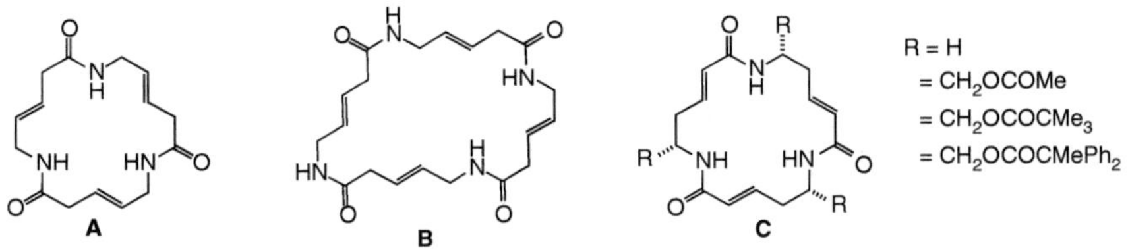

Figure 6.8 Lactams able to form nanotubes developed by Dory's group.

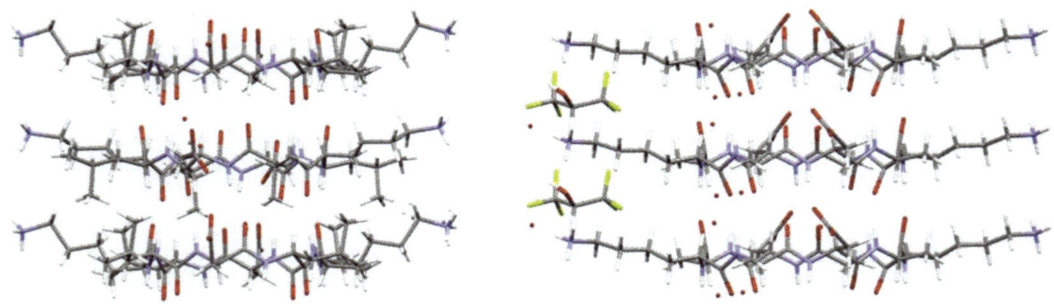

Figure 6.9 Fragments of the X-ray diffraction structures of nanotubes with antiparallel (left) and parallel (right) arrangements obtained by Chalmers.

cyclo-[(*L*-Asp-*D*-Leu-*L*-Lys-*D*-Leu-)$_2$] caused a change in the β-sheet-like structure from antiparallel to parallel stacking. Apparently, side-chain interactions between adjacent nanotubes have a great importance in the packing geometry and overcome the small difference in energy between both sheets.

6.3.3 Microscopy

Microscopy techniques have been widely employed to characterize SCPNs as they provide a very complete picture of the tubular structure, reaching in some cases atomic resolution. However, sometimes, the results can be biased as it is required that the sample is in the solid state or is manipulated to stand it on a solid support. They offer different information, depending on the source of light or energy used to irradiate the sample.

In his seminal work [1], Ghadiri observed by TEM the formation of bundles of hundreds of firmly packed nanotubes 10–30 μm long and 100–500 nm wide. A distance of 14.9 Å between nanotubes was measured by low-dose cryomicroscopy of the crystals. Electron diffraction studies established a trigonal lattice with $a = 4.73$ Å, $b = c = 15.1$ Å, $\alpha = 90°$, and $\beta = \gamma = 99°$, suggesting a highly symmetric nature of the octapeptide in which each nanotube is made up of stacked rings with distances between octapeptides in agreement with antiparallel β-sheet structures.

6.3.3.1 Scanning Electron Microscopy

SEM uses a high energy beam of electrons to irradiate the sample and create a high-resolution image of the sample surface or shape from the reflected electrons. The information that can be obtained from SEM images include the external morphology, aggregation, chemical composition, crystalline structure, and orientation of the crystals. It is also possible to measure the size of the crystals as the spatial resolution ranges from 50 to 100 nm. For example, SEM images were used by Hu and Jiang (Figure 6.10) to prove the self-assembly of in-tether chiral center-induced helical pentapeptides into nanobelts and nanotube structures with rectangular cross sections due to the diffusion-limited crystal growth [43]. The self-assembly process was directed by in-tether groups; chirality directed the helical structures, and aromatic rings provided the driving force for self-assembly by π–π stacking interactions.

6.3.3.2 Transmission Electron Microscopy

TEM also employs a high energy beam of electrons to irradiate the sample, which is then absorbed and scattered through the sample. To function properly, it is necessary

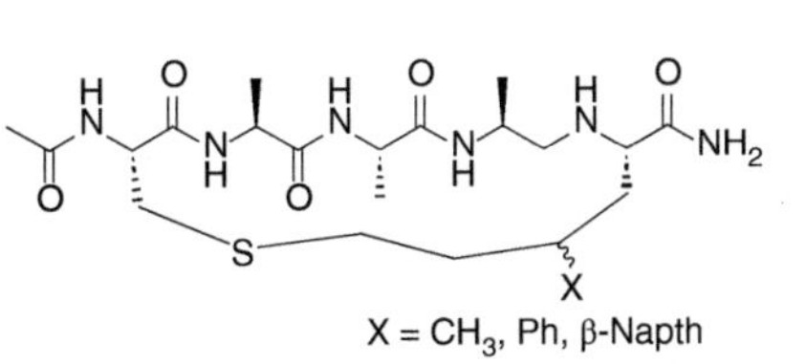

Figure 6.10 Cyclic pentapeptides that assemble into nanotubes based on side chain–side chain interactions of in-tether groups are responsible for the chirality of the resulting helices and whose characterization was carried out by SEM imaging.

to have enough electron density contrast between the sample and the substrate to resolve particles. TEM also allows the collection of images and statistical analysis to obtain information on the lengths, diameter, wall thickness, and width distribution of SCPNs. It provides better resolution than SEM, although sample preparation is more time consuming and expensive.

Scanning transmission electron microscopy (STEM) is a TEM technique in which the electron beam is focused at a large angle and is converged into a focal point which is raster scanned over an electron-transparent sample. Recently, Montenegro's group used this technique to prove the self-assembly of one-dimensional cyclic peptides into transparent two-dimensional giant nanosheet structures, resembling graphene layers (Figure 6.11) [44]. STEM was useful to confirm the presence of aligned nanotubes in areas with fragmented sheets, proving the reversible 1D-to-2D assembly and 2D-to-1D disassembly of the supramolecular system.

Recently, cryo-TEM has emerged as a new TEM technique that does not require the crystallization of the sample, as an aqueous solution of the sample is usually frozen using liquid nitrogen, and it allows to visualize the nanotubes in their natural state [45].

6.3.3.3 Electron Diffraction

Electron diffraction uses an electron beam to irradiate a crystalline sample, and the resulting change in its direction generates diffraction patterns that can be used for structural analysis. This technique was initially used by Ghadiri in his seminal work to characterize the unit cell and angles of SCPNs [1], obtaining highly ordered nanotube crystals, although the observed broad meridional spots in the electron diffraction images (16.1 ± 1.0 Å) suggested a disordered lateral packing of the nanotubes within the crystal lattice [46]. Development of SEM, TEM, and AFM has replaced this technique.

6.3.3.4 Atomic Force Microscopy

AFM technique requires the adsorption of the sample in a flat solid substrate and uses the deflection experienced by a nanoscale cantilever tip to give a surface topography map of the sample. AFM does not require differences in the electron density, but on the height and physicochemical properties of the adsorbed sample. It provides structural information of the nanotube with a high resolution <1 nm. Granja's group employed AFM images to measure the diameter of large-pore SCPNs made of cyclic α,δ-hexadecapeptides able to encapsulate C_{60}. AFM heights of $3.9–4.5 \pm 0.41$ nm were measured after C_{60} encapsulation (Figure 6.12) [11]. The combination of this technique with alternative detection methods can provide further information on the deposited material; for example, current sensing can be implemented to simultaneously obtain the material topography and the electric current flow at the contact point of the tip with the sample. In this alternative, tween nanotubes (carbon and peptide nanotubes, SWCNT/SCPN) were characterized by finding areas in which conducting properties of carbon nanotubes were combined with insulating properties of samples with different heights [47].

6.3.3.5 Fluorescence Microscopy

The functionalization of CPs with fluorescent probes allows SCPN characterization using this microscopy. This optical technique uses the emission properties of a

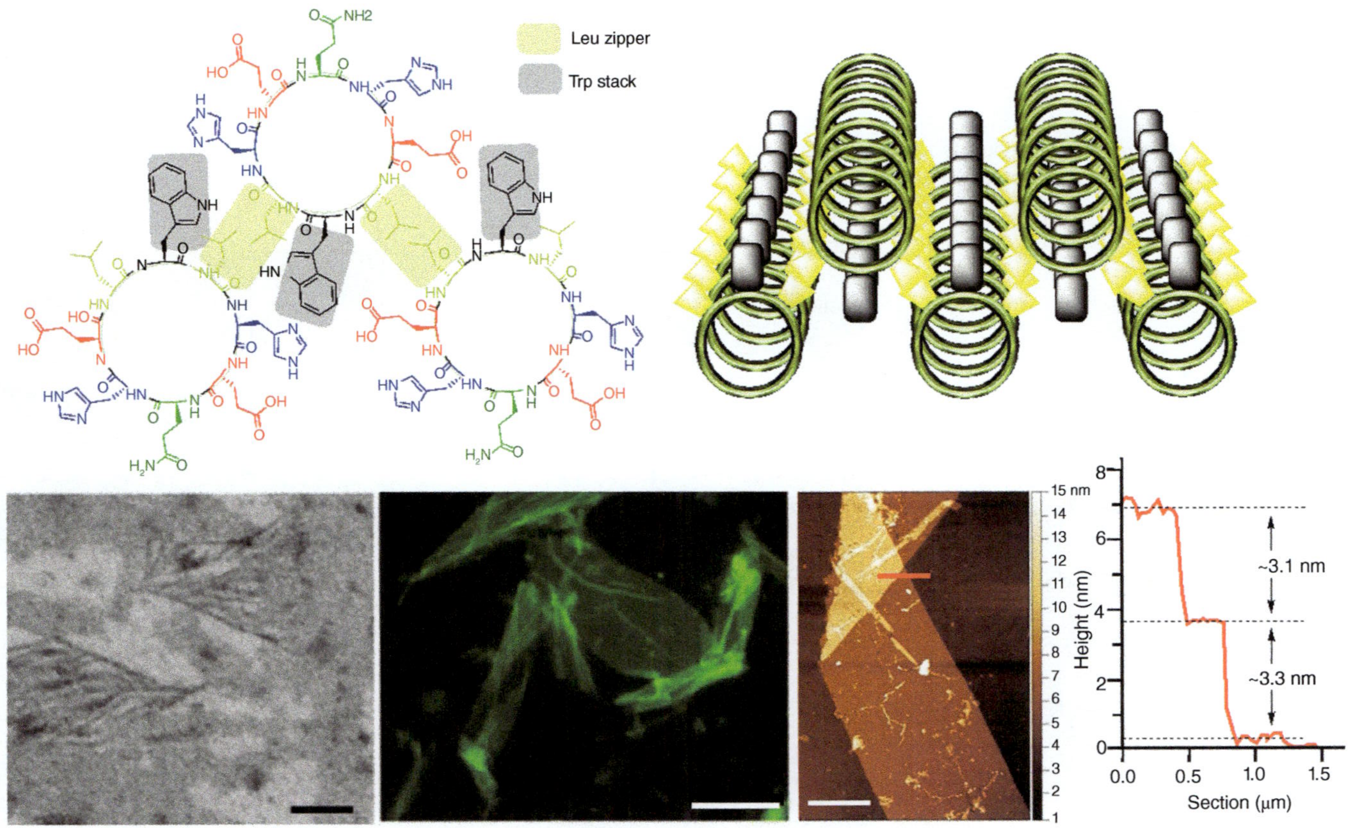

Figure 6.11 Model 2-D nanosheets developed by Montenegro's group based on amphipathic peptides. At the bottom, STEM images (left, scale bar = 100 nm), epifluorescence micrographs (center, scale bar = 50 μm), and AFM images (right, scale bar 2 μm), showing the approximately 3 nm height for the sheet. Images taken in part from reference [44b].

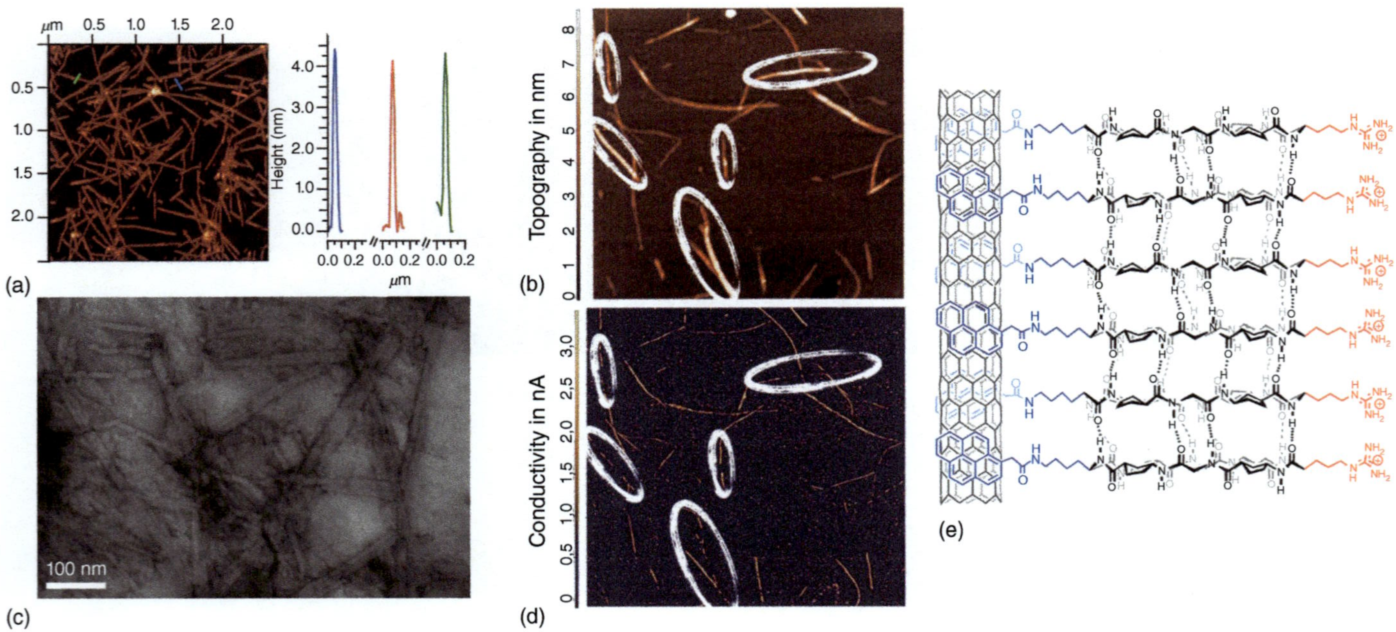

Figure 6.12 (a) AFM and (b) STEM images of α,δ-SCPNs. *Source:* Ref. [11]/Royal Society of Chemistry. (c) Topography image of a hybrid dispersion showing nanotubes. (d) Current mapping showing reduction in conductivity in high height segments. (e) Model of SWCNT/SCPN hybrids.
Source: Adapted from Ref. [47].

sample to create an image, instead of a beam of electrons as SEM or TEM does. Although less resolution could be expected, stochastic optical reconstruction microscopy (STORM) has recently emerged as a super-resolved fluorescence microscopy technique that provides super-resolution information (up to 20–30 nm) about SCPNs. STORM manipulates fluorophores to emit lights at different times by activating and deactivating photoswitchable fluorophores or by combining dye pairs with distinctive wavelength emissions [48]. This method was used by Rho to study the aggregation mechanism of two types of self-assembling cyclic peptide–diblock polymer conjugates with a hydrophobic core and hydrophilic corona decorated with two different dyes [49]. While premix showed colocalization of both peptides, coinjection showed the aggregation of the conjugates into separate nanotubes (Figure 6.13).

6.3.4 Fourier-transform Infrared Spectroscopy

FT-IR uses infrared light to irradiate a solution of the sample. When the chemical bonds interact with infrared light, their vibrational modes change, and this change

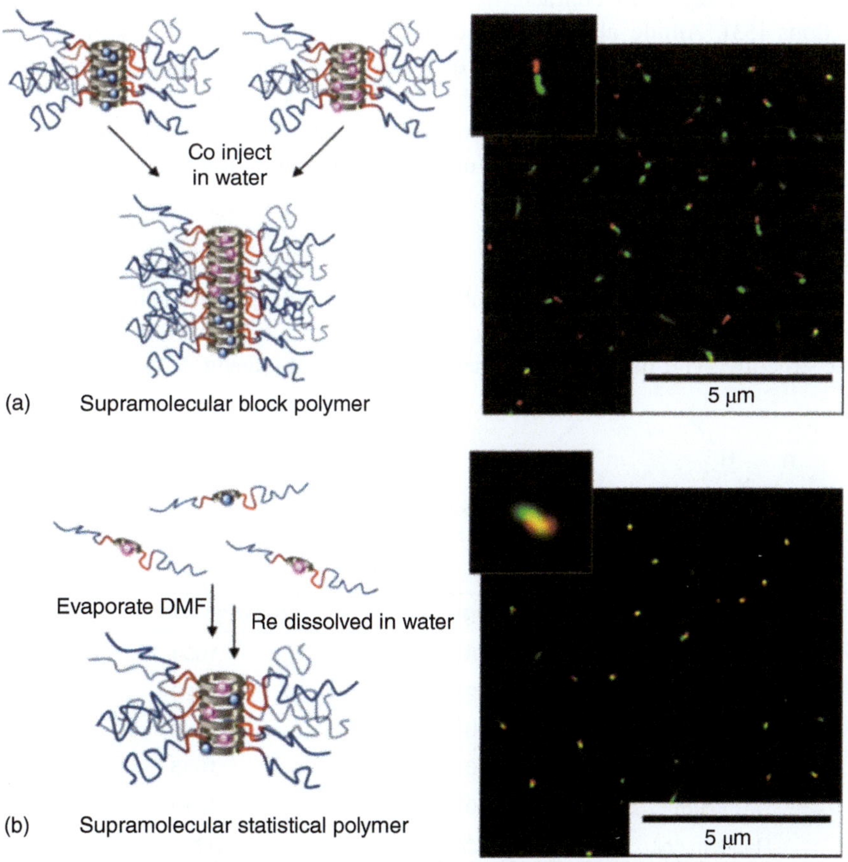

Figure 6.13 STORM images of SCPNs conjugated with a polymer. *Source:* Reproduced with permission from Ref. [49]/Springer Nature/CC BY 4.0.

is different if the atoms are involved in hydrogen bonds or not. Thus, this technique has been used to probe the stacking of CPs by hydrogen bonds, especially to explore the formation of hydrogen bonds between the amide protons and carbonyl groups, as typical C=O stretching amide and NH bending amide bands of tightly packed hydrogen bonds are shifted from non-hydrogen-bonded bonds [50]. This technique is a powerful tool for obtaining information about the secondary structure of peptides (Figure 6.14).

Infrared absorption experiments of SCPNs derived from cyclo-[(*D*-Ala-*L*-Glu-*D*-Ala-*L*-Gln-)$_2$] synthesized by Ghadiri [1] reported a frequency of 3277 cm^{-1} for the N—H stretch mode and a range of frequencies between 1628 and 1736 cm^{-1} with large peaks at 1628 and 1688 cm^{-1}. This is in good agreement with values reported for antiparallel β-sheet secondary structure [51]. Similar results were obtained with other SCPNs, these bands being the fingerprints that denote the formation of this type of structure. In fact, polarized attenuated total reflectance (ATR), grazing angle reflection absorption, and FTIR were used to determine the orientation of SCPN in phospholipid bilayers [52], supporting that hydrophobic CPs form transport-competent, membrane-spanning orientation of the SCPNs. Recently, Bastos group used similar strategies to characterize antimicrobial cyclic peptide–membrane interactions [53]. Amide absorptions adopt characteristic frequency and shape bands, which depends on the strength of hydrogen bonds between CO and NH groups. By combining polarized lens with ATR-FTIR, it is possible to study the orientation of peptides when interacting with lipid membranes, probing the formation of tubular structures that are enhanced upon interaction with the lipid membrane.

6.3.5 Fluorescence Spectroscopy

This technique is usually employed to probe the self-assembly process using a fluorescent probe that changes its fluorescence emission when it is entrapped inside a β-sheet structure. The most used fluorescent probe is thioflavin T, as it is barely

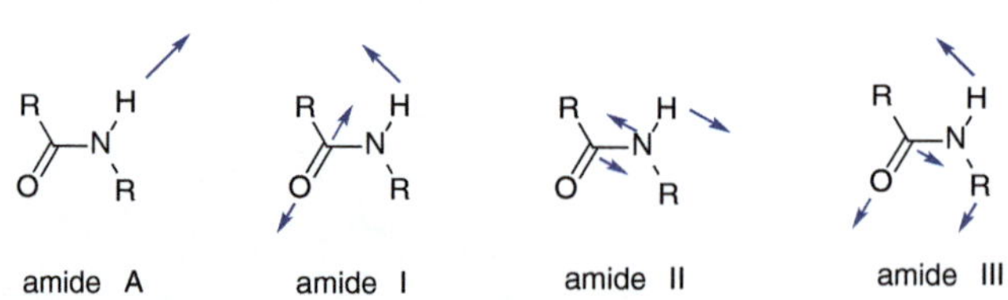

Secondary structure	Range (cm^{-1})	Average (cm^{-1})
α-helix	1648–1657	1654
Anti parallel β-sheet	1612–1641	1625
	1670–1694	1682
Parallel β-sheet	1626–1640	1633
β-turn	1660–1684	1673
Random coil	1640–1650	1645

Figure 6.14 Amide I vibration modes in different secondary structures.

fluorescent in its normal state but experiences an increase in its emission when planarized due to its embedding within an organized structure [54]. Alternatively, the covalent attachment of a fluorophore to the CP, which upon stacking is brought in proximity, generates a bathochromic shift of the fluorescence emission due to excimer formation. In this respect, CPs equipped with a pyrene molecule have been used to determine the association constant of these derivatives [55]. Also, the attachment of FRET pairs to two types of CPs has allowed to study their dynamic behavior. Granja's group has functionalized α,γ-CPs with pyrene and Dapoxyl moieties to study the excimer/FRET effects of the corresponding homo- and heterodimers. These studies would be relevant for the development of biosensors and biological imaging studies.

In recent studies, it has been proposed that the polycyclic aromatic moieties are not only good reporters of assembling properties of the CPs but also part of a hierarchical assembling process to provide hydrogels [56]. The selection of aromatic pendants (number and size) allows tuning the hydrogel properties [57], whose formation can be spatiotemporally controlled in confined spaces [58]. Also, the nanotube bending of CPs bearing an anthracene moiety could be followed in a real-time study, thanks to the fluorescence properties of this pendant [59]. Finally, another example of hierarchical assembling was developed by Perrier group by incorporating an aggregation-induced emission effect CP/polymer hybrids, permitting the *in situ* imaging of drug release upon SCPN disassembly [60].

6.3.6 Scattering Techniques

Some of the techniques described so far (such as TEM or AFM) require the immobilization of the sample in a solid surface. This requirement may modify the original structure of the nanotubular ensemble in solution, providing an unrealistic picture of what is taking place in the bulk solution. They also analyze a portion of the sample, which may not be representative of the complete sample. Nevertheless, scattering techniques allow the analysis of a large population of structures in solution, providing information about the structure and composition of the sample with high resolution. On the contrary, they require data fitting and modeling.

Static light scattering (SLS) has been used to provide information about the number of cyclic peptides that make the SCPNs. It measures the angular dependence of scattered light when a visible laser goes through a solution of the sample. It requires to measure different sample concentrations to construct a plot from which the "zero-angle" intensity of a sample at infinite dilution can be extrapolated, giving the molecular weight of the SCPNs. Additionally, dynamic light scattering (DLS) provides information about the size distribution profile of SCPNs in solution [49].

Small-angle X-ray scattering (SAXS) provides information on the size, shape, and composition of CP aggregates and covers length scales from 1 to 100 nm. It uses 0.07–0.2 nm X-rays to measure the intensity of monochromatic X-rays scattered at small angles by a sample as a function of the momentum transfer. This technique requires a substantial difference between the scattering light density of the particle and the solvent. CP-derived hydrogels were characterized using this technique to complement via electron-microscopy studies [61].

Similarly, small-angle neutron scattering (SANS) employs a neutron ray that scatters from nuclear density and provides information of samples where SCPNs have similar atomic composition to the solvents frequently employed. Also, deuteration of the sample or the solvent allows to enhance the scattering contrast.

SANS has been employed to characterize CP–polymer conjugates, as the scattering contrast is sufficient to differentiate the cyclic peptide core and the solvated polymer corona. SANS has also been used to study the influence of polymer grafting density and solvent polarity on the structure of CP–PEG conjugate nanotubes. For example, Perrier's group has employed this technique to study the influence of the solvent nature in nanotube formation from CP conjugates: while solvents accepting hydrogen bonds restricted the formation of nanotubes to monomeric or oligomeric assemblies, longer nanotubes were obtained in nonpolar solvents [62]. This technique has also been used to prove the different structures acquired by stimuli-responsive polymer–SCPN conjugates as it is able to differentiate 3D morphologies such as nanotubes, micelles, and tubisomes [63].

6.3.7 Size-exclusion Chromatography

This technique allows to measure the dispersity and size/molar mass distribution of an SCPN sample by giving the representative molar mass or size values for every sample fraction. It usually requires the coupling of a chromatographic device (usually an HPLC with a size-exclusion column) with an appropriate detector (light scattering-refractive index-UV–vis). Size-exclusion chromatography relies on the different ability of SCPNs to permeate through a heterogenous porous column due to the different hydrodynamic volume. It is a very sensitive, robust, and reproducible technique.

As a drawback, the interaction of the SCPNs with the column can bias the results as degradation induced by shear forces can modify the structure of SCPNs. This problem can be solved using asymmetric flow-field-flow fractionation in which the separation takes place in an open channel [64].

6.3.8 Computational Calculations

While much information can be extracted from the different techniques described in the previous sections, it is still challenging to get information of the dynamic behavior of SCPNs in solution as well as their variability with solvents, dynamics, deformability, energetic landscapes, the nature of the intra- and intermolecular interactions stabilizing the nanotube, and their properties as a molecular transporter or container [65]. Computational techniques can help to fill this gap. Also, they provide a level of atomic resolution difficult to be obtained by other techniques and allow to study the interactions between SCPNs themselves and with other supramolecular structures such as membranes. Thus, computational methods offer an attractive way to complement the information obtained from experimental methods to get a better understanding of SCPN structure and interactions [10].

The first idea that SCPNs could be constructed by stacking of cyclic peptides came from a theoretical model from DeSantis in 1974 [66]. He predicted that a cyclic derivative with an even number of amino acids of alternating *D*- and *L*-configuration should have a β-type conformation in which hydrogen bonds between perpendicularly pointing carbonyl groups and NHs would favor the cyclic component stacking to form nanotubes. However, it was not until 1993 that Ghadiri synthesized the first SCPNs and verified DeSantis hypothesis [1]. While empirical methods were initially used to study the energetics, geometrical structures, and vibrational modes of cyclic peptides, quantum mechanical methods have expanded nowadays because they give a more complete overview of their properties, such as the electronic structure [67].

6.3.8.1 Molecular Modeling

Molecular modeling of SCPNs can provide information about the arrangement of the CPs that make up the aggregates: structure, CP conformation, possible steric hindrance effect between groups, stability, etc. One of the first features studied was focused to elucidate the antiparallel or parallel arrangement of the β-sheet structures of the cyclic peptides within the nanotube.

Initial semiempirical AM1 calculations made by Gailer and Feigel in 1997 showed very small difference in energy between parallel and antiparallel β-sheets [68]. Lewis studied the cyclic peptide *cyclo*-[(*D*-Ala-*L*-Glu-*D*-Ala-*L*-Gln-)$_2$] and modeled the strong intramolecular interactions with DFT and the weak intermolecular interactions with a simple theory that considered Coulombic, exchange, and hopping interactions between the CPs. He predicted a binding energy per dimer of 58.6 kcal/mol and interaction energy of −7.3 kcal/mol per hydrogen bond [67]. Semiempirical AM1 and DFT calculations (with the B3LYP functional) on CPs of different sizes showed a preference for the antiparallel arrangement (−5.7 kcal/mol per hydrogen bond) [69]. It was shown that antiparallel stacking produces a stronger intermolecular attraction, CO and NH groups being better positioned, more perpendicular to the ring plane to establish better hydrogen bonds. Other groups have also investigated the structures of monomers and dimers of cyclic peptides [70].

García-Fandiño and coworkers have used B3LYP and M05-2X DFT functionals to characterize the molecular structures of dimers of hexameric and octameric γ-Ach- and γ-Acp-based α,γ-cyclopeptides, the association constants, and the preference for heterodimerization [71]. Full geometry optimizations were made at the B3LYP/6-31G(d) level, and single point calculations with the B3LYP and M05- 2X functionals and the 6–31 + G(d,p) basis set were carried out to improve the energetics of α,γ-CPs. They found slow SCF convergence for M05-2X method when compared with B3LYP and showed that hydrogen-bonding interactions were stronger in the α–α stacking for the hexapeptide dimers, probably because the NH groups of the α-amino acids are more polarized than those of the γ-amino acids. α–α dimers showed interaction energies 2 kcal/mol stronger that those determined by the γ–γ species, mainly due to deformation energies in the monomers. Also, α–α heterodimerization was energetically favored over α–α homodimerization, although the differences in energy were very close.

To date, both computational and experimental studies propose a thermodynamic preference for the antiparallel grouping of CPs made up of α-amino acids, which probably minimizes the steric interactions between cross-strand side chains [72]. However, recent experimental and DFT-based calculations have shown that cyclic peptides made up of α,δ-amino acids have a preference for a parallel arrangement [10], probably due to the disposition of the axial hydrogen atoms, and a similar conclusion could be established for α,γ-amino acid-based cyclic peptides.

6.3.8.2 Molecular Dynamics Simulations

Molecular dynamics simulations of SCPNs provide insights into the structural characteristics of the tubular systems. In addition, cyclic peptides usually adopt different conformations in solution, which are difficult to resolve using experimental techniques. Molecular dynamics simulations offer a way to get understanding of sequence-structure–function relationship of CPs. Lin has recently reviewed novel computational methods able to relate the CP sequence and structure with biological activity [73].

It is also possible to use basic logic prediction methods based on computation simulations to model the interactions of CPs with microbial membranes. Understanding these interactions is key to design the appropriate sequence of amino acids to build CPs that self-assemble to form SCPNs able to interact selectively with bacterial membranes [74].

Molecular dynamics has also been used to study the carpet-like mode of interaction between the SCPNs and lipidic bilayers to explain how these materials interfere with the cell membranes of microbes, although initial results have shown that SCPNs retain the tubular structure when interacting with the membrane [75]. Tarek group has found that the interaction of the nanotube formed by eight *cyclo*-[*D*-Arg-*L*-Arg-*D*-Lys-*L*-Trp-*D*-Leu-*L*-Trp-*D*-Leu-*L*-Trp-] subunits with the surface of a hydrated dimyristoylphosphatidylcholine bilayer may partially disassemble before insertion into the membrane [76].

Moreover, dynamic simulations were already used to study the transport of water and other molecules through the SCPNs. Ghadiri already envisioned the potential of these structures to be used as transmembrane channels, as a diffusion constant of $4.4 \cdot 10^{-6}\ cm^2 s^{-1}$ was measured [75a], 17% slower than the diffusion constant of water, but more than twice as fast as the diffusion constant of water in gramicidin [77]. His group was also able to obtain diffusion rates for K^+ and Na^+ almost three times faster with *cyclo*-[*L*-Gln-*D*-Leu-(*L*-Trp-*D*-Leu-)$_3$] than gramicidin A [17a].

Also, the possibility to tune the diameter of the SCPNs makes possible to gain selectivity in the transport of ions such Na^+, K^+, Ca^{2+} [78], water [79], ammonia [80], or glutamic acid (Figure 6.15) [81, 82]. Additionally, it is possible to tune the pore type across the membrane by modifying the CP composition: large pores (bundles of SCPNs) are produced using some amino acids with charged side chains, while monomeric pores are generated using amino acids with neutral side chains [17d].

Recently, the development of machine learning methods has been applied to predict the structure of cyclic peptides in combination with molecular dynamics [83]. The authors have developed a method called StrEAMM (Structural Ensembles Achieved by Molecular Dynamics and Machine Learning) to predict structural ensembles of cyclic peptides with a seven-order-of-magnitude increase in speed

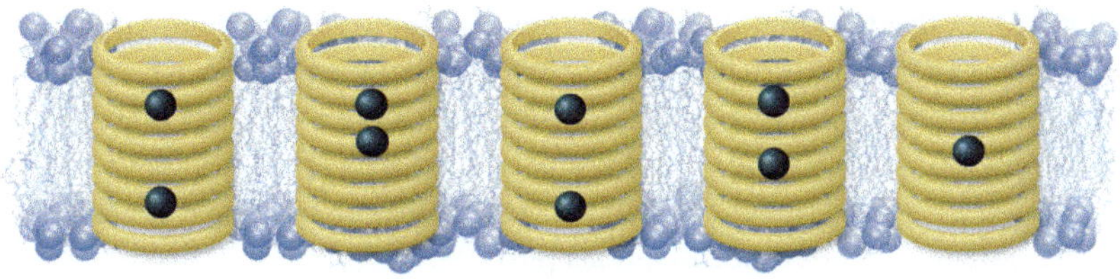

Figure 6.15 Cartoon representation of molecular modeling of α,γ-SCPNs in different electrolyte solutions. *Source:* Adapted from Ref. [81].

(the prediction for each individual CP can be made using less than 1 second of computation time) with respect to explicit-solvent simulations at the same speed.

6.4 Recent Applications

The application fields of SCPNs are immense [2]. Cyclic peptide nanotube crystals have shown to be extraordinarily stable in both highly acidic (pH 1) and highly basic solutions (pH 14) and even in boiling water. They are also stable in most common polar and nonpolar organic solvents. Joshi has recently studied the nanomechanical characterization of *D,L*-CP assemblies of *cyclo*-[(*L*-Gln-*D*-Leu-)$_4$], showing that the peptide crystals are among the most robust known proteinaceous micro- and nanofibers, with an elastic modulus of 11.3 ± 3.3 GPa, hardness of 387 ± 136 MPa, and strength (bending) of 98 ± 19 MPa [84]. These properties make SCPNs ideal to be used as composites.

One of the most extended applications of SCPNs is as synthetic channels that allow transport of small molecules through the lumen across lipid membranes [81]. The tubular structure of SCPNs allows them to interact with lipid bilayers, and consequently, they have found important applications as antibacterial agents and artificial transmembrane ion channels but also in biosensing, pharmacology, catalysis, electronics, molecular recognition, drug delivery, smart materials, molecular motors, and separation technologies [2e].

As antimicrobial agents, SCPNs can specifically insert into bacterial membranes, disrupt transmembrane ion balance, and consequently perturb the membrane insulating properties, increasing the permeability through the membrane and promoting cell death through a mechanism called "carpet-like" [85]. By selecting the appropriate amino-acid sequence, it is possible to get selectivity for Gram-negative or -positive bacteria. Also, the rigid ring scaffolds give stability for proteolysis [86]. As they are easy to synthesize, it is relatively straightforward to construct libraries of CPs to make high-throughput screening of SCPNs as antibacterial agents. Recently, Granja's group has reported a double orthogonal clicking strategy to prepare antimicrobial peptides with increased aqueous solubility and reduced hemolytic properties [87]. The basic component is a CP with only two hydrophobic residues, one of them is a long alkyl chain, incorporated by a copper-catalyzed azide–alkyne cycloaddition. To reduce the hemolytic properties, a saccharide can be incorporated at the hydrophilic core without reducing the cytotoxic properties [86c, 87].

The powerful antimicrobial properties of nanotube forming cyclic peptides were further developed by Ghadiri using a combinatorial strategy, where they have shown the ability of these CPs to work at the strong acidic conditions of the gut to remodel mouse microbiome to prevent atherosclerosis [88].

The ability to encapsulate molecules in the inner part of the nanotubes has found important applications of SCPNs in drug delivery systems. SCPNs make possible to construct nanoparticle ensembles in which size, morphology, and surface can be totally controlled. In this way, they can be used to improve the uptake, bioavailability, and the time of circulation and ensure better cellular uptake of hydrophobic and poorly soluble drugs. Also, their ability to be intercalated into membranes makes it possible to transport drugs through the cell membrane and release drugs in intracellular compartments. Following this strategy, Chen used SCPNs to transport anticancer drug 5-fluorouracil [18c].

Additionally, the possibility to modify the external surface of the nanotube with a polymer permits to tune their size, solubility, aggregation, etc. and enhance the delivery properties [89]. Studies carried out comparing molecular bottle-brush polymers and the supramolecular derivative (SCPN) have shown a reduced plasma circulation time; SCPNs were also able to mostly circumvent organ accumulation [90]. This was attributed to the supramolecular behavior of SCPNs; the weak intermolecular hydrogen bonds should favor the gradual disassembly in the aqueous media, facilitating their metabolism by the body, while a common limitation of nanoparticles is their accumulation in different organs. The attachment of specific drugs at the polymeric components was found to enhance cytotoxicity and, in some cases, implying different mechanisms of action. For example, metal-loaded hybrids, such as ruthenium, showed improved cellular uptake and higher cytotoxicity than free drug [18d].

A recent application of SCPNs has been found as gene vehicles for plasmids, DNA, shRNA, and siRNA transport [91]. Apparently, SCPNs behave as competent shield to protect nucleic material from the immune response and also from DNases and RNases, acidic conditions, and bile digestions.

Cyclic peptides have also shown antiviral activity, preventing the acidification of endosomes by transmembrane channel ion transport, inhibiting the lysis of the membrane and the viral escape from endosomes [92].

Recent biomedical applications include the binding and stabilization of SCPNs with the nonamyloid-β region of α-synucleins, inhibiting aggregation and the downstream pathways that lead to amyloidogenic diseases such as Parkinson's or Alzheimer's [93] and the employment of SCPNs as antiatherosclerotic agents able to regulate HDLs transport [94].

By modifying the side chains of the amino acids pointing outward in the SCPNs with aromatic molecules, such as naphthalene diimides, fullerenes, pyrenes, or tetrathiafulvalene, it is possible to employ SCPNs as organic semiconducting and conduction materials, as the tubular structure helps to align the aromatic rings, with future applications such as optical and electronic devices [95]. Also, the

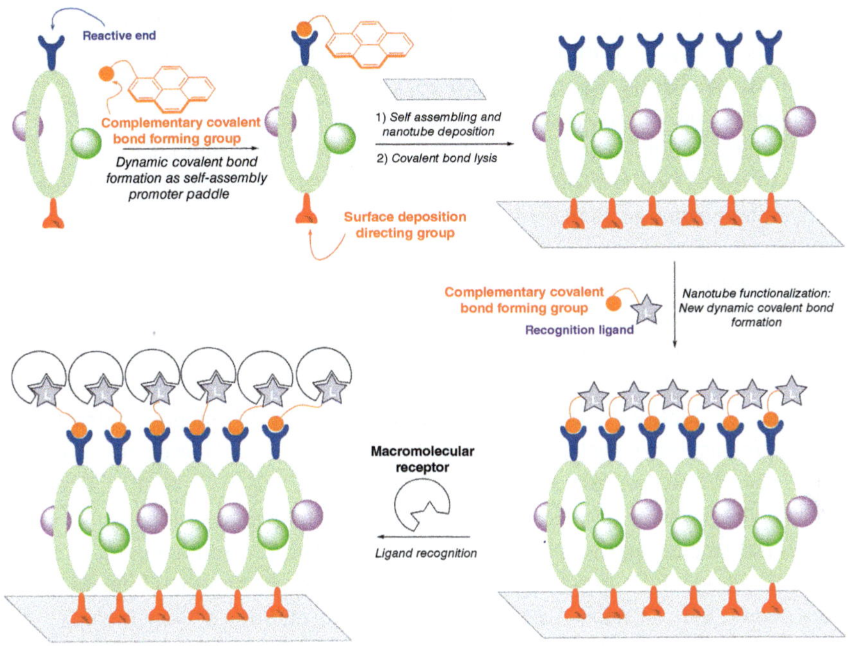

Figure 6.16 Dynamic deposition strategy on nanotube surface developed by Granja's group.

possibility to form heterodimers by combination of cyclic peptides with a different design has been used to create photosystems for artificial light harvesting and light conversion [96].

The use of dynamic covalent bonds allows the exchange of different groups at the nanotube surface changing their properties [97]. In this respect, Granja's group has shown that precisely designed cyclic peptides having one dynamic covalent group can be used in the alignment of different macromolecules on the surface of the nanotube once deposited on a surface. The use of the hydrazine/hydrazone conversion allows to exchange different aldehydes in a controlled way. In this way, small dendrimers or proteins such as concanavalin A were deposited and aligned guided by the nanotube structure (Figure 6.16).

References

1 Ghadiri, M.R., Granja, J.R., Milligan, R.A. et al. (1993). Self-assembling organic nanotubes based on a cyclic peptide architecture. *Nature* 366: 324–327.

2 (a) Gao, X. and Matsui, H. (2005). Peptide-based nanotubes and their applications in bionanotechnology. *Adv. Mater.* 17: 2037–2050. (b) Hamley, I.W. (2014). Peptide Nanotubes. *Angew. Chem. Int. Ed.* 53: 6866–6881. (c) Hsieh, W.-H. and Liaw, J. (2019). Applications of cyclic peptide nanotubes (cPNTs). *J. Food Drug Anal.* 27: 32–47. (d) Shimizu, T., Ding, W., and Kameta, N. (2020). Soft-matter nanotubes:

a platform for diverse functions and applications. *Chem. Rev.* 120: 2347–2407. (e) Song, Q., Cheng, Z., Kariuki, M. et al. (2021). Molecular self-assembly and supramolecular chemistry of cyclic peptides. *Chem. Rev.* 121: 13936–13995.

3 (a) Tomasic, L. and Lorenzi, G.P. (1987). Some cyclic oligopeptide with S_{2n} symmetry. *Helv. Chim. Acta* 70: 1012–1016. (b) Pavone, V., Benedetti, E., Blasio, B.D. et al. (1989). Regularly alternating LD-peptides. III. Hexacyclic peptides from valine or phenylalanine. *Biopolymers* 28: 215–223.

4 (a) Khazanovich, N., Granja, J.R., McRee, D.E. et al. (1994). Nanoscale tubular ensembles with specified internal diameters. design of a self-assembled nanotube with a 13 Å pore. *J. Am. Chem. Soc.* 116: 6011–6012. (b) Clark, T.D., Buriak, J.M., Kobayashi, K. et al. (1998). Cylindrical β-sheet peptide assemblies. *J. Am. Chem. Soc.* 120: 8949–8962.

5 (a) Seebach, D., Matthews, J.L., Meden, A. et al. (1997). Cyclo-β-peptides: structure and tubular stacking of cyclic tetramers of 3-aminobutanoic acid as determined from powder diffraction data. *Helv. Chim. Acta* 80: 173–182. (b) Clark, T.D., Buehler, L.K., and Ghadiri, M.R. (1998). Self-assembling cyclic β3-peptide nanotubes as artificial transmembrane ion channels. *J. Am. Chem. Soc.* 120: 651–656.

6 Karle, I.L., Handa, B.K., and Hassall, C.H. (1975). The conformation of the cyclic tetrapeptide L-Ser(O-t-Bu)-$^{\beta}$-Ala-Gly-L-$^{\beta}$-Asp(OMe) containing a 14-membered ring. *Acta Crystallogr. Sect. B* 31: 555–560.

7 (a) Tabata, Y., Uji, H., Imai, T., and Kimura, S. (2018). Two one-dimensional arrays of naphthyl and anthryl groups along peptide nanotubes prepared from cyclic peptides comprising α- and β-amino acids. *Soft Matter* 14: 7597–7604. (b) Ohmura, H., Tabata, Y., Kimura, S., and Uji, H. (2021). Piezoelectric properties reflecting nanostructures of tetrathiafulvalene and chloranil complexes using cyclic peptide nanotube scaffolds. *Peptide Sci.* 113: e24192.

8 Amorín, M., Castedo, L., and Granja, J.R. (2003). New cyclic peptide assemblies with hydrophobic cavities: the structural and thermodynamic basis of a new class of peptide nanotubes. *J. Am. Chem. Soc.* 125: 2844–2845.

9 Brea, R.J., Castedo, L., and Granja, J.R. (2007). Large-diameter self-assembled dimers of α,γ-cyclic peptides, with the nanotubular solid-state structure of cyclo-[(L-Leu-D-MeN-γ-Acp-)$_4$]·4$CHCl_2COOH$. *Chem. Commun.* 3267–3269.

10 Calvelo, M., Lamas, A., Guerra, A. et al. (2020). Parallel versus antiparallel β-sheet structure in cyclic peptide hybrids containing γ- or δ-cyclic amino acids. *Chem. Eur. J.* 26: 5846–5858.

11 Lamas, A., Guerra, A., Amorín, M., and Granja, J.R. (2018). New self-assembling peptide nanotubes of large diameter using δ-amino acids. *Chem. Sci.* 9: 8228–8233.

12 (a) Fujimura, F., Hirata, T., Morita, T. et al. (2006). Columnar assembly of cyclic β-amino acid functionalized with pyranose rings. *Biomacromolecules* 7: 2394–2400. (b) Hirata, T., Fujimura, F., Horikawa, Y. et al. (2007). Molecular assembly formation of cyclic Hexa-β-peptide composed of acetylated Glycosamino acids. *Biopolymers* 88: 150–156. (c) Fujimura, F., Horikawa, Y., Morita, T. et al. (2007). Double assembly composed of lectin association with columnar molecular assembly of cyclic tri-β-peptide having sugar units. *Biomacromolecules* 8: 611–616.

13 Ishihara, Y. and Kimura, S. (2010). Nanofiber formation of amphiphilic cyclic tri-β-peptide. *J. Pept. Sci.* 16: 110–114.

14 (a) Ghadiri, M.R., Kobayashi, K., Granja, J.R. et al. (1995). The structural and thermodynamic basis for the formation of self-assembled peptide nanotubes. *Angew. Chem. Int. Ed. Engl.* 34: 93–95; (b) Tabata, Y., Kamano, Y., Kimura, S., and Uji, H. (2020). Engineering pH-responsive switching of donor–π–acceptor chromophore alignments along a peptide nanotube scaffold. *RSC Adv.* 10: 3588–3592.

15 Couet, J., Samuel, J.D.J.S., Kopyshev, A. et al. (2005). Peptide-polymer hybrid nanotubes. *Angew. Chem. Int. Ed.* 44: 3297–3301.

16 Silva, A.R.P., Guimarães, M.S., Rabelo, J. et al. (2022). Recent advances in the design of antimicrobial peptide conjugates. *J. Mater. Chem. B* 10: 3587–3600.

17 (a) Ghadiri, M.R., Granja, J.R., and Buehler, L.K. (1994). Artificial transmembrane ion channels from self-assembling peptide nanotubes. *Nature* 369: 301–304. (b) Danial, M., My-Nhi Tran, C., Young, P.G. et al. (2013). Janus cyclic peptide-polymer nanotubes. *Nat. Commun.* 4: 2780. (c) Danial, M., Tran, C.M.N., Jolliffe, K.A., and Perrier, S. (2014). Thermal gating in lipid membranes using Thermoresponsive cyclic peptide- polymer conjugates. *J. Am. Chem. Soc.* 136: 8018–8026. (d) Danial, M., Perrier, S., and Jolliffe, K.A. (2015). Effect of the amino acid composition of cyclic peptides on their self-assembly in lipid bilayers. *Org. Biomol. Chem.* 13: 2464–2473.

18 (a) Blunden, B.M., Chapman, R., Danial, M. et al. (2014). Drug conjugation to cyclic peptide-polymer self-assembling nanotubes. *Chem. Eur. J.* 20: 12745–12749. (b) Wang, Y., Yi, S., Sun, L. et al. (2014). Doxorubicin-loaded cyclic peptide nanotube bundles overcome chemoresistance in breast cancer cells. *J. Biomed. Nanotechnol.* 10: 445–454. (c) Chen, J., Zhang, B., Xia, F. et al. (2016). Transmembrane delivery of anticancer drugs through self-assembly of cyclic peptide nanotubes. *Nanoscale* 8: 7127–7136. (d) Larnaudie, S.C., Brendel, J.C., Romero-Canelon, I. et al. (2018). Cyclic peptide-polymer nanotubes as efficient and highly potent drug delivery systems for organometallic anticancer complexes. *Biomacromolecules* 19: 239–247. (e) Yang, J., Song, J.-I., Song, Q. et al. (2020). Hierarchical self- assembled photo-responsive Tubisomes from a cyclic peptide-bridged amphiphilic block copolymer. *Angew. Chem. Int. Ed.* 59: 8860–8863.

19 (a) Gil, E.S. and Hudson, S.M. (2004). Stimuli-Reponsive polymers and their bioconjugates. *Prog. Polym. Sci.* 29: 1173–1222. (b) Cobo, I., Li, M., Sumerlin, B.S., and Perrier, S. (2015). Smart hybrid materials by conjugation of responsive polymers to biomacromolecules. *Nat. Mater.* 14: 143–159.

20 Amorín, M., Brea, R.J., Castedo, L., and Granja, J.R. (2005). The smallest α,γ-peptide nanotubulet segments: cyclic α,γ-tetrapeptide dimers. *Org. Lett.* 7: 4681–4684.

21 Blanco-González, A., Calvelo, M., Garrido, P.F. et al. (2021). Transmembrane self-assembled cyclic peptide nanotubes based on α-residues and cyclic δ-amino acids: a computational study. *Front. Chem.* 9: 704160.

22 (a) Horne, W.S., Stout, C.D., and Ghadiri, M.R. (2003). A heterocyclic peptide nanotube. *J. Am. Chem. Soc.* 125: 9372–9376. (b) van Maarseveen, J.H., Horne, W.S., and Ghadiri, M.R. (2005). Efficient route to C2 symmetric heterocyclic backbone modified cyclic peptides. *Org. Lett.* 7: 4503–4506.

23 Wang, D., Guo, L., Zhang, J. et al. (2001). Artificial ion channels formed by a synthetic cyclic peptide. *J. Pept. Res.* 57: 301–306.

24 Rodríguez-Vázquez, N., García-Fandiño, R., Aldegunde, M.J. et al. (2017). Cis-platinum complex encapsulated in self-assembling cyclic peptide dimers. *Org. Lett.* 19: 2560–2563.

25 Rodríguez-Vázquez, N., García-Fandiño, R., Amorín, M., and Granja, J.R. (2016). Self-assembling α,γ-cyclic peptides that generate cavities with tunable properties. *Chem. Sci.* 7: 183–187.

26 Burade, S.S., Saha, T., Bhuma, N. et al. (2017). Self-assembly of fluorinated sugar amino acid derived α,γ-cyclic peptides into transmembrane anion transport. *Org. Lett.* 19: 5948–5951.

27 (a) Darnall, S.M., Li, C., Dunbar, M. et al. (2019). Organic nanotube with subnanometer, pH-responsive lumen. *J. Am. Chem. Soc.* 141: 10953–10957. (b) Hourani, R., Zhang, C., van der Weegen, R. et al. (2011). Processable cyclic peptide nanotubes with tunable interiors. *J. Am. Chem. Soc.* 133: 15296–15299.

28 Coin, I., Beyermann, M., and Bienert, M. (2007). Solid-phase peptide synthesis: from standard procedures to the synthesis of difficult sequences. *Nat. Protoc.* 2: 3247–3256.

29 Coppock, M.B. and Winton, A.L. (2022). *Peptide Macrocycles, Methods and Protocols*. Eds. 2 vol, 371. Human Press.

30 Chow, H.Y., Zhang, Y., Matheson, E., and Li, X. (2019). Ligation technologies for the synthesis of cyclic peptides. *Chem. Rev.* 119: 9971–10001.

31 Ösapay, G., Profit, A., and Taylor, J.W. (1990). Synthesis of tyrocidine a: Use of oxime resin for peptide chain assembly and cyclization. *Tetrahedron Lett.* 31: 6121–6124.

32 El-Faham, A. and Albericio, F. (2011). Peptide coupling reagents, more than a letter soup. *Chem. Rev.* 111: 6557–6602.

33 Günther, H. (2013). *NMR Spectroscopy, Basic Principles, Concepts and Applications in Chemistry*, 3rde. Weinheim: Wiley-VCH.

34 Silk, M.R., Price, J.R., Mohanty, B. et al. (2021). Side-chain interactions in d/l peptide nanotubes: studies by crystallography, NMR spectroscopy and molecular dynamics. *Chem. Eur. J.* 27: 14489–14500.

35 Harwood, L.M. and Claridge, T.D.W. (1996). *Introduction to Organic Spectroscopy*. Oxford: Oxford University Press.

36 Brea, R.J., Amorín, M., Castedo, L., and Granja, J.R. (2005). Methyl-blocked dimeric α,γ-peptide nanotube segments: formation of a peptide heterodimer through backbone–backbone interactions. *Angew. Chem.* 117: 5856–5859.

37 Panciera, M., Amorín, M., Castedo, L., and Granja, J.R. (2013). Design of Stable β-sheet-based cyclic peptide assemblies assisted by metal coordination: selective homo- and heterodimer formation. *Chem. Eur. J.* 19: 4826–4834.

38 Panciera, M., Gónzalez-Freire, E., Calvelo, M. et al. (2020). Induced α,γ-cyclic peptide rotodimer recognition by nucleobase scaffolds. *Pep. Sci.* 112: e24132.

39 Fuertes, A., Amorín, M., and Granja, J.R. (2020). Versatile symport transporters based on cyclic peptide dimers. *Chem. Commun.* 56: 46–49.

40 (a) Gauthier, D., Baillargeon, P., Drouin, M., and Dory, Y.L. (2001). Self-assembly of cyclic peptides into nanotubes and then into highly anisotropic crystalline materials. *Angew. Chem. Int. Ed.* 40: 4635–4638. (b) Leclair, S., Baillargeon, P., Skouta, R. et al. (2004). Micrometer-sized hexagonal tubes self-assembled by a cyclic peptide in a liquid crystal. *Angew. Chem. Int. Ed.* 43: 349–353.

41 Nazari, N., Bernard, S., Fortin, D. et al. (2023). Triple Thorpe-Ingold effect in the synthesis of 18-membered C3 symmetric lactams stacking as endless supramolecular tubes. *Chem. Eur. J.* 29: e202203717.

42 Silk, M.R., Newman, J., Ratcliffe, J.C. et al. (2017). Parallel and antiparallel cyclic D/L-peptide nanotubes. *Chem. Commun.* 53: 6613–6616.

43 Hu, K., Jiang, Y.W.X., Li, H. et al. (2018). Tuning peptide self-assembly by an in-tether chiral center. *Sci. Adv.* 4: eaar5907.

44 (a) Insua, I. and Montenegro, J. (2020). 1D to 2D self assembly of cyclic peptides. *J. Am. Chem. Soc.* 142: 300–307. (b) Díaz, S., Insua, I., Bhak, G., and Montenegro, J. (2020). Sequence decoding of 1D to 2D self-assembling cyclic peptides. *Chem. Eur. J.* 26: 14765–14770.

45 Doerr, A. (2016). Single-particle cryo-electron microscopy. *Nat Methods* 13: 23.

46 Hartgerink, J., Granja, J.R., Milligan, R.A., and Ghadiri, M.R. (1996). Self-assembling peptide nanotubes. *J. Am. Chem. Soc.* 118: 43–50.

47 Montenegro, J., Vázquez-Vázquez, C., Kalinin, A. et al. (2014). Coupling of carbon and peptide nanotubes. *J. Am. Chem. Soc.* 136: 2484–2491.

48 (a) Rust, M.J., Bates, M., and Zhuang, X. (2006). Sub-diffraction-limit imaging by stochastic optical reconstruction microscopy (STORM). *Nat. Methods* 3: 793–796. (b) Yuste, R. (2011). *Imaging: A Laboratory Manual*. New York: Cold Spring Harbor Laboratory Press. (c) Tam, J. and Merino, D. (2015). Stochastic optical reconstruction microscopy (STORM) in comparison with stimulated emission depletion (STED) and other imaging methods. *J. Neurochem.* 135: 643–658. (d) Xu, J., Ma, H., and Liu, Y. (2017). Stochastic optical reconstruction microscopy (STORM). *Curr. Protoc. Cytom.* 81: 12.46.1–12.46.27.

49 Rho, J.Y., Cox, H., Mansfield, E.D.H. et al. (2019). Dual self- assembly of supramolecular peptide nanotubes to provide stabilization in water. *Nat. Commun.* 10: 4708.

50 Yang, H., Yang, S., Kong, J. et al. (2015). Obtaining information about protein secondary structures in aqueous solution using Fourier transform IR spectroscopy. *Nat. Protoc.* 10: 382–396.

51 Yan, Z., Li, Q., and Zhang, P. (2017). Soy protein isolate and glycerol hydrogen bonding using two-dimensional correlation (2D-COS) attenuated Total reflection Fourier transform infrared (ATR FT-IR) spectroscopy. *Appl. Spectrosc.* 71: 2437–2445.

52 Kim, H.S., Hartgerink, J.D., and Ghadiri, M.R. (1998). Oriented self-assembly of cyclic peptide nanotubes in lipid membranes. *J. Am. Chem. Soc.* 120: 4417–4424.

53 Claro, B., Goormaghtigh, E., and Bastos, M. (2021). Attenuated total reflection-Fourier transform infrared spectroscopy: a tool to characterize antimicrobial cyclic peptide-membrane interactions. *Eur. Biophys. J.* 50: 629–639.

54 Wolfe, L.S., Calabrese, M.F., Nath, A. et al. (2010). Protein-induced photophysical changes to the amyloid indicator dye thioflavin T. *Proc. Natl. Acad. Sci. U. S. A.* 107: 16863–16868.

55 Brea, R.J., Vázquez, M.E., Mosquera, M. et al. (2007). Controlling multiple fluorescent signal output in cyclic peptide-based supramolecular systems. *J. Am. Chem. Soc.* 129: 1653–1657.

56 Méndez-Ardoy, A., Granja, J.R., and Montenegro, J. (2018). pH-triggered self-assembly and hydrogelation of cyclic peptide nanotubes confined in water micro-droplets. *Nanoscale Horiz.* 3: 391–396.

57 Bayón-Fernández, A., Méndez-Ardoy, A., Alvarez-Lorenzo, C. et al. (2023). Self-healing cyclic peptide hydrogels. *J. Mater. Chem. B* 11: 606–617.

58 Méndez-Ardoy, A., Bayón-Fernández, A., Yu, Z. et al. (2020). Spatially controlled supramolecular polymerization of peptide nanotubes by microfluidics. *Angew. Chem. Int. Ed.* 59: 6902–6908.

59 Novelli, F., Vilela, M., Pazó, A. et al. (2021). Molecular plumbing to bend self-assembling peptide nanotubes. *Angew. Chem. Int. Ed.* 60: 18838–18844.

60 Yang, J., Yu, X., Song, J.-I. et al. (2022). Aggregation-induced emission featured supramolecular Tubisomes for imaging-guided drug delivery. *Angew. Chem. Int. Ed.* 61: e202115208.

61 Shaikh, H., Rho, J.Y., Macdougall, L.J. et al. (2018). Hydrogel and organogel formation by hierarchical self-assembly of cyclic peptides nanotubes. *Chem. Eur. J.* 24: 19066–19074.

62 Mansfield, E.D.H., Hartlieb, M., Catrouillet, S. et al. (2018). Systematic study of the structural parameters affecting the self-assembly of cyclic peptide-poly(ethylene glycol) conjugates. *Soft Matter* 14: 6320–6326.

63 (a) Chapman, R., Warr, G.G., Perrier, S., and Jolliffe, K.A. (2013). Water-soluble and pH-responsive polymeric nanotubes from cyclic peptide templates. *Chem. Eur. J.* 19: 1955–1961. (b) Koh, M.L., FitzGerald, P.A., Warr, G.G. et al. (2016). Study of (cyclic peptide)–polymer conjugate assemblies by small-angle neutron scattering. *Chem. Eur. J.* 22: 18419–18428.

64 (a) Schimpf, M.E., Caldwell, K., and Giddings, J.C. (2000). *Field-Flow Fractionation Handbook*. Canada: John Wiley & Sons Inc. (b) Wahlund, K.G. and Nilsson, L. (2011). Flow FFF-basics and key applications. In: *Field-Flow Fractionation in Biopolymer Analysis* (ed. S.K.R. Williams and K.D. Caldwell), 1–21. Springer: Vienna.

65 García-Fandiño, R., Granja, J.R., D'Abramo, M., and Orozco, M. (2009). Theoretical characterization of the dynamical behavior and transport properties of α,γ-peptide nanotubes in solution. *J. Am. Chem. Soc.* 131: 15678–15686.

66 De Santis, P., Morosetti, S., and Rizzo, R. (1974). Conformational analysis of regular enantiomeric sequences. *Macromolecules* 7: 52–58.

67 Lewis, J.P., Pawley, N.H., and Sankey, O.F. (1997). Theoretical investigation of the cyclic peptide system Cyclo[(D-Ala-Glu-D-Ala-Gln)$_{m = 1\text{-}4}$]. *J. Phys. Chem. B* 101: 10576–10583.

68 Gailer, C. and Figel, M. (1997). Is the parallel or antiparallel β-sheet more stable? A semiempirical study. *J. Comput.-Aided Mol. Des.* 11: 273–277.

69 Chen, G., Su, S., and Liu, R.J. (2002). Theoretical studies of monomer and dimer of Cyclo[(−L-Phe^1-D-Ala^{2-})$_n$] and Cyclo[(−L-Phe^1-D-^{Me}N-Ala^{2-})$_n$] ($n = 3-6$). *Phys. Chem. B* 106: 1570–1575.

70 (a) Carloni, P., Andreoni, W., and Parrinello, M. (1997). Self-assembled peptide nanotubes from first principles. *Phys. Rev. Lett.* 79: 761–764. (b) Jishi, R.A., Flores, R.M., Valderrame, M. et al. (1998). Equilibrium geometry and properties of Cyclo[(Gly-D-Ala)$_4$] and {Cyclo[(Gly-D-Ala)$_4$]}$_2$ from density functional theory. *J. Phys. Chem. A* 102: 9858–9862. (c) Fukasaku, K., Takeda, K., and Shiraishi, K. (1998). First-principles study on electronic structures of protein nanotubes. *J. Phys. Soc. Jpn.* 67: 3751–3760.

71 García-Fandiño, R., Castedo, L., Granja, J.R., and Vázquez, S.A. (2010). Interaction and dimerization energies in methyl-blocked α,γ-peptide nanotube segments. *J. Phys. Chem. B* 114: 4973–4983.

72 Qu, W., Tan, H., Chen, G., and Liu, R. (2010). The self-assembled of cyclic D,L-α-peptide systems: insights into the structure and energetics. *Int. J. Quantum Chem.* 110: 1648–1659.

73 Damjanovic, J., Miao, J., Huang, H., and Lin, Y.-S. (2021). Elucidating solution structures of cyclic peptides using molecular dynamics simulations. *Chem. Rev.* 121: 2292–2324.

74 Claro, B., González-Freire, E., Calvelo, M. et al. (2020). Membrane targeting antimicrobial cyclic peptide nanotubes—an experimental and computational study. *Colloids Surf. B* 196: 111349.

75 (a) Engels, M., Bashford, D., and Ghadiri, M.R. (1995). Structure and dynamics of self-assembling peptide nanotubes and the channel-mediated water organization and self-diffusion. A molecular dynamics study. *J. Am. Chem. Soc.* 117: 9151–9158. (b) Tarek, M., Maigret, B., and Chipot, C. (2003). Molecular dynamics investigation of an oriented cyclic peptide nanotube in DMPC bilayers. *Biophys. J.* 85: 2287–2298.

76 Chipot, C. and Tarek, M. (2006). Interaction of a peptide nanotube with a water–membrane interface. *Phys. Biol.* 3: S20.

77 Dani, J.A. and Levitt, D.G. (1981). Water transport and ion-water interaction in the gramicidin channel. *Biophys. J.* 35: 501–508.

78 (a) Yan, X., Fan, J., Yu, Y. et al. (2015). Transport behavior of a single Ca^{2+}, K^+, and Na^+ in a water-filled transmembrane cyclic peptide nanotube. *J. Chem. Inf. Model.* 55: 998–1011. (b) Carvajal-Diaz, J.A. and Cagin, T. (2016). Electrophoretic transport of Na^+ and K^+ ions within cyclic peptide nanotubes. *J. Phys. Chem. B* 120: 7872–7879.

79 (a) Liu, J., Fan, J., Tang, M. et al. (2010). Water diffusion behaviors and transportation properties in transmembrane cyclic Hexa-, octa- and decapeptide nanotube. *J. Phys. Chem. B* 114: 12183–12192. (b) Liu, J., Fan, J., Cen, M. et al. (2012). Dependences of water permeation through cyclic octa-peptide nanotubes on channel length and membrane thickness. *J. Chem. Inf. Model.* 52: 2132–2138.

80 Zhang, M., Fan, J., Xu, J. et al. (2016). Different transport behaviors of NH_4^+ and NH_3 in transmembrane cyclic peptide nanotubes. *J. Mol. Model.* 22: 233.

81 García-Fandiño, R., Amorín, M., Castedo, L., and Granja, J.R. (2012). Transmembrane ion transport by self-assembling α,γ-peptide nanotubes. *Chem. Sci.* 3: 3280–3285.

82 Sanchez-Quesada, J., Kim, H.S., and Ghadiri, M.R. (2001). A synthetic pore-mediated transmembrane transport of glutamic acid. *Angew. Chem. Int. Ed.* 40: 2503–2506.

83 Miao, J., Descoteaux, M.L., and Lin, Y.-S. (2021). Structure prediction of cyclic peptides by molecular dynamics + machine learning. *Chem. Sci.* 12: 14927–14936.

84 Rubin, D.J., Nia, H.T., Desire, T. et al. (2013). Mechanical reinforcement of polymeric fibers through peptide nanotube incorporation. *Biomacromolecules* 14: 3370–3375.

85 Magana, M. et al. (2020). The value of antimicrobial peptides in the age of resistance. *Lancet Infect. Dis.* 20: e216–e230.

86 (a) Fernandez-Lopez, S., Kim, H.-S., Choi, E.C. et al. (2001). Antibacterial agents based on the cyclic d,l-α-peptide architecture. *Nature* 412: 452–455. (b) Dartois, V.R., Sanchez-Quesada, J., Cabezas, E. et al. (2005). Systemic antibacterial activity of novel synthetic cyclic peptides. *Antimicrob. Agents Chemother.* 49: 3302–3310. (c) Motiei, L., Rahimipour, S., Thayer, D.A. et al. (2009). Antibacterial cyclic D,L-α-glycopeptides. *Chem. Commun.* 3693–3695.

87 González-Freire, E., Novelli, F., Pérez-Estévez, A. et al. (2021). Double orthogonal click reactions for the development of antimicrobial peptide nanotubes. *Chem. Eur. J.* 27: 3029–3038.

88 Chen, P.B., Black, A.S., Sobel, A.L. et al. (2020). Directed remodeling of the mouse gut microbiome inhibits the development of atherosclerosis. *Nat. Biotechnol.* 38: 1288–1297.

89 Larnaudie, S.C., Sanchis, J., Nguyen, T.-H. et al. (2018). Cyclic peptide-poly(HPMA) nanotubes as drug delivery vectors: *in vitro* assessment, pharmacokinetics and biodistribution. *Biomaterials* 178: 570–582.

90 Kerr, A., Sagita, E., Mansfield, E.D.H. et al. (2022). Polymeric nanotubes as drug delivery vectors–comparison of covalently and supramolecularly assembled constructs. *Biomacromolecules* 23: 2315–2328.

91 (a) Hsieh, W.-H., Chang, S.-F., Chen, H.-M. et al. (2012). Oral gene delivery with cyclo-(D-Trp-Tyr) peptide nanotubes. *Mol. Pharmaceutics* 9: 1231–1249. (b) Lee, Y.-H., Chang, S.-F., and Liaw, J. (2015). Anti-apoptotic gene delivery with cyclo-(D-Trp-Tyr) peptide nanotube via eye drop following corneal epithelial debridement. *Pharmaceutics* 7: 122–136. (c) Li, M., Ehlers, M., Schlesiger, S. et al. (2016). Incorporation of a non-natural arginine analogue into a cyclic peptide leads to formation of positively charged nanofibers capable of gene transfection. *Angew. Chem. Int. Ed.* 55: 598–601. (d) Panigrahi, B., Singh, R.K., Mishra, S., and Mandal, D. (2018). Cyclic peptide-based nanostructures as efficient siRNA carriers. *Artif. Cells Nanomed. Biotechnol.* 46: S763–S773.

92 Horne, W.S., Wiethoff, C.M., Cui, C. et al. (2005). Antiviral cyclic D,L-α- peptides: targeting a general biochemical pathway in virus infections. *Bioorg. Med. Chem.* 13: 5145–5153.

93 (a) Richman, M., Wilk, S., Chemerovski, M. et al. (2013). *In vitro* and mechanistic studies of an antiamyloidogenic self-assembled cyclic D,L-α-peptide architecture. *J. Am. Chem. Soc.* 135: 3474–3484. (b) Chemerovski-Glikman, M., Richman, M., and Rahimipour, S. (2014). Structure-based study of antiamyloidogenic cyclic

D,L-α-peptides. *Tetrahedron* 70: 7639–7644. (c) Chemerovski-Glikman, M., Rozentur-Shkop, E., Richman, M. et al. (2016). Self-assembled cyclic D,L-α-peptides as generic conformational inhibitors of the α-synuclein aggregation and toxicity: *in vitro* and mechanistic studies. *Chem. Eur. J.* 22: 14236–14246.

94 Zhao, Y., Leman, L.J., Search, D.J. et al. (2017). Self-assembling cyclic D,L-α-peptides as modulators of plasma HDL function. A supramolecular approach toward antiatherosclerotic agents. *ACS Cent. Sci.* 3: 639–646.

95 (a) Horne, W.S., Ashkenasy, N., and Ghadiri, M.R. (2005). Modulating charge transfer through cyclic D,L-α-peptide self-assembly. *Chem. Eur. J.* 11: 1137–1144. (b) Ashkenasy, N., Horne, W.S., and Ghadiri, M.R. (2006). Design of self-assembling peptide nanotubes with delocalized electronic states. *Small* 2: 99–102. (c) Reiriz, C., Brea, R.J., Arranz, R. et al. (2009). α,γ-Peptide nanotube templating of one-dimensional parallel fullerene arrangements. *J. Am. Chem. Soc.* 131: 11335–11337. (d) Ishihara, Y. and Kimura, S. (2012). Peptide nanotube composed of cyclic tetra-β-peptide having polydiacetylene. *Biopolymers* 98: 155–160. (e) Ishihara, Y. and Kimura, S. (2013). Four-peptide-nanotube bundle formation by self-assembling of cyclic tetra-β-peptide using GQuartet motif. *Biopolymers* 100: 141–147. (f) Uji, H., Kim, H., Imai, T. et al. (2016). Electronic properties of tetrathiafulvalene-modified cyclic-β-peptide nanotube. *Biopolymers* 106: 275–282. (g) Kamano, Y., Tabata, Y., Uji, H., and Kimura, S. (2019). Chiral and random arrangements of flavin chromophores along cyclic peptide nanotubes on gold influencing differently on surface potential and piezoelectricity. *RSC Adv.* 9: 3618–3624. (h) Tabata, Y., Mitani, S., Uji, H. et al. (2019). The effect of macrodipole orientation on the piezoelectric response of cyclic β- peptide nanotube bundles on gold substrates. *Polym. J.* 51: 601–609. (i) Tabata, Y., Kamano, Y., Uji, H. et al. (2019). Electronic properties of cyclic β-peptide nanotube bundles reflecting structural arrangement. *Chem. Lett.* 48: 322–324.

96 (a) Song, Q., Goia, S., Yang, J. et al. (2021). Efficient artificial light-harvesting system based on supramolecular peptide nanotubes in water. *J. Am. Chem. Soc.* 143: 382–389. (b) Brea, R.J., Pérez-Alvite, M.J., Panciera, M. et al. (2011). Highly efficient and directional homo- and heterodimeric energy transfer materials based on fluorescently derivatized α,γ-cyclic octapeptides. *Chem. Asian J.* 6: 110–121.

97 Priegue, J.M., Louzao, I., Gallego, I. et al. (2022). 1D alignment of proteins and other nanoparticles by using reversible covalent bonds on cyclic peptide nanotubes. *Org. Chem. Front.* 9: 1226–1233.

7

Design and Control of Self-sorting Patterns of Supramolecular Peptide Nanofibers

Keisuke Nakamura and Itaru Hamachi

Department of Synthetic Chemistry and Biological Chemistry, Graduate School of Engineering, Kyoto University, Nishikyo-ku, Katsura, Kyoto 615-8510, Japan
JST-ERATO, Hamachi Innovative Molecular Technology for Neuroscience, Nishikyo-ku, Katsura, Kyoto 615-8530, Japan

7.1 Introduction

7.1.1 Self-sorting in a Biological System

Life exhibits a myriad of unique functions, such as extraordinary mechanics, dynamic mobility, adaptiveness, self-healing properties, and self-replication, which are yet to be achieved in the existing synthetic materials. Living systems also execute multiple tasks simultaneously; they recognize their surrounding conditions and adaptively alter their structure and composition. These remarkable features of life have fascinated many chemists and inspired them to develop novel life-like synthetic materials. One of the essential structural aspects for emergent functions of the living systems is the fact that they are a multicomponent system containing a myriad of proteins/peptides, nucleic acids, lipids, saccharides, and other metabolites. Notably, these biomolecules recognize themselves through specific interactions and independently (or cooperatively) assemble into the higher architectures in a spatiotemporally controlled manner: that is *self-sorting*. In living cells, for example, different cytoskeleton proteins, such as actin filament, intermediate fiber, and microtubules, create self-sorted distinct fibrous structures without any interference with each other (Figure 7.1); these self-sorted fibers orthogonally associate and dissociate to play different types of roles in their motility, molecular transportation, and force generation, all of which are essential for structures and functions of living cells [1–4]. Self-sorting also prevails at larger scales. Cells are divided into various self-sorted compartments, termed subcellular organelles: nuclei, mitochondria, endoplasmic reticulum, Golgi apparatus, lysosome, and so forth [5]. These substructures are able to provide various

Peptide Self-Assembly and Engineering: Fundamentals, Structures, and Applications, First Edition.
Edited by Xuehai Yan.

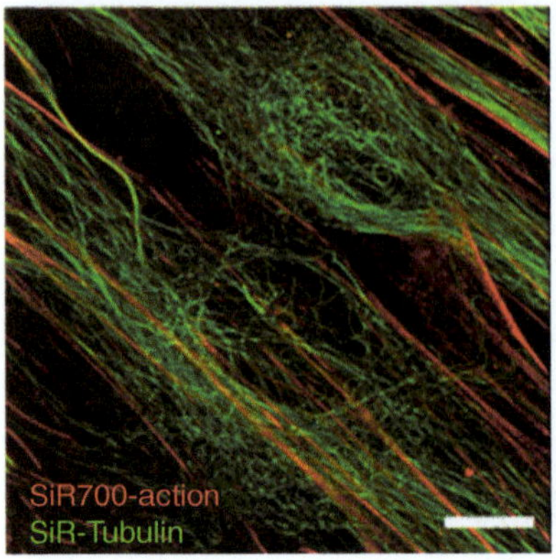

Figure 7.1 Fluorescence images of a cultured fibroblast showing the self-sorting fibrous structures of the actin filament (red) and microtubules (green). Scale bar: 10 μm. Source: From Ref. [1]. Reproduced with permission from American Chemical Society.

environments for transcription, translation, metabolism, and signal transduction inside the cell. At a tissue scale, the self-sorting of cells into distinct domains/compartments is a key process for morphogenesis [6]. It is now apparent that self-sorting events contribute to the emergence of complexity and versatility in well-ordered and flexible living systems. In recent years, there have been many approaches to realize life-like emergent functions using synthetic materials made of multiple self-assembled structures [7–9]. A number of supramolecular chemists now consider that self-sorting should be one of the key factors for developing such sophisticated synthetic materials inspired by living systems.

7.1.2 Self-sorting of Supramolecular Peptide Nanofibers

Supramolecular peptide nanofibers are one of the most attractive materials as alternatives to existing synthetic polymer materials, promising for a wide variety of applications ranging from electronics to biomedicine [10, 11]. Since these are constructed through noncovalent interactions, the nanofibers formation is inherently reversible and responsive to external stimuli such as pH or temperature, producing dynamic and unique materials, such as environmentally friendly recyclable materials and intelligent materials that can alter their structures/functions adaptive to external stimuli. Another important aspect is that these one-dimensional self-assembly nanofibers often form into hydrogels via entanglement and cross-linking of the nanofibers in water [12]. The bulk properties of the hydrogel can be finely tuned by the rational design of the monomeric structure, so-called (hydro)gelators, applicable for tissue engineering, drug release matrix, and sensing device.

Supramolecular nanofibers exhibit features similar to the cytoskeletal protein assembly to some extent, in terms of their dynamic self-assembly. Thus, from the viewpoint of biomimetics, the bioinspired design of supramolecular peptide nanofibers would lead to the construction of novel and sophisticated materials with cell-like performance. Notable roles played by cellular cytoskeletons inspired us to regard that self-sorting should be a powerful platform to rationally integrate multiple properties of individual synthetic nanofibers without loss of them. Indeed, a focus of recent studies on supramolecular self-assembly is gradually shifting from a single-component system to a self-sorting multicomponent system. Several studies have reported that some pairs of synthetic supramolecular nanofibers are self-sorted [13–15]. However, the rational design of synthetic self-sorting systems still remains challenging. There is a huge gap between the biological system and the artificial supramolecular system in the design of self-sorting pairs. Unlike biomacromolecules such as proteins and nucleic acids, where multiple interactions are involved in precise molecular recognition, synthetic molecules are generally small and susceptible to subtle changes in their surroundings because they can employ only limited numbers of weak molecular recognitions. Also, the synthetic self-sorting system requires a delicate balance of noncovalent interactions between the intra- and interfiber components, which is very difficult to control, particularly under aqueous conditions. Indeed, the mixture of different synthetic small monomers in water tends to form a wide variety of self-assembly structures in which co-assembly is often induced rather than self-sorting (Figure 7.2) [16]. At a molecular

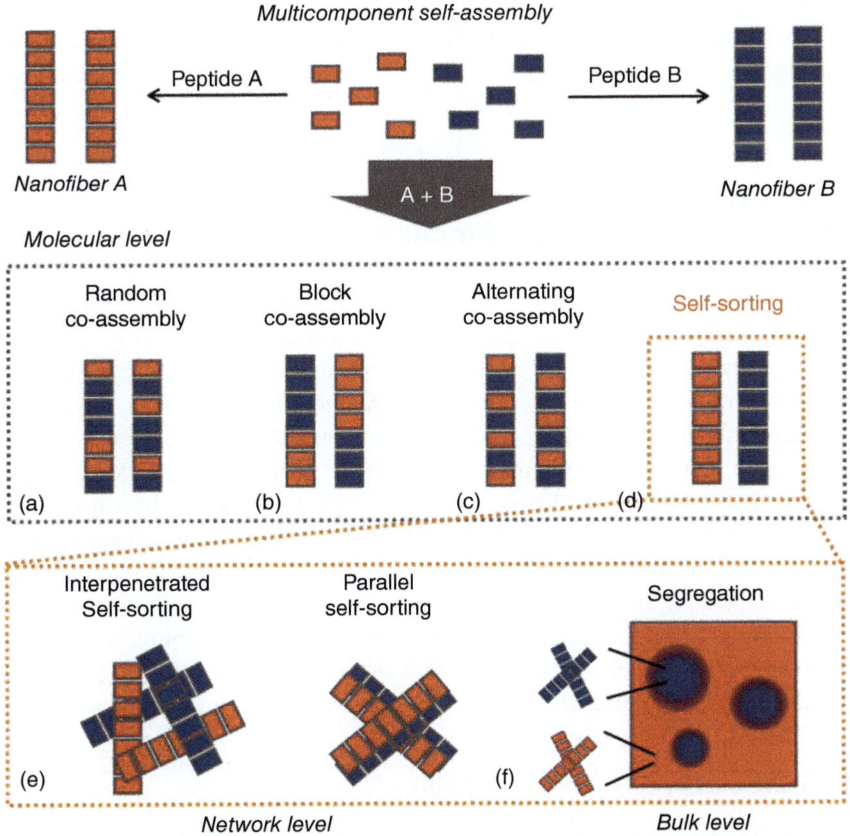

Figure 7.2 Complexity in multicomponent self-assembly. The molecular-level self-assembly structure of two different peptides is divided into (a) random co-assembly, (b) block co-assembly, (c) alternating co-assembly, and (d) self-sorting. The self-sorting nanofibers are further differentiated at (e) network and (f) bulk scales. Source: Keisuke Nakamura.

level, there are at least four possible classes of multicomponent self-assembly: (i) a random co-assembly (Figure 7.2a); (ii) a block co-assembly (Figure 7.2b); (iii) an alternating co-assembly, or a self-assembled fiber of alternating molecules (Figure 7.2c); and (iv) a self-sorting where two monomers orthogonally self-assemble into two distinct nanofibers (Figure 7.2d). In addition to molecular-level self-sorting, self-sorting patterns of nanofibers can be categorized at the network level or bulk scale. For example, two self-sorted nanofibers may well entangle and contact each other via interfiber interactions, or the two different nanofibers interpenetrate each other in the self-sorting system (Figure 7.2e). Also, molecular-level self-sorting sometimes results in segregated domains like phase separation at the bulk scale (Figure 7.2f). Such spatial and hierarchical complexity of multicomponent synthetic self-assembly hampers the precise analysis and rational design of self-sorting nanofibers to date. Chemical and physical controlling factors that lead to a self-sorting structure have been controversial and yet to be explicit.

This chapter focuses on how to create purely synthetic supramolecules achieving self-sorting systems. We describe here valuable methods for precisely analyzing the self-sorting network structures and their controlling factors using a few of state-of-the-art examples. What functions are obtained in self-sorting supramolecular nanofibers are also introduced. To highlight the more general design principle of self-sorting, this chapter discusses not only self-sorted peptide nanofibers but also other types of supramolecular nanofibers.

7.2 How to Characterize Self-sorting Patterns of Supramolecular Peptide Nanofibers

The starting point for the rational design of self-sorting supramolecular materials is how clearly and deeply we can characterize the self-sorting structures [16]. The multicomponent system inevitably enhances the complexity, and the data increased dramatically and require careful interpretation. Though there are many reports on multicomponent supramolecular nanofibers, there are few discussions on how best to characterize the multicomponent self-assembly structure [16]. Spectroscopic methods, such as infrared (IR), UV–Vis, and Circular Dichroism (CD), have been classically used to judge whether a system is self-sorting, but these give only ensembled information. Recent studies have demonstrated that direct imaging of self-assembly structures is powerful for analyses of self-sorting because it provides information on individual fibers. Given that each technique has strong and weak points, a variety of analyses should be rationally combined. In this section, we discuss the pros and cons of several techniques applied to characterize a self-sorting supramolecular system.

7.2.1 Spectroscopy (IR/UV–Vis/CD/NMR)

Conventional spectroscopic analyses, IR spectroscopy, UV–vis spectroscopy, and CD spectroscopy are easily accessible techniques to probe the intermolecular interactions and molecular packing structure. If the spectra of two components are distinguishable, these techniques are useful for investigating whether they form a self-sorting structure by comparing the spectra of single-component nanofibers and the resulting mixture of the two components. If a system is self-sorting, the spectra of the mixture become identical to the overlay of the spectra of each single-component nanofiber. In contrast, if there is a co-assembly structure, the spectra may be altered from the simple summation. For example, Hirst et al. used CD spectroscopy to prove the self-sorting of L- and D-type dendric peptide self-assemblies [13]. The two enantiomers had opposite cotton peaks derived from the chirality of the self-assembly structure with equal intensity. The cotton signal of the mixture linearly altered in correspondence to the ratio of enantiomers and was consistent with the sum of each component, suggesting the self-sorting of two enantiomeric peptides.

Nuclear magnetic resonance (NMR) is also helpful for monitoring the formation process of self-sorting supramolecular nanofibers [17]. Monomers can be detected by NMR, while the signal of the polymeric self-assembled compounds became invisible due to a decrease in relaxation time. Thus, the formation kinetics can be determined by monitoring the disappearance of the NMR signals. If the signals of two different monomers disappear in a stepwise manner, it may suggest that the two monomers are self-sorting to form orthogonal assemblies (as mentioned later in Section 7.3.1, kinetics is one of the key factors for self-sorting).

Although these spectroscopic techniques may be the first choice to characterize the self-sorting structure, it is important to realize that the spectroscopic methods provide only information on the average of nanofibers and cannot address how self-sorted structures are organized at the larger length scale.

7.2.2 Small-angle Scattering

Small-angle X-ray scattering (SAXS) and small-angle neutron scattering (SANS) are recently attracting more and more attention as techniques to analyze the self-assembly structure at the nanoscale [17–19]. The scattering data are fitted to a model to obtain information about the radius of the structure, the length, persistence length, etc. Similar to the spectroscopic analyses, the scattering between the single- and multicomponent systems is compared to probe the self-sorting structure. The contrast matching technique represents another advantage of SANS. The scattering intensity of individual components can be selectively weakened under the solvent with the specific ratio of H_2O and D_2O. By modifying the ratio of the solvent to diminish the contribution of one network, a specific component of the multicomponent system can be focused on, allowing a greater understanding of the self-assembly structure within the self-sorted network. Cross et al. used this technique to characterize a self-sorted network of perylene-bisimide (PBI)- and stilbene-based nanofibers (Figure 7.3) [19]. In 45% D_2O, the stilbene-based nanofibers are contrast-matched, and only PBI-based nanofibers scatter in the multicomponent system. On the other hand, the PBI-based nanofibers are contrast-matched in 60% D_2O. They revealed that these nanofibers were self-sorted based on the collective data of spectroscopic analyses, while SANS suggested that the nanostructure of the stilbene-based nanofibers in the multicomponent system had a larger

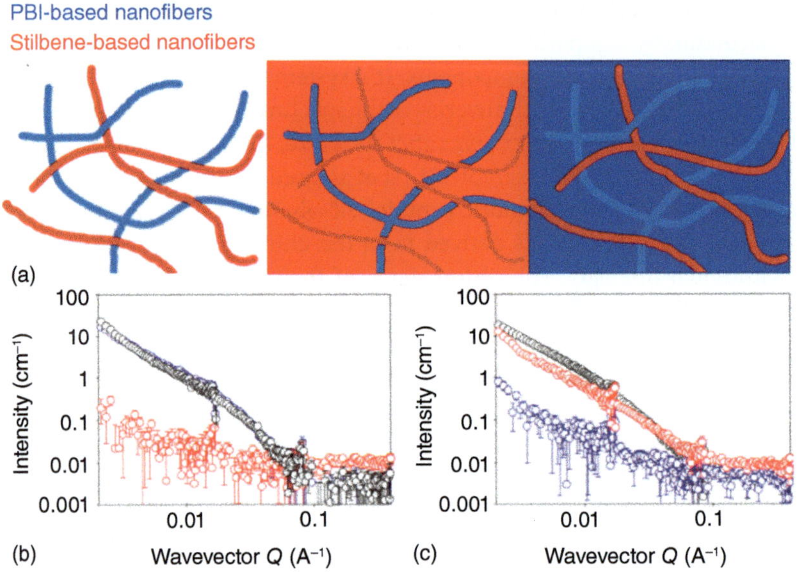

Figure 7.3 (a) Schematic illustration of contrast matching in SANS experiments; (b) scattering data in 45% D_2O, where the PBI-based nanofibers scatter, and the stilbene-based nanofibers are contrast-matched; (c) scattering data in 60% D_2O, where the stilbene-based nanofibers scatter, and the PBI-based nanofibers are contrast-matched. (blue) single-component PBI-based nanofiber, (red) single-component stilbene-based nanofiber, and (black) multicomponent system. Source: Ref. [19]. Reproduced with permission from American Chemical Society.

Kuhn length than a single-component system did. SAXS and SANS provide detailed structure in a self-sorting network, but these tools require large-scale facilities and expensive deuterated compounds. Besides, the data fitting of the model should always be optimized.

7.2.3 Electron Microscopy/Atomic Force Microscopy

Direct imaging of supramolecular nanofibers is powerful because it can provide information on the spatiotemporal distribution of the individual nanofibers [20]. Indeed, structural analysis of many supramolecular polymers has been mainly conducted by transmittance electron microscopy (TEM) and scanning electron microscopy (SEM), which visualize the nanoscale structure of the network. The resolution of SEM/TEM reaches ca. 1 nm order, by which a single nanofiber can be visualized. On the other hand, it is generally hard for conventional SEM/TEM imaging to discriminate individual nanofibers in the multicomponent system because most of the nanofibers form similar morphology and diameter. However, EM imaging can give reliable data in the case that morphologies of two supramolecular nanofibers are completely different and discernable [21–23]. For example, Moffat and Smith visualized the self-assembly structure of two different types of dendric peptide nanofibers with SEM imaging [21]. The two fibers with very different diameters coexisted in a multicomponent system. Liu et al. characterized a self-sorting network of two amino-acid-type nanofibers on the basis of the distinct helical morphology obtained by SEM/TEM imaging

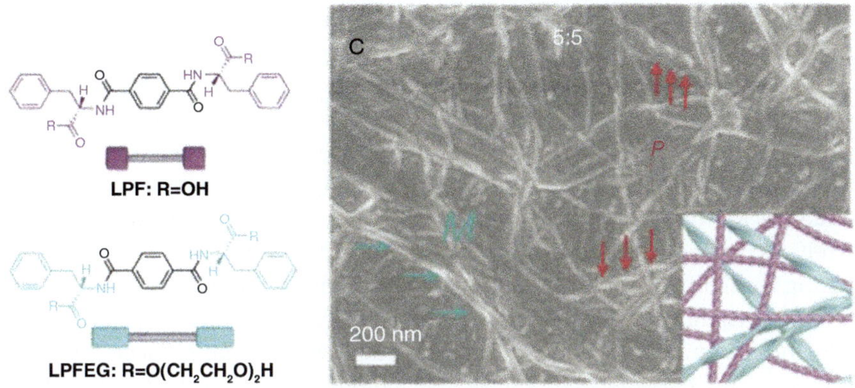

Figure 7.4 SEM images of the self-sorted structures of two different amino-acid-type nanofibers. Source: Ref. [22]. Reproduced with permission from John Wiley & Sons.

(Figure 7.4) [22]. Similarly, atomic force microscopy (AFM) is used to image the self-sorting network of two nanofibers with a distinct morphology [24].

It must be cautious in mind that TEM/SEM/AFM imaging usually requires drying during the sample preparation process, which may induce any artifacts, such as aggregation or structural changes in the self-assembly. Indeed, it is reported that drying a dipeptide-based monomer leads to the fibers being thicker than those without drying [25]. It is also noteworthy that high-magnification SEM/TEM images visualize only a limited region within a sample, which may cause a biased interpretation of the self-assembly structure.

7.2.4 Confocal Laser Scanning Microscopy

It recently turned out that confocal laser scanning microscopy (CLSM) is quite powerful for studying multicomponent and self-sorting supramolecular systems [20, 26]. In CLSM imaging, only fluorescence close to a focal plane is detected, resulting in better resolution, particularly in the depth section, than the wide-field microscopies. CLSM has several advantages in analyzing self-sorting structures over TEM/SEM/AFM imaging. First, CLSM imaging can noninvasively evaluate the intact self-assembly structure without a freezing/drying process of samples. This enables the analysis of the dynamic features of nanofibers, such as their formation process or fluidity. Besides, three-dimensional structures of supramolecular nanofibers can be visualized through sequential images in the different z-section. More importantly, multiple components can be distinguished by selective staining: this is quite useful for analyzing self-sorting structures. The drawback of CLSM is its lower spatial resolution (c. 200–300 nm) than EM/AFM imaging due to Abbe's diffraction limit. Thus, it is important to point out that CLSM gives information on bundled fibers rather than a single nanofiber at the nanoscale. However, the recent development of super-resolution microscopic technologies, such as stimulated emission depletion (STED) microscopy, stochastic optical reconstruction microscopy (STORM), structured illumination microscopy (SIM), and Airyscan imaging, has improved the resolution to reach less than 100 nm [27].

For analyses of self-sorting supramolecular fibers, selective staining of individual components is critical. The most straightforward approach is to use intrinsically fluorescent self-assembly components with different excitation wavelengths [28]. However, this approach is strictly limited to intrinsically emissive nanofibers. A more general one is desirable, which is relying on the usage of a small amount of fluorescent additive which is selectively incorporated into one nanofiber. For example, a fluorescent dye conjugated with a self-assembling motif via linker can be selectively incorporated into a corresponding nanofiber bearing the identical self-assembling motif. In this case, the fluorescent dye should be carefully selected in the points of their chemical properties (hydrophobicity, hydrophilicity, and charges) to achieve efficient and selective fluorescent staining. Onogi et al. pioneered *in situ* CLSM imaging of self-sorted supramolecular nanofibers using orthogonally selective probes (Figure 7.5) [15]. To visualize the self-sorting nanofibers made of peptide- and lipid-type monomers, they designed the Oregon Green-modified peptide monomers and Alexa Fluor 546-modified lipid monomers that can selectively co-assemble with peptide nanofibers and lipid nanofibers, respectively. CLSM study clearly revealed that the peptide- and lipid-type nanofibers are not spatially overlapped with each other, and that the super-resolution STED imaging allows the visualization of distinct nanofibers with a diameter of 80–100 nm. In order to image self-sorted peptide self-assembly, fluorescent proteins fused to peptides can be employed for selective staining [29, 30]. For example, two fluorescent proteins,

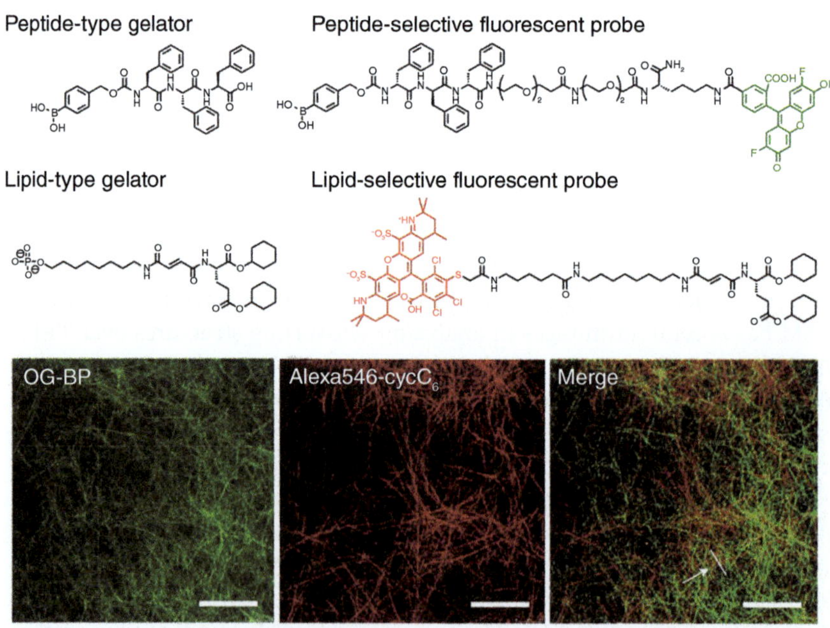

Figure 7.5 *In situ* CLSM imaging of self-sorted network of peptide- and lipid-type nanofibers. Two fluorescent probes are used for the selective staining of individual nanofibers. Scale bar: 5 μm. Source: Ref. [15]. Reproduced with permission from Springer Nature.

EGFP and mCherry, are introduced to two different amyloid-like peptide nanofibers by genetic fusion and a tag system using SpyTag/SpyCatcher [29].

Dynamic features of self-sorting nanofibers can be quantitatively analyzed, thanks to the *in situ* CLSM observation. For example, the formation and degradation process of each nanofiber during construction of a self-sorting system can be monitored by real-time imaging. One can also determine the fluidity of self-sorting nanofibers by fluorescence recovery after photobleaching (FRAP), where the rate of influx diffusion of a fluorescent probe into the photobleached region is monitored to estimate the diffusion coefficient. The diffusion coefficient is a parameter to identify individual nanofibers in a self-sorting system. Onogi et al. visualized the degradation process of self-sorting peptide- and lipid-type nanofibers, where only peptide nanofibers disappeared in response to external stimuli [15]. The same group also reported that the diffusion coefficient of the lipid-type nanofiber in the self-sorting system is identical to that in the single-component system. These clearly suggest that the chemical and physical properties of the nanofibers remain intact even after mixing up.

When the staining selectivity is strictly confirmed, *in situ* CLSM imaging undoubtedly provides invaluable information on the spatiotemporal distribution of nanofibers that other microscopic techniques otherwise cannot. It is now recognized that CLSM is an indispensable tool for analyzing various types of multicomponent self-assembly structures, including not only self-sorting at the molecular level but also hierarchical self-sorting structures at a larger scale and other multicomponent self-assembly structures such as block copolymers [20, 26].

7.3 How to Design and Control Self-sorting Patterns of Supramolecular Peptide Nanofibers

Although there is an increasing number of reports about self-sorting nanofibers, control factors essential for self-sorting are still controversial. The kinetics of self-assembly is one of the essential parameters because most supramolecular nanofibers are under kinetic control. Indeed, recent studies suggested that in many self-sorting systems, the individual nanofiber forms stepwise with distinct kinetics [14, 15, 31–35]. Another control factor to be considered is how the chemical structure of monomers affects self-sorting behavior. Several researchers reported that various parameters of the component molecules, such as hydrophobicity, net charge, and chirality, may give impacts on the self-sorting events [36–42]. This section discusses the control factors for self-sorting supramolecular fibers at the monomer level, focusing on kinetics and chemical structures by enumerating several examples.

7.3.1 Kinetics

Although it is not perfectly clear and is system dependent whether the self-sorting process is under thermodynamic or kinetic control, several studies suggest that the kinetic parameter often involves the self-sorting process. Early examples of self-sorted supramolecular nanofibers are typically obtained by a heating/cooling process. Sugiyasu et al. demonstrated that the kinetic nature of nanofiber formation is important for the thermally induced self-sorting process of polythiophene- and perylene-based nanofibers. In this system, the dissociation and association of individual nanofibers independently occurred during a heating/cooling process. The formation of thiophene-based nanofibers is faster than that of perylene-one [14]. Onogi et al. visualized the formation process of self-sorted peptide- and lipid-type nanofibers during cooling by real-time CLSM imaging [15]. They observed that the lipid-type nanofibers self-assembled at higher temperatures, and their formation was immediately completed at the beginning of imaging. On the other hand, the peptide-type nanofibers gradually appeared with an induction time of 15 minutes and elongated. These findings suggest that the distinct formation kinetic may contribute to the self-sorting event. It is recently accepted that the stepwise formation of two distinct nanofibers is common in various self-sorting systems.

In addition to thermally controlled self-sorting, chemical stimuli are used to induce stepwise nanofiber formation leading to self-sorting in a highly controlled manner [17, 31–34]. Adams and coworkers pioneered the programmable design of self-sorting nanofibers of two different peptides with similar chemical structures using a pH control with the gradual hydrolysis of glucono-δ-lactone (Figure 7.6) [17]. These two peptides comprise a carboxyl group with distinct pK_a values and undergo self-assembly into nanofibers through protonation at the corresponding pHs. The glucono-δ-lactone is spontaneously hydrolyzed to decrease pH gradually; the stepwise formation of each supramolecular network around the respective pH value was archived, resulting in self-sorting. Singh et al. also used chemical fuels to generate self-sorting nanofibers from the co-assembled structure of three monomers bearing aldehyde groups [33]. The heat-cooling process of the three monomers yielded a co-assembly structure. Sodium dithionite ($Na_2S_2O_4$) and formaldehyde were used as fuels to induce kinetic-controlled self-assembly: Dithionite was firstly added to induce dissociation of the co-assembly through the conversion of the aldehyde to hydroxysulfonate; then, the sequential addition of formaldehyde reverted the hydroxysulfonate group back to the aldehyde group. In such a fuel-driven self-assembly process, where the formation kinetics of nanofibers is stepwise, the three

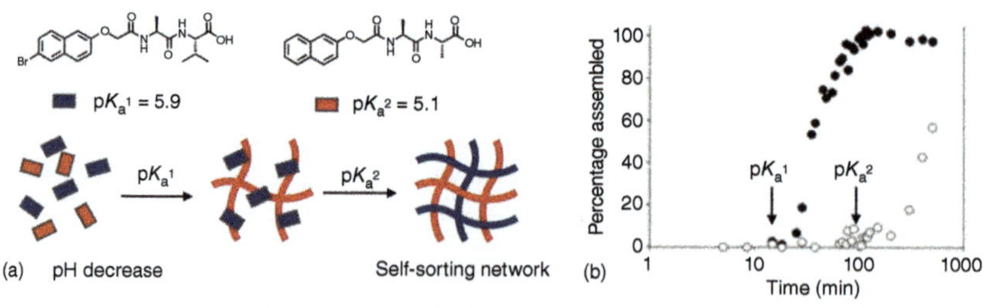

Figure 7.6 (a) Chemically programmed self-sorting network of two peptide hydrogelators with distinct pK_a values; (b) 1H NMR analysis shows the stepwise formation of two self-assembly structures during a pH decrease. Source: Ref. [35]. Reproduced with permission from The Royal Society of Chemistry.

components are self-sorted. The precise control of kinetics using chemical stimuli would be the key to the rational design of self-sorting nanofibers.

7.3.2 Chemical Structure

The chemical structure of component molecules (monomers) is another important aspect in the design of artificial self-sorting systems. Of course, the chemical structure should affect the kinetics of fiber formation. However, it is reported that some peptides tend to co-assemble even if they have different formation kinetics [35]. This fact suggests that the chemical structure/interactions are more elaborately involved in the self-sorting process.

As a pioneering work, Boekhoven et al. examined the self-assembly structure of various kinds of supramolecular nanofibers and a phospholipid liposome [36]. They revealed that the orthogonal self-assembly between phospholipids and nanofibers could occur only in the following cases: (i) the self-assembly is driven by strong intermolecular interactions, and (ii) both self-assembled structures are based on distinct sets of interactions. These must be applied to a self-sorting system between two different supramolecular nanofibers. In addition to the distinct driving force for self-assembly, small differences in each chemical structure also impact the self-sorting structure. Kubota et al. investigated the self-assembly properties between the peptide- and lipid-type hydrogelators using a small library of them bearing varied hydrophobicity and net charge [37]. The CLSM imaging-based assay revealed that a pair of peptide- and lipid-type peptide with the identical net charge preferentially leads to a self-sorted network, while the oppositely charged pair tends to form co-assembly. It is also suggested that moderate hydrophobicity (calculated logP value) of the peptide-type monomers would be required to achieve the self-sorting. Self-sorting behavior is affected by not only hydrophobicity but also immiscibility or incompatibility that may induce demixing. Wakabayashi et al. evaluated the self-sorting property of two peptide amphiphiles bearing alkyl chains (H-PAs) and fluoroalkyl chains as immiscible hydrophobic substituents (F-PAs) [38]. A mixture of H-PA and F-PA bearing short-chain lengths provides a self-sorting network after the cooling process and a high-temperature incubation. On the other hand, in the case of H-PA and F-PA with the longer chain length, a co-assembly is obtained. The hydrophobic effect is more susceptible to the chain length than the immiscibility. Therefore, in the case of the longer chain, the hydrophobic interaction, which leads to co-assembly, has a larger contribution to assembly formation, relative to immiscibility, which leads to self-sorting. Self-sorting should be controlled by the balance between the hydrophobic interaction and immiscibility between alkyl and fluoroalkyl chains. The abovementioned results may imply that repulsive interaction between two monomers affects the self-assembly structure. However, a few of the conflicted examples were reported for certain systems. Regarding the net charge, for example, positively charged peptide DWDW and negatively charged peptide KWKW form self-sorting nanofibers [39]. These interactions may work in concert with other factors, such as kinetics.

Chirality may be another important factor that affords a distinct set of molecular interactions. Peptides are a suitable scaffold for the design of chiral nanofibers because they have multiple chiral points. In fact, the chiral differentiation of peptide nanofibers leads to both co-assembly and self-sorting, which depend on a system [40–42]. For example, Sasmal et al. directly visualized the chirality-driven self-sorting of enantiomeric oligopeptide pairs with super-resolution SIM imaging [40]. On the other hand, Swanekamp et al. demonstrated that L- and D-Ac-$(EKFE)_2$-NH_2 peptides co-assembled into a ripped β-sheet structure that contains alternating L- and D-peptides rather than self-sorted fibrous structure [41].

Many other factors still need to be unveiled. In addition, multiple factors are involved in complex interactions that affect the self-sorting behavior. Although the *de novo* design of chemical structures to undergo self-sorting is yet far from realized, the above examples could help us design a synthetic self-sorting system.

7.4 Multiscale Self-sorting Pattern

In the previous section, we focused on the self-sorting network structure at the molecular level. On the other hand, a self-sorting structure is classified into a further different category of a larger scale, as mentioned in Section 7.1.2. For example, the self-sorted nanofibers possibly entangle via interfiber interactions, leading to the different localization of each nanofiber. At bulk scale, compartmented domains of discrete nanofibers, such as phase-separated structures, are often observed. In addition to molecular-level self-sorting, such multiscaled self-sorting should be considered for the precise design of multicomponent hierarchical supramolecular materials.

7.4.1 Network-level Self-sorting

Understanding network-level self-sorting remains challenging in artificial materials due to the limited methods available for distinguishing them clearly [42–46]. However, the biological system takes advantage of such a hierarchical (network-level) structure. For example, it is now unveiled that interplay through physical contact among self-sorted cytoskeletal fibers is essential for cellular functions; bundled actin filaments guide microtubule polymerization to regulate the turnover of stress fibers and cell motility [4]. Therefore, the design and control of self-sorting at the network scale would be potentially crucial for creating sophisticated functions of artificial life-like systems.

Cross et al. compared two self-sorting hydrogel structures of electron-rich (*p*-type) and electron-deficient (*n*-type) hydrogelators prepared at different temperatures (25 and 30 °C) (Figure 7.7a) [19]. Spectroscopic and SANS experiments suggested that self-sorting structures at the molecular level formed at 30 °C are essentially identical to those formed at 25 °C. In contrast, NMR experiments suggested that gelation at 25 °C results in a stepwise formation, while at 30 °C, the formation of *p*- and *n*-type nanofibers occurred simultaneously, suggesting the differences in the

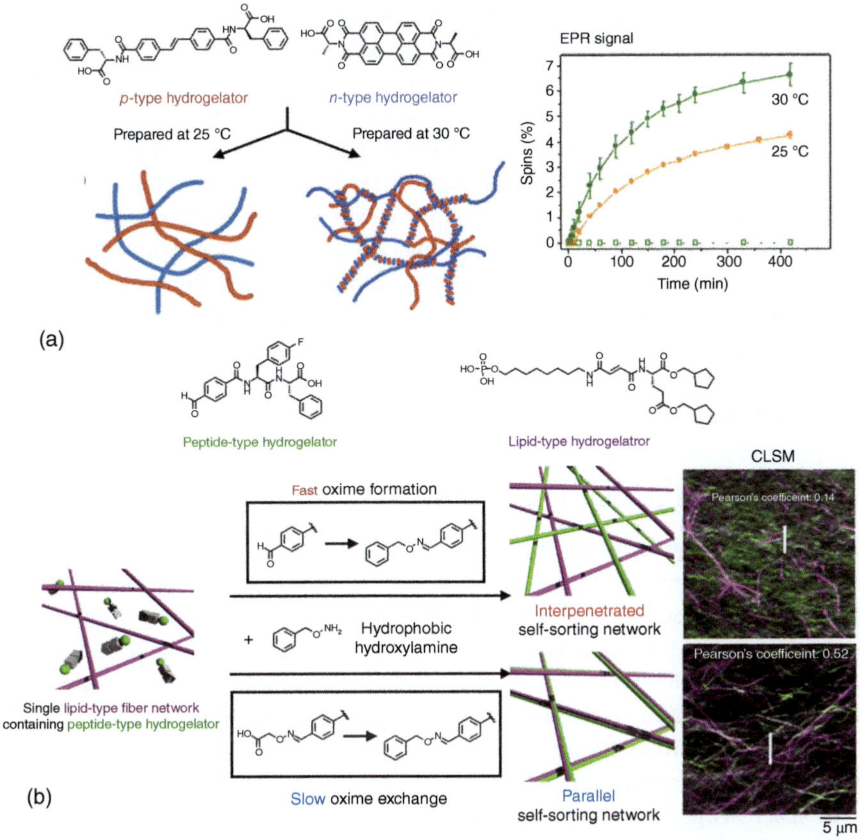

Figure 7.7 Network-level self-sorting patterns. (a) Distinct patterns of the self-sorted network comprising p- and n-type nanofibers. EPR signal suggests the difference in the network structure between the system prepared at 25 and 30 °C. Source: [19]. Reproduced with permission from American Chemical Society. (b) Control of seed formation allows the distinct network patterns of self-sorted nanofibers (interpenetrated and parallel). CLSM shows (green) peptide-type nanofibers and (magenta) lipid-type nanofibers. Scale bar: 5 μm. Source: Ref. [45]. Reproduced with permission from Ref [45]/ Springer Nature/ CC BY 4.0.

network. The difference in the energy transfer efficiency also implied that the network-level patterns were different though both are self-sorted.

Recent studies indicated that CLSM imaging can effectively visualize the network-level spatial distribution of self-sorting structures. For example, Kubota et al. utilized CLSM imaging and distinguished two different network patterns in self-sorted nanofibers of peptide- and lipid-type nanofibers (Figure 7.7b) [45]. They designed an aldehyde-tethered peptide that forms into nanofibers via oxime formation (or exchange) upon the addition of hydrophobic hydroxylamine. When the peptide nanofibers were prepared by the rapid oxime formation in the presence of lipid-type nanofibers, the obtained self-sorted nanofibers were interpenetrated with each other at the molecular level. On the other hand, the slower formation of peptide nanofibers through the oxime exchange (from hydrophilic to hydrophobic) reaction provided a parallel self-sorting network, where the peptide nanofibers preferentially localized on the surface of lipid-type nanofibers. The slow kinetics condition allows the selective nucleation of peptide nanofibers on the energetically favorable lipid nanofiber surface rather than in the water space, while the faster oxime formation produces a supersaturated state within a short time that induces the nucleation in both sites. Higashi et al. also visualized the hierarchical self-assembly structure of peptide nanofibers and DNA tile nanotubes by CLSM imaging [46]. These nanofibers produce parallel self-sorting by a heating–cooling one-pot process, where the DNA tile nanofibers are mainly localized on the peptide nanofibers. On the other hand, the system is an interpenetrated self-sorting when it is prepared by mixing two nanofibers independently obtained by the heating–cooling process. They suggested that the attractive reversible interaction between the self-sorted supramolecular assemblies is crucial for forming integrative self-sorted nanostructures.

7.4.2 Macroscopic Segregation

Self-sorted nanofibers generate macroscopic patterns at micrometer to millimeter scales like phase-separated structures in some cases. Wang et al. reported phase-separated compartmentalized structure in supramolecular hydrogels of tris-hydrazone gelators that formed from a soluble hydrazide, neutral and anionic aldehydes (Figure 7.8A) [47]. Mixing three building blocks in a phosphate buffer produced a turbid gel through imine formation. CLSM imaging of the hydrogel stained with an anionic fluorescein probe showed a sheet-like phase-separated structure with a width of 100 μm, composed of segregated domains rich in anionic fibers and neutral nanofibers. The macroscopically segregated structure is also reported in self-sorted peptide amphiphile (PA) nanofibers. Chen et al. prepared negatively charged PA molecules containing only a tiny fraction of the two PAs conjugated with fluoresceinisothiocyanate (FTIC) or tetramethylrhodamine (TAMRA) [49]. CLSM imaging revealed the

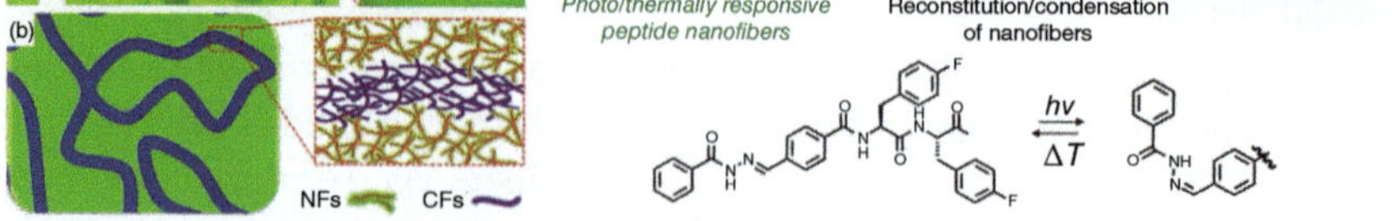

Figure 7.8 (A) Sheet-like phase-separated structure of neutral nanofibers (NFs) and charged nanofibers (CFs). Source: Ref. [47]. Reproduced with permission from American Chemical Society. (B) Out-of-equilibrium pattern formation in self-sorted peptide- and lipid-type nanofibers. Source: Ref. [48]. Reproduced with permission from American Chemical Society.

microscale structure of the two fluorescent PA molecules, where TAMRA-PA and FITC-PA were segregated at the micron-scale domains. They concluded that conjugation of the negatively charged FITC to PA disrupted its co-assembly of the majority of unlabeled PAs, inducing the formation of micron-scale domains. In contrast, conjugation of the zwitterionic TAMRA caused no such disruption.

A complicated macroscopic structure can also be created under an out-of-equilibrium state controlled by the diffusion of peptide monomers in a self-sorting network. Nakamura et al. describe phototriggered macroscopic patterns of peptide nanofibers in a self-sorting supramolecular network composed of lipid- and peptide-type nanofibers modified with acylhydrazone as a photoswitch moiety (Figure 7.8B) [48]. A spatially limited photoirradiation using a photomask induces the degradation of the peptide nanofibers via Z-to-E isomerization, whereas the network of lipid-type nanofibers is intact. Thereafter, the thermal incubation caused the spatial condensation of peptide-type nanofibers in the photoirradiated areas and concurrent nanofiber depletion in the nonirradiated areas due to the diffusion controlled by this self-sorting structure. They also revealed that the reconstructed peptide nanofibers were thermodynamically stable, and the supply of monomers from the metastable nanofibers promoted seed-driven nanofiber condensation. Complex macroscopic self-sorted patterns at the microscale were produced from a simple photomask, that is two-line patterns from a one-wide-line photomask and grid-like patterns from consecutive photoirradiation using the one-line photomask.

7.5 Functions of Self-sorted Nanofibers

A few of the unique functions of self-sorted fibrous materials have been investigated so far. In a self-sorting system, multiple functions, such as stimuli-responsiveness or optoelectronic property, should be integrated without losing their original property. Self-sorting nanofiber is also utilized as a template to provide spatially separated distinct environments where various functional materials, such as catalysts and bioactive molecules, are individually aligned.

7.5.1 Multiple Stimuli-responsive Hydrogels

Stimuli-responsive hydrogels are promising soft materials. They undergo a gel–sol transition in response to external stimuli, which may be utilized for controlled drug release matrix, biomarker detection, and degradable biomaterials. Stimuli-responsive self-sorting double network (SDN) hydrogels show unique physical properties and sophisticated responses to external stimuli. Adams et al. pioneered the responsive behavior of SDN hydrogel. Photoresponsive stilbene-based nanofibers were prepared in an SDN hydrogel through the pH decrease with glucono-δ-lactone [31]. The stilbene-based nanofibers can be selectively decomposed through E-to-Z isomerization by LED light irradiation, while the other fibers remain intact. Spatial patterning of SDN hydrogel was achieved by photoirradiation with a photomask, where one network was selectively removed in the irradiated area. A similar example using photoacids was also reported [50, 51]. Multiple stimuli responsiveness is also rationally integrated into an SDN hydrogel. Shigemitsu et al. demonstrated the bidirectional control of the rheological property of the SDN hydrogel [52]. The peptide-fiber network decomposes in response to $Na_2S_2O_4$, while the phosphate-lipid fiber network stiffens via dephosphorylation catalyzed by bovine alkaline phosphatase (BAP) (Figure 7.9a). The SDN hydrogel softened in response to $Na_2S_2O_4$, while it became rigid by the presence of BAP. Such a unique feature allowed us to tune the release rate of proteins encapsulated in this SDN hydrogel in a different direction. $Na_2S_2O_4$ promotes protein release coupled with the degradation of peptide nanofibers, whereas BAP suppresses the release of protein by the dense network of lipid nanofiber (Figure 7.9b). Interestingly, this SDN material can recognize the order of applied stimuli following the network's change in the mesh size (Figure 7.9c).

The incorporation of stimuli-responsive moiety into nanofibers often impairs self-sorting in multicomponent systems. In order to sidestep this challenge, it turned out that postassembly fabrication (PAF) is an effective strategy for functionalizing the SDN hydrogel. In the PAF protocol, a key molecule that can modify already existing nanofibers is added after the formation of a self-sorting network. For example, Tanaka et al. modified an SDN hydrogel composed of hydrogen peroxide-degradative peptide nanofibers and phospholipid-type nanofibers by adding sarcosine oxidase (SO_x) and Ca^{2+} ion. SOx oxidizes sarcosine and produces hydrogen peroxide as a

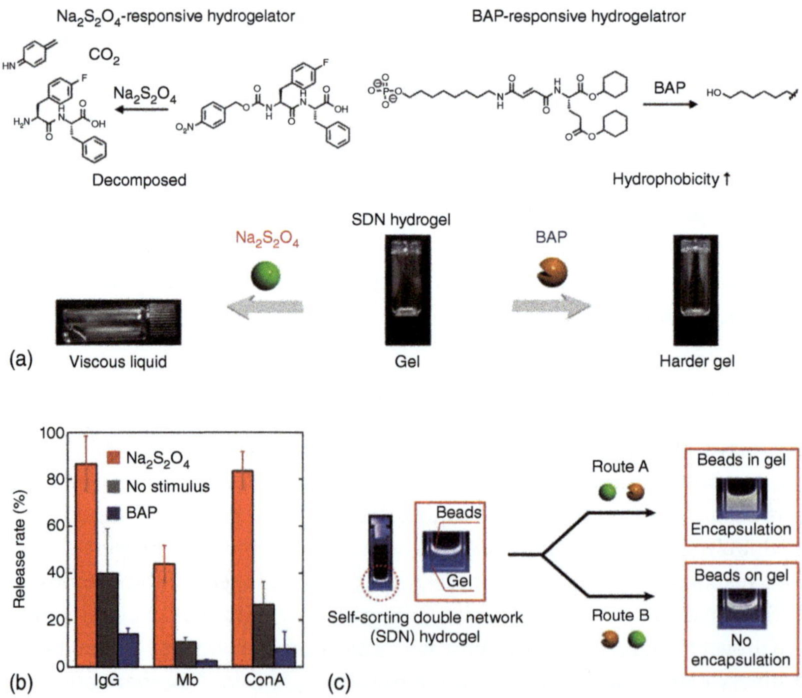

Figure 7.9 Bidirectional response of SDN hydrogels. (a, b) Bidirectionally controlled release of encapsulated proteins in response to stimuli; (c) recognition of the order of applied stimuli. Source: Ref. [52]. Reproduced with permission from Springer Nature.

by-product which reacts with the peptide gelator, resulting in the decomposition of nanofibers [53]. Ca^{2+} ions, on the other hand, cross-link the phospholipid nanofibers through the coordination to the phosphate groups, and the removal of Ca^{2+} ions by a metal chelator such as ATP results in the weakened hydrogel. They confirmed that the self-sorting structure remained intact after adding SOx and Ca^{2+} ions. This SDN hydrogel containing SOx and Ca^{2+} ions was able to respond to two biomolecules, sarcosine and ATP, in an AND logic gate mode. It is expected that the PAF strategy should facilitate the development of stimuli-responsive SDN hydrogels because it does not require the daunting redesign of self-sorting gelator pairs from the beginning.

7.5.2 Cascade Reaction

A spatially separated environment in a self-sorting system is used for a tandem cascade reaction. For example, two different catalytic sites can be placed in each network to realize a cascade reaction that is otherwise difficult in a homogeneous system. In Singh's report, two nanofibers containing a carboxyl group and proline at the N-terminus, which catalyze deprotection of the acetal group and aldol reaction, respectively, are mixed into the self-sorting nanofibers [54]. When benzaldehyde dimethyl acetal and cyclohexanone were simultaneously dissolved in the self-sorting system, more than 85% of the final product (aldol) was obtained. On the other hand, in the co-assembly system, no final product was obtained. In this case, a salt bridge between the acidic and proline groups of the two molecules rendered proline catalytically inactive through protonation. Self-sorting would provide subcompartments where chemical reactions are confined in a system, analogous to subcellular organelles in a cell.

7.5.3 Optoelectronics

Self-sorting is an attractive bottom-up approach to realizing efficient optoelectronic properties of π-conjugated supramolecular polymers. Heterojunctions of a doner and an accepter are preferable to achieve efficient charge separation and charge-carrier transport in an efficient photovoltaic system. A self-sorting structure of p- and n-type nanofibers can create p–n heterojunctions where two nanofibers meet. Draper and coworkers examined the photoconductivity of self-sorted nanofibers prepared from p-type stilbene-based nanofibers and p-type PBI-type nanofibers [55]. The conductivity is obtained in the presence of the PBI radical anion. The stilbene-based nanofibers act as an electron donor leading to photoconductivity at longer wavelengths and long-lived radical anion, indicating self-sorting system improved electronic properties over the single-component system.

The formation of p–n heterojunction largely depends on the extent of the hierarchical self-sorting structure. Cross et al. demonstrated by electron paramagnetic resonance (EPR) that electron transfer efficiency differs between hierarchical self-sorting structures from stilbene-based and PBI-type nanofibers (Figure 7.7a) [19]. A more radical anion was generated from a self-sorting structure on visible light irradiation, where two fibers were supposed to be sufficiently close in space than the other to allow electron transfer. These findings may be significant for the delicate design of electronic supramolecular materials.

7.5.4 Biological Application of Self-sorting

Self-sorting nanofibers may provide useful matrices to control cellular functions for regenerative medicine and tissue engineering. Okesola et al. applied SDN hydrogels comprising peptide amphiphilic and biocompatible sugar-derived gelators to cell culture [56]. The mechanical property and cell compatibility of this self-sorting gel can be tuned by altering the ratio of each component. Wu et al. recently reported a PAF strategy to introduce different bioactive molecules into the distinct network through photoinduced covalent modification [57]. Compared to a covalently cross-linked polymeric network, this system would provide a promising platform for cell engineering because they well resemble a dynamic network of extracellular matrix, where the bioactive ligands are specifically anchored to a type of fiber network rather than mixed on both fibers.

In situ formation of a self-sorting network in the biological milieu is challenging but promising for manipulating cell functions [58, 59]. Liu et al. realized the self-sorting of two amphiphilic peptide self-assemblies *in vivo* system (Figure 7.10a) [58]. They developed a lipid-inspired peptide interdigitating amphiphiles bearing selenium oxide group and a bola-amphiphilic peptide bearing disulfide bond. These two peptides self-assemble into twisted and flat nanoribbons, respectively, through reduction by glutathione (GSH). The GSH-responsive *in situ* self-sorting process of the two peptides inside of HeLa cells was directly confirmed by super-resolution SIM (Figure 7.10b). These peptides were localized in ER and the Golgi apparatus and formed self-sorted assemblies, leading to combinatorial organelle dysfunction and cell death.

7.6 Conclusion

The bioinspired design of supramolecular materials has strongly attracted many chemists, and such materials are now becoming more and more complicated toward life-like structures and functions. In this context, the self-sorting of multiple components can be recognized as an essential aspect of rational design of synthetic materials. Compared to life, synthetic self-sorting systems are so far quite simple. Nevertheless, solid guidelines for the design and control of its structure are yet to be established. Further accumulation of experimental data and new technologies for analysis are keenly needed for the *de novo* design of such fascinating materials. Given that self-sorting phenomena are complicated, evaluation of the self-assembly structure ranging all the scales should be carefully examined, and thus, technologies that can seamlessly connect a variety of results obtained by

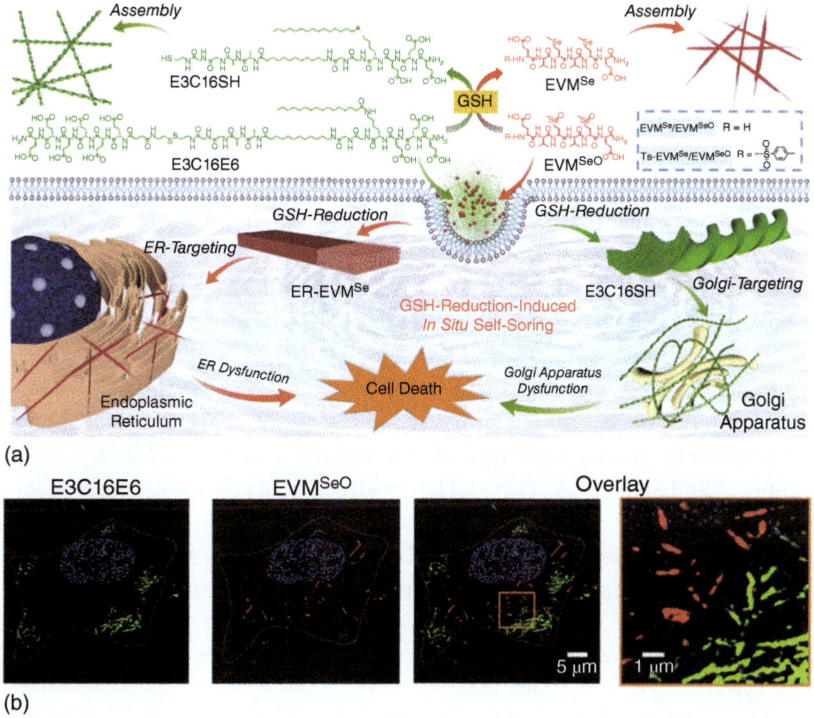

Figure 7.10 (a) *In situ* self-sorting of peptides E3C16E6 and EVMSeO induced by GSH reduction for the formation of peptide assemblies simultaneously targeting the ER and Golgi apparatus to promote organelle dysfunction and subsequent cell death; (b) SIM images of self-sorted nanofibers in a living cell. Scale bar: 5 μm. Source: Ref. [58]. Reproduced with permission from American Chemical Society.

spectroscopic and microscopic experiments are keenly desirable. Besides, the functions of self-sorting nanofibers should be further explored in various directions. Multicomponent systems not only provide a powerful platform to integrate the functions of each component rationally but also enable to create more sophisticated synergic functions through efficient cross-talks, compared to the simple summation. In biology, far-from-equilibrium phenomena, such as actin wave, oscillatory signal transduction, and skin pattern formation, arise from reaction–diffusion processes where two diffusible molecules interact [60]. If interactions between two supramolecular nanofibers are appropriately designed, such a far-from-equilibrium state is achieved with a synthetic system, leading to remarkably dynamic features [61]. In addition to the self-sorting of supramolecular nanofibers with similar structures, self-sorting behaviors of supramolecular peptide assembly and other functional materials, such as polymer, inorganic materials, and biomolecules, should also be explored for next-generation composite materials [62, 63]. We envisage that further studies on self-sorting phenomena would create more complex life-like synthetic materials showing unprecedented functions.

References

1 Lukinavičius, G., Reymond, L., Umezawa, K. et al. (2016). Fluorogenic probes for multicolor imaging in living cells. *J. Am. Chem. Soc.* 138: 9365–9368.

2 Alberts, B., Johnson, A., Lewis, J. et al. (2008). *Molecular Biology of the Cell*, 5e. New York: Garland Science.

3 Blanchoin, L., Boujemaa-Paterski, R., Sykes, C., and Plastino, J.A. (2014). Dynamics, architecture, and mechanics in cell motility. *Physiol. Rev.* 94: 235–263.

4 Dogterom, M. and Koenderink, G.H. (2019). Actin-microtubule crosstalk in cell biology. *Nat. Rev. Mol. Cell Biol.* 20: 38–54.

5 van Bergeijk, P., Hoogenraad, C.C., and Kapitein, L.C. (2016). Right time, right place: probing the functions of organelle positioning. *Trends Cell Biol.* 26: 121–134.

6 Gómez-Gálvez, P., Anbari, S., Escudero, L.M., and Buceta, J. (2021). Mechanics and self-organization in tissue development. *Semin. Cell Dev. Biol.* 120: 147–159.

7 Vantomme, G. and Meijer, E.W. (2019). The construction of supramolecular system. *Science* 363: 1396–1397.

8 Goor, O.J.G.M., Hendrikse, S.I.S., Dankers, P.Y.W., and Meijer, E.W. (2017). From supramolecular polymers to multi-component biomaterials. *Chem. Soc. Rev.* 46: 6621–6637.

9 Miljanić, O. and Osowska, K. (2011). Kinetic and thermodynamic self-sorting in synthetic systems. *Synlett* 12: 1643–1648.

10 Aida, T., Meijer, E.W., and Stupp, S.I. (2012). Functional supramolecular polymers. *Science* 335: 813–817.

11 Hauser, C.A.E. and Zhang, S. (2010). Designer self-assembling peptide nanofiber biological materials. *Chem. Soc. Rev.* 39: 2780–2790.

12 Draper, E.R. and Adams, D.J. (2017). Low-molecular-weight gels: the state of the art. *Chem.* 3: 390–410.

13 Hirst, A.R., Huang, B., Castelletto, V. et al. (2007). Self-organisation in the assembly of gels from mixtures of different dendritic peptide building blocks. *Chem. Eur. J.* 13: 2180–2188.

14 Sugiyasu, K., Kawano, S., Fujita, N., and Shinkai, S. (2008). Self-sorting organogels with p–n heterojunction points. *Chem. Mater.* 20: 2863.

15 Onogi, S., Shigemitsu, H., Yoshii, T. et al. (2016). In situ real-time imaging of self-sorted supramolecular nanofibres. *Nat. Chem.* 8: 743–752.

16 Draper, E.R. and Adams, D.J. (2018). How should multicomponent supramolecular gels be characterised? *Chem. Soc. Rev.* 47: 3395–3405.

17 Morris, K.L., Chen, L., Raeburn, J. et al. (2013). Chemically programmed self-sorting of Gelator networks. *Nat. Commun.* 4: 1480.

18 McDowall, D., Adams, D.J., and Seddon, A.M. (2022). Using small angle scattering to understand low molecular weight gels. *Soft Matter* 18: 1577–1590.

19 Cross, E.R., Sproules, S., Schweins, R. et al. (2018). Controlled tuning of the properties in optoelectronic self-sorted gels. *J. Am. Chem. Soc.* 140: 8667–8670.

20 Kubota, R., Tanaka, W., and Hamachi, I. (2021). Microscopic imaging techniques for molecular assemblies: electron, atomic force, and confocal microscopies. *Chem. Rev.* 121: 14281–14347.

21 Moffat, J.R. and Smith, D.K. (2009). Controlled self-sorting in the assembly of 'multi-Gelator' gels. *Chem. Commun.* 3: 316–318.

22 Liu, G., Zhou, C., Teo, W.T. et al. (2019). Self-sorting double-network hydrogels with tunable supramolecular handedness and mechanical properties. *Angew. Chem. Int. Ed.* 58: 9366–9372.

23 Ji, W., Zhang, S., Yukawa, S. et al. (2018). Regulating higher-order organization through the synergy of two self-sorted assemblies. *Angew. Chem. Int. Ed.* 57: 3636–3640.

24 Adhikari, B., Nanda, J., and Banerjee, A. (2011). Multicomponent hydrogels from enantiomeric amino acid derivatives: helical nanofibers, handedness and self-sorting. *Soft Matter* 7: 8913–8922.

25 Mears, L.L.E., Draper, E.R., Castilla, A.M. et al. (2017). Drying affects the fiber network in low molecular weight hydrogels. *Biomacromolecules* 18: 3531–3540.

26 Kubota, R., Nakamura, K., Torigoe, S., and Hamachi, I. (2020). The power of confocal laser scanning microscopy in supramolecular chemistry: in situ real-time imaging of stimuli-responsive multicomponent supramolecular hydrogels. *ChemistryOpen* 9: 67–79.

27 Pujals, S., Feiner-Gracia, N., Delcanale, P. et al. (2019). Super-resolution microscopy as a powerful tool to study complex synthetic materials. *Nat. Rev. Chem.* 3: 68–84.

28 Sarkar, A., Sasmal, R., Empereur-Mot, C. et al. (2020). Self-sorted, random, and block supramolecular copolymers via sequence controlled, multicomponent self-assembly. *J. Am. Chem. Soc.* 142: 7606–7617.

29 An, B., Wang, X., Cui, M. et al. (2017). Diverse supramolecular nanofiber networks assembled by functional low-complexity domains. *ACS Nano* 11: 6985–6995.

30 Cox, H., Georgiades, P., Xu, H. et al. (2017). Self-assembly of mesoscopic peptide surfactant fibrils investigated by STORM super-resolution fluorescence microscopy. *Biomacromolecules* 18: 3481–3491.

31 Draper, E.R., Eden, E.G.B., McDonald, T.O., and Adams, D.J. (2015). Spatially resolved multicomponent gels. *Nat. Chem.* 7: 848–852.

32 Ravarino, P., Panja, S., Bianco, S. et al. (2023). Controlled annealing in adaptive multicomponent gels. *Angew. Chem. Int. Ed.* e202215813.

33 Singh, N., Lopez-Acosta, A., Formon, G.J.M., and Hermans, T.M. (2022). Chemically fueled self-sorted hydrogels. *J. Am. Chem. Soc.* 144: 410–415.

34 Liu, B., Beatty, M.A., Pappas, C.G. et al. (2021). Self-sorting in dynamic combinatorial libraries leads to the co-existence of foldamers and self-replicators. *Angew. Chem. Int. Ed.* 133: 13681–13685.

35 Colquhoun, C., Draper, E.R., Eden, E.G.B. et al. (2014). The effect of self-sorting and co-assembly on the mechanical properties of low molecular weight hydrogels. *Nanoscale* 6: 13719–13725.

36 Boekhoven, J., Brizard, A.M., Stuart, M.C.A. et al. (2016). Bio-inspired supramolecular materials by orthogonal self-assembly of hydrogelators and phospholipids. *Chem. Sci.* 7: 6021–6031.

37 Kubota, R., Liu, S., Shigemitsu, H. et al. (2018). Imaging-based study on control factors over self-sorting of supramolecular nanofibers formed from peptide- and lipid-type hydrogelators. *Bioconjugate Chem.* 29: 2058–2067.

38 Wakabayashi, R., Imatani, R., Katsuya, M. et al. (2022). Hydrophobic immiscibility controls self-sorting or co-assembly of peptide amphiphiles. *Chem. Commun.* 58: 585–588.

39 Sahoo, J.K., VandenBerg, M.A., Ruiz Bello, E.E. et al. (2019). Electrostatic-driven self-sorting and nanostructure speciation in self-assembling tetrapeptides. *Nanoscale* 11: 16534–16543.

40 Gupta, D., Sasmal, R., Singh, A. et al. (2020). Enzyme-responsive chiral self-sorting in amyloid-inspired minimalistic peptide amphiphiles. *Nanoscale* 12: 18692–18700.

41 Swanekamp, R.J., DiMaio, J.T.M., Bowerman, C.J., and Nilsson, B.L. (2012). Coassembly of enantiomeric amphipathic peptides into amyloid-inspired rippled β-sheet fibrils. *J. Am. Chem. Soc.* 134: 5556–5559.

42 Nagy-Smith, K., Beltramo, P.J., Moore, E. et al. (2017). Molecular, local, and network-level basis for the enhanced stiffness of hydrogel networks formed from Coassembled racemic peptides: predictions from Pauling and Corey. *ACS Cent. Sci.* 3: 586–597.

43 Singh, N., Maity, C., Zhang, K. et al. (2017). Synthesis of a double-network supramolecular hydrogel by having one network catalyse the formation of the second. *Chem. Eur. J.* 23: 2018–2021.

44 Piras, C.C. and Smith, D.K. (2019). Sequential assembly of mutually interactive supramolecular hydrogels and fabrication of multi-domain materials. *Chem. Eur. J.* 25: 11318–11326.

45 Kubota, R., Nagao, K., Tanaka, W. et al. (2020). Control of seed formation allows two distinct self-sorting patterns of supramolecular nanofibers. *Nat. Commun.* 11: 4100.

46 Higashi, S.L., Hirosawa, K.M., Suzuki, K.G.N. et al. (2020). One-pot construction of multicomponent supramolecular materials comprising self-sorted supramolecular architectures of DNA and semi-artificial glycopeptides. *ACS Appl. Bio Mater.* 3: 9082–9092.

47 Wang, Y., Lovrak, M., Liu, Q. et al. (2019). Hierarchically compartmentalized supramolecular gels through multilevel self-sorting. *J. Am. Chem. Soc.* 141: 2847–2451.

48 Nakamura, K., Tanaka, W., Sada, K. et al. (2021). Phototriggered spatially controlled out-of-equilibrium patterns of peptide nanofibers in a self-sorting double network hydrogel. *J. Am. Chem. Soc.* 143: 19532–19541.

49 Chen, C.H., Palmer, L.C., and Stupp, S.I. (2021). Self-sorting in supramolecular assemblies. *Soft Matter* 17: 3902–3912.

50 Cornwell, D.J., Daubney, O.J., and Smith, D.K. (2015). Photopatterned multidomain gels: multi-component self-assembled hydrogels based on partially self-sorting 1,3:2,4-dibenzylidene-d-sorbitol derivatives. *J. Am. Chem. Soc.* 137: 15486–15492.

51 Schlichter, L., Piras, C.C., and Smith, D.K. (2021). Spatial and temporal diffusion-control of dynamic multi-domain self-assembled gels. *Chem. Sci.* 12: 4162–4172.

52 Shigemitsu, H., Fujisaku, T., Tanaka, W. et al. (2018). An adaptive supramolecular hydrogel comprising self-sorting double nanofibre networks. *Nat. Nanotechnol.* 13: 165–172.

53 Tanaka, W., Shigemitsu, H., Fujisaku, T. et al. (2019). Post-assembly fabrication of a functional multicomponent supramolecular hydrogel based on a self-sorting double network. *J. Am. Chem. Soc.* 141: 4997–5004.

54 Singh, N., Zhang, K., Angulo-Pachón, C.A. et al. (2016). Tandem reactions in self-sorted catalytic molecular hydrogels. *Chem. Sci.* 7: 5568–5572.

55 Ardoña, H.A.M., Draper, E.R., Citossi, F. et al. (2017). Kinetically controlled coassembly of multichromophoric peptide hydrogelators and the impacts on energy transport. *J. Am. Chem. Soc.* 139: 8685–8692.

56 Okesola, B.O., Wu, Y., Derkus, B. et al. (2019). Supramolecular self-assembly to control structural and biological properties of multicomponent hydrogels. *Chem. Mater.* 31: 7883–7897.

57 Wu, D., Lei, H., Xie, X. et al. (2022). Self-sorting double network hydrogels with photo-definable biochemical cues as artificial synthetic extracellular matrix. *Nano Research* 15: 4294–4301.

58 Liu, X., Li, M., Liu, J. et al. (2022). In situ self-sorting peptide assemblies in living cells for simultaneous organelle targeting. *J. Am. Chem. Soc.* 144: 9312–9323.

59 Yang, X., Lu, H., Wu, B., and Wang, H. (2022). Triggered self-sorting of peptides to form higher-order assemblies in a living system. *ACS Nano* 16: 18244–18252.

60 Kondo, S. and Miura, T. (2010). Reaction-diffusion model as a framework for understanding biological pattern formation. *Science* 329: 1616–1620.

61 Mattia, E. and Otto, S. (2015). Supramolecular systems chemistry. *Nature Nanotechnol.* 10: 111–119.

62 Shigemitsu, H. and Hamachi, I. (2017). Design strategies of stimuli-responsive supramolecular hydrogels relying on structural analyses and cell-mimicking approaches. *Acc. Chem. Res.* 50: 740–750.

63 Shigemitsu, H., Kubota, R., Nakamura, K. et al. (2020). Protein-responsive protein release of supramolecular/polymer hydrogel composite integrating enzyme activation systems. *Nat. Commun.* 11: 3859.

8

Protein-decorated Artificial Viral Capsids Self-assembled from β-Annulus Peptides

Kazunori Matsuura

Tottori University, Graduate School of Engineering, Department of Chemistry and Biotechnology, Koyama-Minami 4-101, Tottori, Japan
Tottori University, Centre for Research on Green Sustainable Chemistry, Koyama-Minami 4-101, Tottori, Japan

8.1 Introduction

A spherical virus is a supramolecular complex several tens of nanometers in size consisting of genome nucleic acids encapsulated in self-assembled coat proteins called capsids. Many spherical viral capsids can self-assemble from proteins without nucleic acids to form nanocapsules with a discrete internal space [1]. Therefore, spherical viral capsids, also referred to as virus-like particles, can be used as nanoreactors [1, 2], drug delivery carriers [3, 4], and base materials for vaccines [5]. Although the artificial construction of spherical viral capsids using genetic or protein engineering has been reported [6], the construction of viral capsids from chemically synthesized peptides remains challenging. If it was possible to chemically synthesize viral capsids with customized functionalities, they could serve as tools for analyzing the capsid formation mechanism, genome packaging into the capsid, cell infection, and intracellular dynamics.

Over the past 20 years, there have been significant advances in the construction of molecular assemblies from rationally designed biomolecules such as DNAs, lipids, proteins, and peptides [7]. In particular, peptides are promising building blocks for nanomaterials because of their structural diversity, high designability, and ease of synthesis [8, 9]. It has become possible to predict higher-order structures from the amino-acid sequences of proteins and peptides and to rationally design their assembled structures [9–13]. The artificial design of symmetrical assembling protein/peptide subunits has enabled the construction of virus-like nanocapsules that exhibit dodecahedral and icosahedral symmetry [14]. The ability to molecularly design and construct artificial protein/peptide nanocapsules could enhance the potential of such materials for applications in a variety of areas and could contribute to advances in the field of bionanotechnology.

Peptide Self-Assembly and Engineering: Fundamentals, Structures, and Applications, First Edition.
Edited by Xuehai Yan.

This chapter describes the chemical strategies for constructing artificial viral capsids by the self-assembly of designed peptides and their decoration with functional proteins. Section 8.2 describes the construction of artificial viral capsids by the self-assembly of β-annulus peptides from tomato bushy stunt virus (TBSV) and the encapsulation of various guest molecules into the artificial viral capsid. In Section 8.3, decoration of the artificial viral capsid with functional proteins by the C-terminal modification of β-annulus peptides is described. Finally, Section 8.4 describes the construction of enveloped viral replicas by complexing cationic lipid bilayers on anionic capsids and embedding membrane proteins in the envelope.

8.2 Artificial Viral Capsids Self-assembled from β-Annulus Peptides

8.2.1 Self-assembly of β-Annulus Peptides

The TBSV capsid self-assembles from 180 quasi-equivalent protein subunits, each containing 388 amino acids. The capsid subunit includes a trimeric annular *β*-structure, the "*β*-annulus motif," which participates in the formation of a dodecahedral internal skeleton [14]. Matsuura and coworkers pioneered self-assembled artificial viral capsids from the 24-residue β-annulus peptide of TBSV in 2010 [15]. They synthesized a β-annulus peptide (INHVGGTGGAIMAPVAVTRQLVGS) self-assembled spontaneously into hollow spherical nanocapsules 30–50 nm in diameter in water without the formation of nanofibers or unimolecular folding structures (Figure 8.1a–c). The existence of a hollow interior in the artificial viral

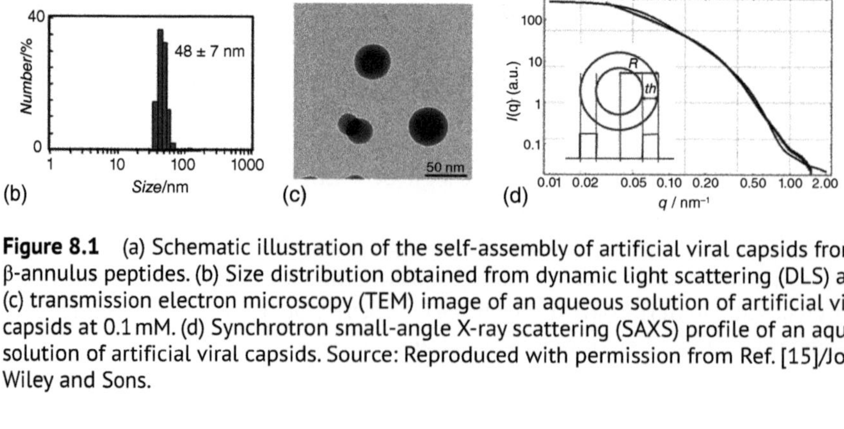

Figure 8.1 (a) Schematic illustration of the self-assembly of artificial viral capsids from β-annulus peptides. (b) Size distribution obtained from dynamic light scattering (DLS) and (c) transmission electron microscopy (TEM) image of an aqueous solution of artificial viral capsids at 0.1 mM. (d) Synchrotron small-angle X-ray scattering (SAXS) profile of an aqueous solution of artificial viral capsids. Source: Reproduced with permission from Ref. [15]/John Wiley and Sons.

capsid was confirmed by transmission electron microscopy (TEM) and synchrotron small-angle X-ray scattering (SAXS) methods (Figure 8.1c, d). The concentration dependence of dynamic light scattering (DLS) indicated that the critical aggregation concentration (CAC) of the β-annulus peptide was 25 μM at 25 °C, and that the size and morphology of the artificial viral capsids were minimally affected at concentrations above the CAC. The dissociation constant of the fluorescently labeled artificial viral capsid was determined by fluorescence correlation spectroscopy (FCS) to be 45 μM in Tris–HCl buffer, pH 7.0, at 25 °C [16]. The formation of artificial viral capsids can be controlled by the concentration and temperature, which is suitable for the encapsulation and release of guest molecules.

8.2.2 Encapsulation of Guest Molecules into Artificial Viral Capsids

The pH dependence of the surface ζ-potential of the artificial viral capsids indicated that the N-terminal segment of the peptide is directed toward the interior of the artificial capsid, and its C-terminal is directed toward the outer surface [17], which corresponds to the terminal direction of the natural TBSV capsid. Since the N-terminus of the peptide is cationic at neutral pH, the interior of the capsid is presumed to be cationic. Therefore, anionic dyes, DNAs [17], and anionic quantum dots [18] can be encapsulated in artificial viral capsids through electrostatic interactions. Modification of the N-terminal amino acid of the β-annulus peptide with Ni-NTA enabled the selective encapsulation of EGFP (enhanced green fluorescent protein) with a His tag into the artificial viral capsids (Figure 8.2a) [19]. Fluorescent zinc oxide (ZnO) nanoparticles were encapsulated in artificial viral capsids that were formed from β-annulus peptides bearing a ZnO-binding sequence (HCVAHR) at the N-terminus [20]. Single-stranded DNAs could be encapsulated in artificial viral capsids via disulfide bonds at the N-terminus of the β-annulus peptide and

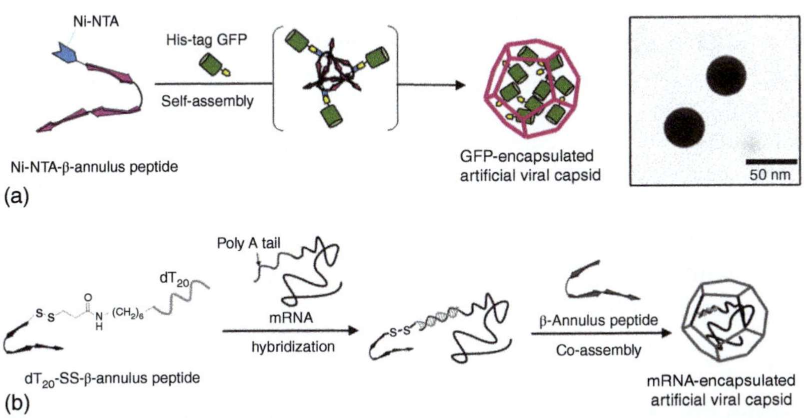

Figure 8.2 (a) Schematic illustration of the encapsulation of His-tagged GFPs into artificial viral capsids and a TEM image. Source: Reproduced with permission from Ref. [19]/ the Royal Society of Chemistry. (b) Schematic illustration of the encapsulation of mRNAs bearing a poly(A) tail into artificial viral capsids. Source: Reproduced and modified from Ref. [22] with permission from MDPI.

then released under reducing conditions [21]. Eukaryotic mRNAs bearing poly(A) tails were selectively encapsulated in the artificial viral capsids that were formed by co-assembly of β-annulus peptides bearing a dT_{20} at the N-terminal and unmodified β-annulus peptides via hybridization of dT_{20} and poly(A) (Figure 8.2b) [22]. The mRNA-encapsulating artificial viral capsid exhibited nuclease resistance. The capsid was internalized in human hepatoma (HepG2) cells by modification of cell-penetration peptides on the surface, and the intracellular expression of the fluorescent protein mCherry that was encoded by the mRNA was confirmed. Based on these results, it may be expected that artificial viral capsids can be useful as delivery devices for protein and nucleic acid drugs.

8.3 Protein-decorated Artificial Viral Capsids

8.3.1 Decoration of Functional Molecules on Artificial Viral Capsids

Considering that the C-terminus of the β-annulus peptide is directed to the outer surface of the artificial viral capsid as mentioned above, a C-terminal modification enables the construction of artificial viral capsids that are decorated with functional molecules such as proteins [23]. This means replacing the outer coat proteins that are decorated on the internal skeleton consisting of the β-annulus motif of natural TBSV by other functional molecules. Matsuura and coworkers constructed gold nanoparticle (AuNP)-decorated artificial viral capsids by self-assembling AuNP-conjugated β-annulus peptides via Cys–Au interactions at the C-terminus of the peptide (Figure 8.3a) [24]. The TEM image showed the formation of spherical assemblies of 5 nm AuNPs on the artificial viral capsid. The ζ-potential of the AuNP-decorated artificial viral capsids shifted more toward a negative potential than did the unmodified capsids. This demonstrates that AuNPs were decorated on the outer surfaces of the artificial viral capsids.

Artificial viral capsids with peptide spikes that mimic natural viruses that display protein spikes (e.g. adenovirus) were also constructed by self-assembly of β-annulus peptides bearing a coiled-coil-forming sequence at the C-terminus (Figure 8.3b) [25]. Addition of a complementary coiled-coil-forming peptide to the capsid showed attachment of small spheres of approximately 5 nm that correspond to dimeric coiled-coil spikes on spherical assemblies 40–60 nm in size. Using a similar strategy, artificial viral capsids decorated with single-stranded DNA were constructed by self-assembly of β-annulus peptide conjugates bearing dA_{20} at the C-terminus [26]. The addition of a complementary poly(dT) to the dA_{20}-decorated artificial viral capsids led to the formation of aggregates, whereas the addition of noncomplementary poly(dA) did not produce such aggregates. These results indicate that dA_{20} on artificial viral capsids hybridize with the complementary poly(dT) to form aggregates by cross-linking of capsids with polynucleotide chains. Recently, Futaki and coworkers demonstrated that artificial viral capsids decorated with the amphiphilic cationic lytic peptide HAad efficiently delivered encapsulated proteins to the cytosol [27]. Efficient cytosolic delivery of encapsulated His-tagged EGFP and anti-mCherry

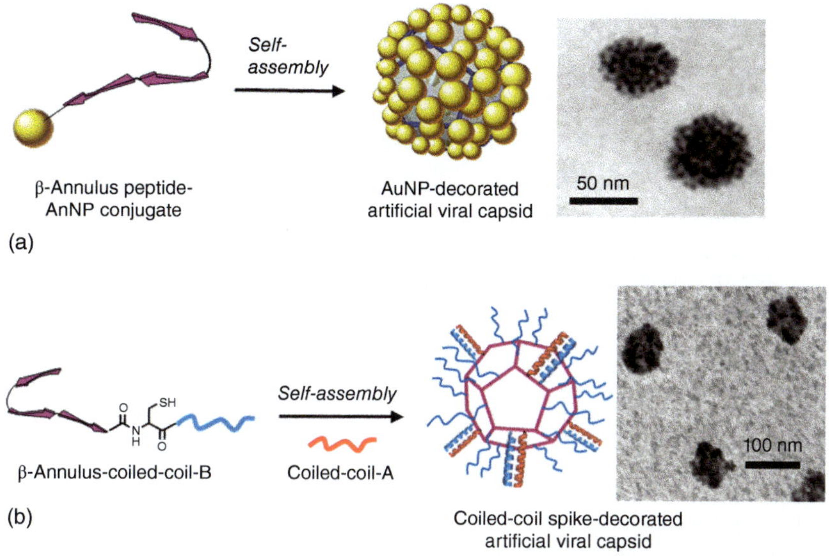

Figure 8.3 Schematic illustration of the construction of artificial viral capsids decorated with (a) gold nanoparticles (AuNPs) and (b) coiled-coil peptide spikes, along with TEM images. Source: Reproduced with permission from Refs. [24, 25]/Springer Nature and Royal Society of Chemistry.

nanobody-EGFP fusion protein was achieved using the HAad-decorated capsid. The targetability of anti-mCherry nanobody-EGFP to Lifeact-mCherry was demonstrated by costaining the cytoskeletal F-actin structure.

8.3.2 Human Serum Albumin-decorated Artificial Viral Capsids

Since human serum albumin (HSA) is a blood-compatible protein and is minimally immunogenic, it has attracted attention as a biocompatible drug delivery material. If artificial vial capsids decorated with HSA can be constructed by self-assembly, the nanocapsules would be promising candidates as less toxic designer drug carriers. Matsuura and coworkers constructed HSA-decorated artificial viral capsids by self-assembly of HSA-β-annulus peptide conjugates (Figure 8.4a) [28]. The reduced Cys at position 34 in HSA was linked with a Cys at the C-terminal position of the β-annulus peptide via a bismaleimide linker. A TEM image of a solution of HSA-β-annulus conjugate (0.1 μM) in Tris–HCl buffer (pH 7.0) showed the formation of spherical structures of approximately 50–70 nm in size (Figure 8.4b). Remarkably, the CAC of the conjugate at pH 7 was approximately 0.01 μM (Figure 8.4c), which is 1/2500 lower than that of the unmodified β-annulus peptides (25 μM). In fact, the conjugate formed spherical capsid structures of 84 ± 26 nm, even at 0.01 μM. These results suggested that artificial viral capsids were stabilized by modification with HSA, possibly due to lateral interactions between dense HSAs on the capsid and interactions between HSA and the capsid surface. The ζ-potential of the HSA-decorated capsid was -43 ± 15 mV at pH 7.0, whereas those at pH 3.5 and 4.8

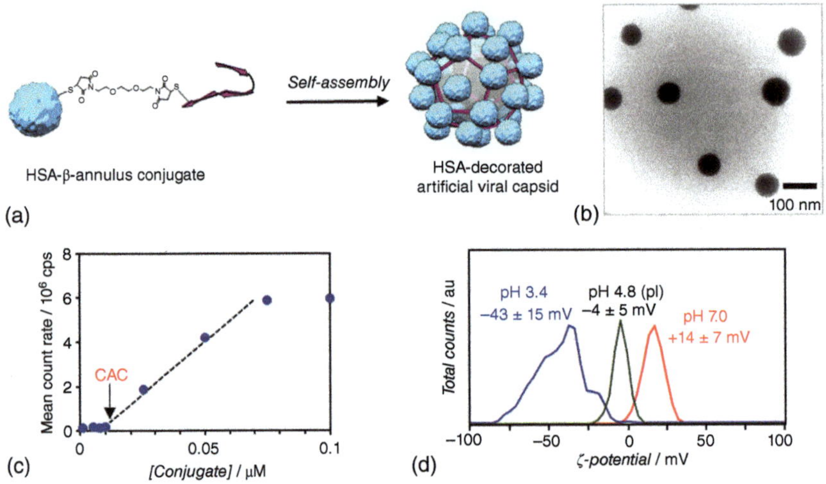

Figure 8.4 (a) Schematic illustration of the self-assembly of artificial viral capsids decorated with HSA. (b) TEM image of an HSA-β-annulus peptide conjugate at 0.1 μM. (c) The critical aggregation concentration for the HSA-β-annulus peptide conjugate. (d) ζ-Potentials of artificial viral capsids decorated with HSA at pHs 3.4, 4.8, and 7.5. Source: Reproduced with permission from Ref. [28]/American Chemical Society.

(the isoelectric point of HSA) were +14±7 mV and −4.4±5 mV, respectively (Figure 8.4d). Considering that the ζ-potentials of the unmodified artificial viral capsids were almost neutral at pH 7.0, the negative charges on the assemblies of the conjugate at pH 7 could be due to the presence of HSA on the surface.

8.3.3 Horseradish Peroxidase-decorated Artificial Viral Capsids

The strategy for decorating artificial viral capsids using a bismaleimide linker described above would be limited to proteins bearing a reduced Cys, such as HSA. Thus, Matsuura and coworkers developed a simple, general method for decorating proteins bearing His tags on the artificial viral capsids (Figure 8.5a) [29]. An Ni-NTA-modified artificial viral capsid was constructed by self-assembly of β-annulus peptides that were modified with Ni-NTA at the C-terminal. Then, horseradish peroxidase (HRP), which catalyzes the oxidation of various organic substrates by hydrogen peroxide, bearing a His tag was decorated on the surface of the Ni-NTA-modified capsid. The DLS of an equimolar mixture of β-annulus-Ni-NTA peptides and His-tagged HRP had average hydrodynamic diameters of approximately 30 nm. A TEM image of the mixture showed the formation of spherical assemblies with diameters of 30–90 nm. The ζ-potential revealed that His-tagged HRP was decorated on the outer surface of the Ni-NTA-modified artificial viral capsid. The specific enzyme activity of intact His-tagged HRP in potassium phosphate buffer (pH 7.0) at 25 °C was 1.63 units/mg (Figure 8.5b), whereas that of the HRP-decorated capsid was 0.238 units/mg under the same conditions (Figure 8.5c). Unfortunately, very dense packing of HRP on the surface of the nanoassembly could result in partial thermal denaturation and a decrease in its activity.

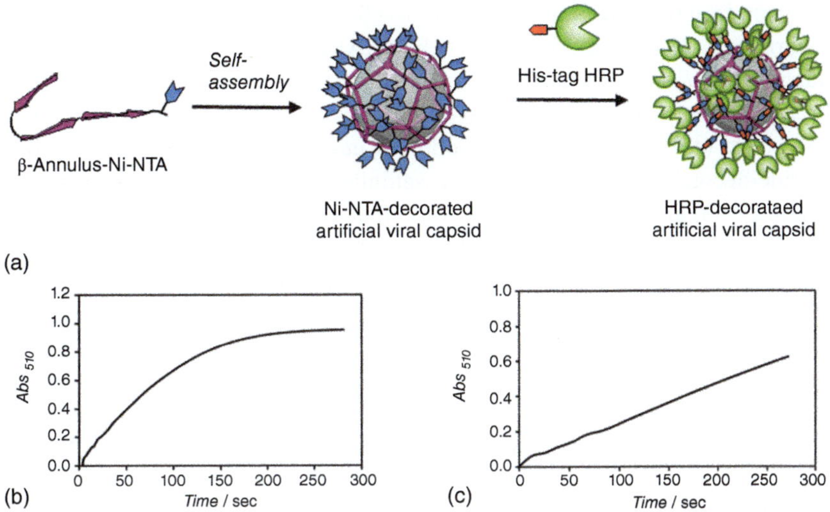

Figure 8.5 (a) Schematic illustration of the construction of artificial viral capsids decorated with horseradish peroxidase (HRP). (b, c) Time courses of absorbance increases at 510 nm due to the enzymatic reaction of (b) His-tagged HRP (11.4 μM) and (c) HRP-decorated artificial viral capsids (25 μM) using 4-aminoantipyrine, phenol, and hydrogen peroxide as substrates in phosphate buffer (pH 7.0) at 25 °C. Source: Reproduced with permission from Ref. [29]/MDPI.

8.3.4 Ribonuclease S-decorated Artificial Viral Capsids

Ribonuclease (RNase) is an enzyme that catalyzes the degradation of RNA into nucleotides or oligonucleotides and exhibits anticancer activity by degrading RNAs within cancer cells [30–33]. The degradation of intracellular RNA with RNase induces apoptosis of cancer cells by caspases which are proteolytic enzymes, regulation of DNA transcription, and cytokine production [31]. Some cancer cells have specific ligands for RNase on their surface; thus, RNases selectively accumulate on cancer cells and are not toxic to normal cells [32, 33]. However, the low efficiency of cellular uptake of RNase by cancer cells is an issue that needs to be improved. It is known that the cellular uptake efficiency of nanoparticles depends on their size, and that approximately 50 nm nanoparticles are efficiently taken up into cancer cells [34, 35]. In order to transport RNase efficiently into cells, it is necessary to design a carrier of approximately 50 nm in size.

Ribonuclease S (RNase S) is a split-type enzyme that is produced by digesting RNase A with subtilisin and is reconstructed from S-protein and S-peptide with an association constant of 7×10^6 M^{-1} at 25 °C [30]. Thus, supramolecular assemblies possessing ribonuclease activity can be constructed from S-proteins and S-peptides as reversible building blocks [36–38]. Matsuura and coworkers successfully constructed an RNase S-decorated artificial viral capsid by reconstructing the S-protein on the surface of an S-peptide-decorated artificial viral capsid (Figure 8.6a) [39, 40]. A β-annulus peptide bearing an S-peptide sequence at the C-terminus (β-annulus-S-peptide) was synthesized by native chemical ligation of the benzylthioesterified β-annulus peptide at the C-terminus (β-annulus-SBz) with an S-peptide bearing Cys at the N-terminus (Cys-S-peptide: CGGGKETAAAKFERQHMDS) [39]. DLS and TEM analyses demonstrated that the β-annulus-S-peptide self-assembled at 25 μM into spherical structures of approximately 50 nm in both acetate buffer (pH 4.5) and phosphate buffer (pH 7.4). The reconstruction of equimolar S-protein (25 μM) on the S-peptide-decorated capsid afforded assemblies with sizes of 51 ± 32 nm at pH 4.5 (Figure 8.6b) and 71 ± 43 nm at pH 7.4 (Figure 8.6c). A TEM image at pH 4.5 showed the formation of spherical assemblies of 50–65 nm attached small spheres of 5–10 nm on the surface (Figure 8.6d) [39]. In contrast, at pH 7.4, hollow spherical structures with rugged walls were observed (Figure 8.6e) [40]. The small spheres at pH 4.5 and the rugged wall at pH 7.4 may be ascribed to RNase S on the capsid surface. It appears that more RNase S may be reconstituted at pH 7.4 compared with that at pH 4.5. The reconstruction of RNase S on the surface of the capsid was also confirmed by CD spectroscopy (Figure 8.7a). The negative molar ellipticity of RNase S on the capsid was larger than that of S-protein alone, suggesting that reconstitution of RNase S on the surface of the capsid was achieved. The ribonuclease activity of the RNase S-decorated capsid was evaluated using a fluorescently labeled RNA substrate with quencher to be approximately 85% of the activity of RNase S alone at pH 7.4 (Figure 8.7b). These results indicate that an RNase S-decorated artificial viral capsid with high ribonuclease activity was constructed successfully.

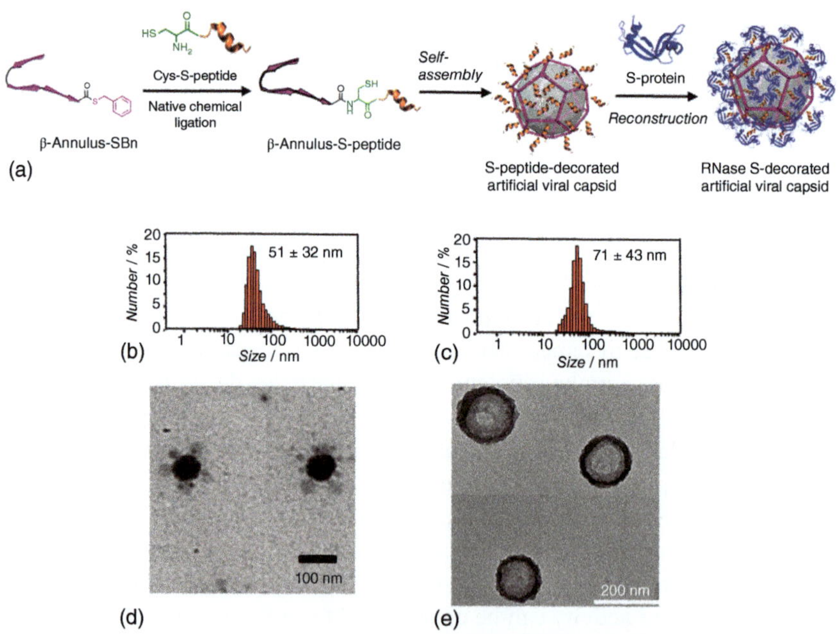

Figure 8.6 (a) Schematic illustration of the construction of artificial viral capsids decorated with ribonuclease S (RNase S). (b–e) Size distributions obtained from DLS (b, c) and TEM images (d, e) in solutions of equimolar mixtures of β-annulus-S-peptide and S-protein (25 μM) in 10 mM ammonium acetate buffer (pH 4.5, b, d) and 10 mM phosphate buffer (pH 7.4, c, e) at 25 °C. Source: Reproduced with permission from Refs. [39, 40]/ American Chemical Society and John Wiley and Sons.

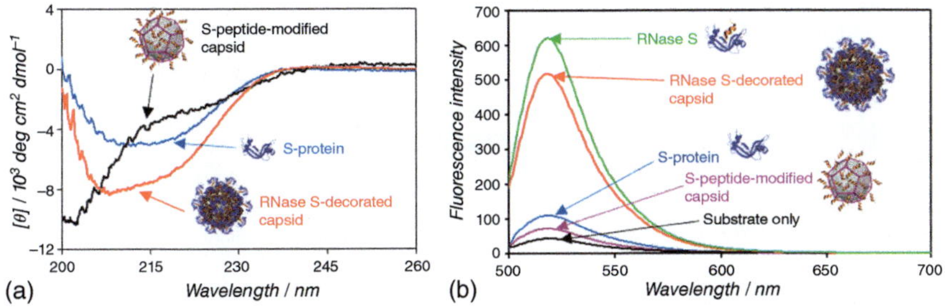

Figure 8.7 (a) CD spectra of S-peptide-modified artificial viral capsids ([β-annulus-S-peptide] = 25 μM), 25 μM S-protein, and RNase S-decorated artificial viral capsids ([β-annulus-S-peptide] = [S-protein] = 25 μM) in 10 mM phosphate buffer (pH 7.4) at 25 °C. (b) Fluorescence spectra of reaction mixtures using the RNaseAlert Lab Test in the presence of 25 μM RNase S, RNase S-decorated artificial viral capsids ([β-annulus-S-peptide] = [S-protein] = 25 μM), 25 μM S-protein, S-peptide-modified artificial viral capsids ([β-annulus-S-peptide] = 25 μM), and substrate alone in 10 mM phosphate buffer (pH 7.4) at 25 °C. The excitation wavelength was 490 nm. Source: Reproduced with permission from Ref. [40]/John Wiley and Sons.

The RNase S-decorated artificial viral capsid was spontaneously taken up by human hepatoma HepG2 cells more efficiently than RNase S alone, and it degraded the intracellular RNA (Figure 8.8a) [40]. Confocal laser scanning microscopy (CLSM) images produced by staining with an RNA-specific fluorescent dye (Strand Brite™ RNA Green) revealed degradation of some of the intracellular RNAs (Figure 8.8b). The intracellular RNAs in HepG2 cells were extracted and quantified to estimate the extent of RNA degradation by the RNase S-decorated capsid. The amount of RNA in HepG2 cells was decreased to 56% of the original amount by the addition of RNase S-decorated capsid, which was more effective than RNase S or RNase A alone at the same concentration. As a result, the cell viability of the untreated HepG2 cells was decreased by 52% by adding RNase S-decorated capsids, which is comparable to that obtained with the commercial anticancer drugs taxol and actinomycin D (Figure 8.8c). The anticancer activity of RNase S-decorated capsids was higher than that of RNase S or RNase A alone, which might be caused by higher intracellular uptake efficiency and intracellular RNA degradation activity of RNase S-decorated capsids.

8.4 Enveloped Viral Replicas Embedded with Membrane Proteins

8.4.1 Enveloped Artificial Viral Capsids

The surfaces of the nucleocapsids of natural envelope-type viruses such as influenza virus, human immunodeficiency virus, and coronavirus are covered with a lipid bilayer membrane which embed membrane proteins. The membrane proteins on enveloped viruses play important roles in interactions with host cells such as adhesion, infection, and budding. For example, hemagglutinin and neuraminidase

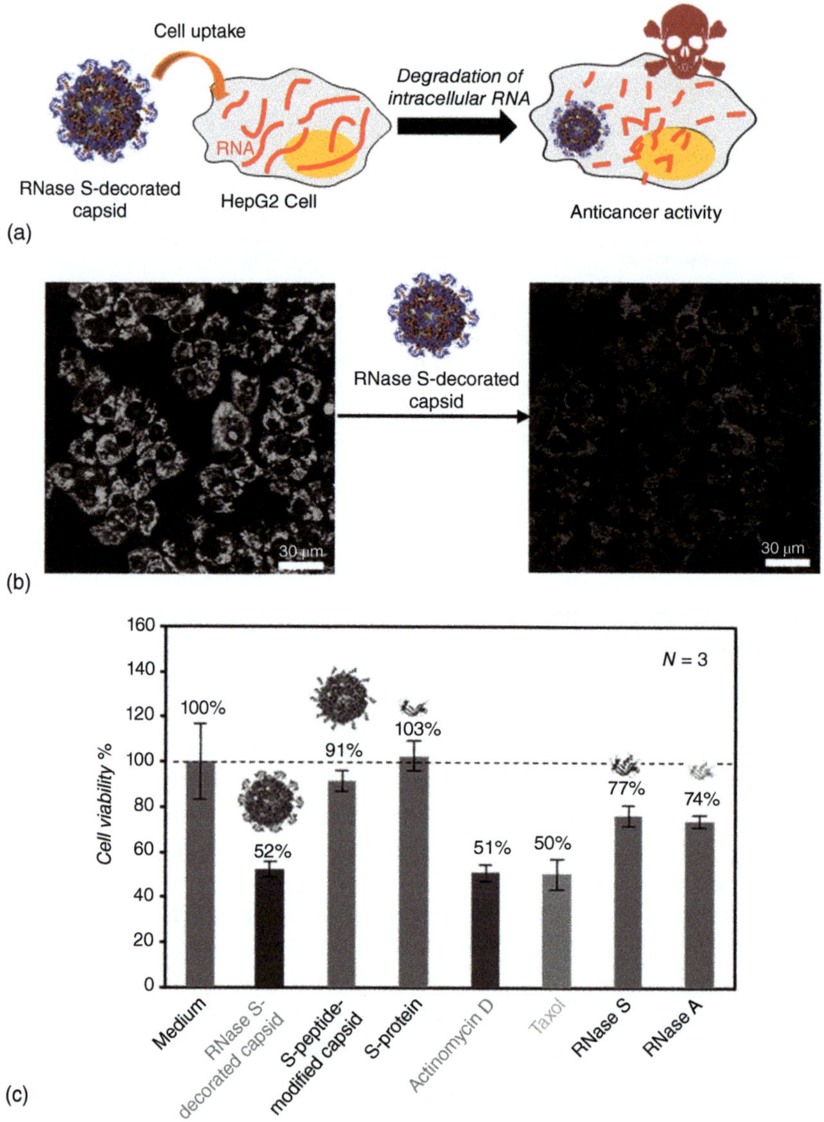

Figure 8.8 (a) Schematic illustration of cell death by degradation of intracellular RNAs by RNase S-decorated artificial viral capsids. (b) CLSM images of HepG2 cells stained with Strand Brite™ RNA Green in the absence and presence of RNase S-decorated artificial viral capsids ([β-annulus-S-peptide] = [S-protein] = 25 μM). (c) Cytotoxicity of HepG2 cells incubated in the presence of each sample for 48 h at 37 °C under 5% CO_2. RNase S-decorated artificial viral capsids ([β-annulus-S-peptide] = [S-protein] = 25 μM), S-peptide-modified artificial viral capsids ([β-annulus-S-peptide] = 25 μM), 25 μM S-protein, 1.9 μM actinomycin D, 25 μM taxol, 25 μM RNase S, and 25 μM RNase A). Source: Reproduced with permission from Ref. [40]/John Wiley and Sons.

embedded in the envelope of the influenza virus are involved in infection and budding, respectively. To date, several artificial nanomaterials that mimic enveloped viruses have been developed. For example, Kostarelos and coworkers reported that coating the spikes of adenovirus with cationic lipids improves their cell permeability [41]. Shih and coworkers reported that a spherical assembly constructed by DNA origami is structurally stabilized by complexing it with a lipid bilayer [42]. However, progress on a chemical strategy for constructing artificial enveloped virus-like structures has been slow.

Matsuura and coworkers created an enveloped artificial viral capsid by complexing an anionic artificial viral capsid with a cationic lipid bilayer (Figure 8.9a) [43]. An anionic artificial viral capsid was constructed by self-assembly of β-annulus-EE peptides having two anionic Glu residues at the C-terminus which directs it to the outside of the capsid. The anionic artificial viral capsid was then complexed with a lipid

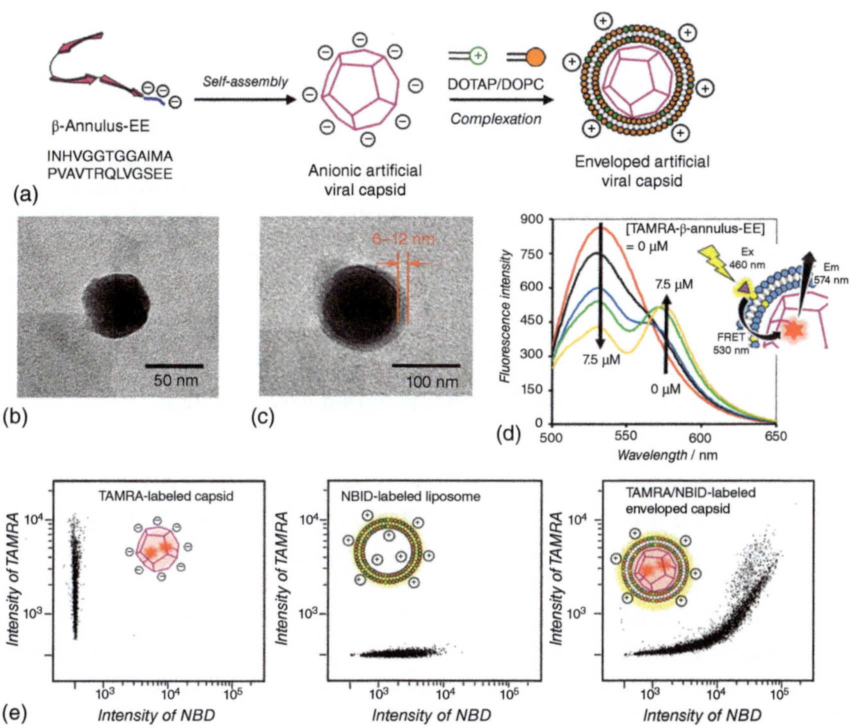

Figure 8.9 (a) Schematic illustration of the construction of an enveloped artificial viral capsid via electrostatic interactions. (b, c) TEM images of an artificial viral capsid bearing an anionic surface (b) and an enveloped artificial viral capsid (c) stained with EM stainer. (d) Fluorescence spectra of NBD-PE (3.75 μM) and TAMRA-β-annulus-EE (0–7.50 μM) excited at 460 nm in 10 mM Tris–HCl buffer (pH 7.0) at 25 °C. (e) Flow cytometry analyses of TAMRA-labeled artificial viral capsids, NBD-labeled liposomes, and TAMRA/NBD-labeled enveloped artificial viral capsids. Source: Reproduced with permission from Ref. [43]/Royal Society of Chemistry.

bilayer consisting of a cationic lipid (DOTAP; 1,2-dioleoyl-3-trimethylammonium propane) and a zwitterionic lipid (DOPC; 1,2-dioleoyl-*sn*-glycero-3-phosphocholine) at a 1:1 charge ratio. The DLS measurements indicated that the size distribution of the complex with a cationic lipid bilayer was 84 ± 23 nm, which is slightly larger than that of the anionic artificial viral capsid (67 ± 21 nm). TEM images of the anionic capsid showed the formation of spherical assemblies with diameters of approximately 50 nm (Figure 8.9b), whereas that of the complex revealed spherical structures of about 100 nm that were covered with lightly stained fringe structures of thickness 6–12 nm (Figure 8.9c), which is comparable to one or two lipid bilayers. The ζ-potential of the anionic capsid at pH 7.0 was -29 ± 7 mV, while that of the complex was $+72 \pm 10$ mV. These results indicate that complexation of the anionic artificial viral capsid with a DOTAP/DOPC lipid bilayer formed an artificial viral capsid with a cationic envelope.

In order to confirm the formation of an enveloped artificial viral capsid further, a fluorescence-labeled enveloped capsid was constructed by complexing a tetramethylrhodamine (TAMRA)-labeled anionic capsid with a nitrobenzoxadiazole (NBD)-labeled lipid bilayer. When the NBD on the enveloped capsid was excited in the presence of different concentrations of TAMRA-β-annulus-EE, fluorescence resonance energy transfer (FRET) from NBD to TAMRA was observed (Figure 8.9d), indicating that the lipid bilayer is in close proximity to the artificial viral capsid. In addition, imaging flow cytometry analysis of the TAMRA/NBD-labeled enveloped artificial viral capsids indicated a broad fluorescence intensity distribution for both TAMRA and NBD (Fig. 8.9e). These results confirm the coexistence of the β-annulus-EE peptide and the lipid bilayer on the enveloped artificial viral capsids.

8.4.2 Embedding of Connexin-43 onto Enveloped Artificial Viral Capsids

Over the past two decades, embedding membrane proteins onto liposomes has been made possible by the use of a cell-free protein expression system (PURE system) in the presence of liposomes. Using the PURE system, Akiyoshi and coworkers constructed liposomes embedded with connexin-43 (Cx43), which is a membrane protein that forms a tubular gap junction with a hole of approximately 1.6 nm by associating between the cells, enabling molecular transport between the cells [44–46]. However, the artificial construction of enveloped viral capsids embedded with membrane proteins remains a challenge.

Recently, Matsuura and coworkers embedded Cx43 onto the enveloped artificial viral capsids described above (Figure 8.10a) [47]. Western blot analysis demonstrated that Cx43 was expressed from the Cx43-encoding plasmid DNA using the PURE system in the presence of cationic enveloped artificial viral capsids. FCS indicated that fluorescence-labeled anti-Cx43 antibody was bound to the Cx43-embedd viral replicas. TEM images of the Cx43-embedded enveloped viral replicas with anti-Cx43 primary antibody and gold nanoparticle-labeled secondary antibody revealed many gold nanoparticles on the surface of the spherical structures, indicating the existence

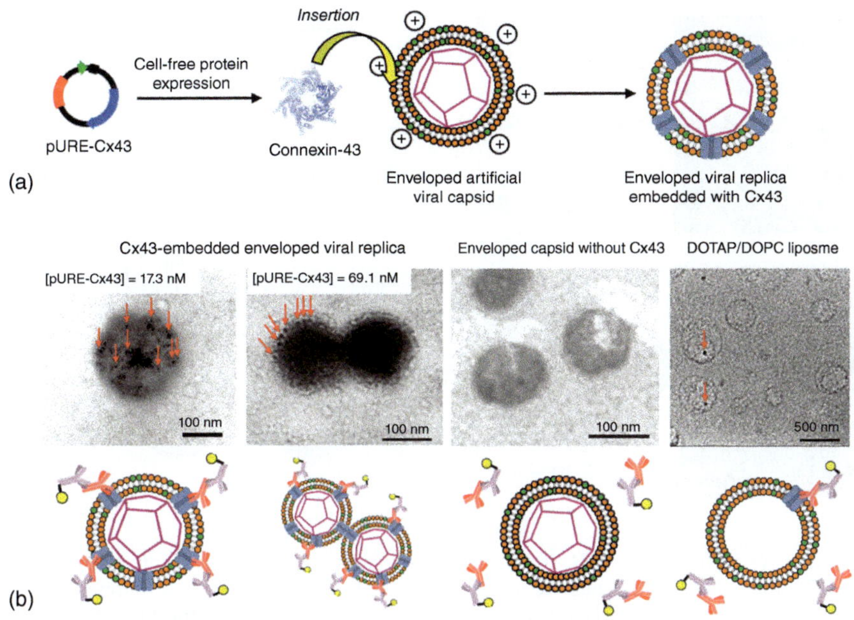

Figure 8.10 (a) Schematic illustration of the construction of enveloped viral replicas embedded with Connexin-43 (Cx43). (b) TEM images of Cx43-embedded viral replicas, the enveloped viral capsid without Cx43, and Cx43-expressing liposomes in the presence of anti-Cx43 antibody and gold nanoparticle-labeled secondary antibody in 10 mM Tris–HCl buffer (pH 7.0). Cx43 was expressed from 17.3 nM or 69.1 nM pURE-Cx43 using the PURE system. TEM samples were stained with EM stainer.
Source: Reproduced with permission from ref. [47]/ Royal Society of Chemistry / CC BY 3.0.

of many Cx43 protein molecules on the envelope (Figure 8.10b). In contrast, gold nanoparticles were not observed on the enveloped artificial viral capsids that did not contain Cx43. On Cx43-expressing liposomes consisting of DOTAP/DOPC, only a few gold nanoparticles were observed. Interestingly, when the concentration of the Cx43-encoding plasmid increased, spherical structures that were connected to each other were observed, suggesting the formation of gap junction structures among Cx43-embedded viral replicas.

Furthermore, it was demonstrated that molecular transport from Cx43-embedded enveloped viral replicas into Cx43-expressing cells through the gap junction formation was enabled [47]. The CLSM images showed that fluorescence of TAMRA was observed inside of Cx43-expressing HepG2 cells after the addition of TAMRA-encapsulated Cx43-embedded viral replicas (Figure 8.11a), whereas the transport of TAMRA from the enveloped capsid without Cx43 to the cells was minimal (Figure 8.11b). Quantitative analyses revealed that the fluorescence intensity in cells transported from Cx43-embedded viral replicas was significantly stronger than that from the enveloped capsids without Cx43 (Figure 8.11c). These results indicate that TAMRA can be transported from the Cx43-embedded viral replica into Cx43-expressing cells through the gap junctions.

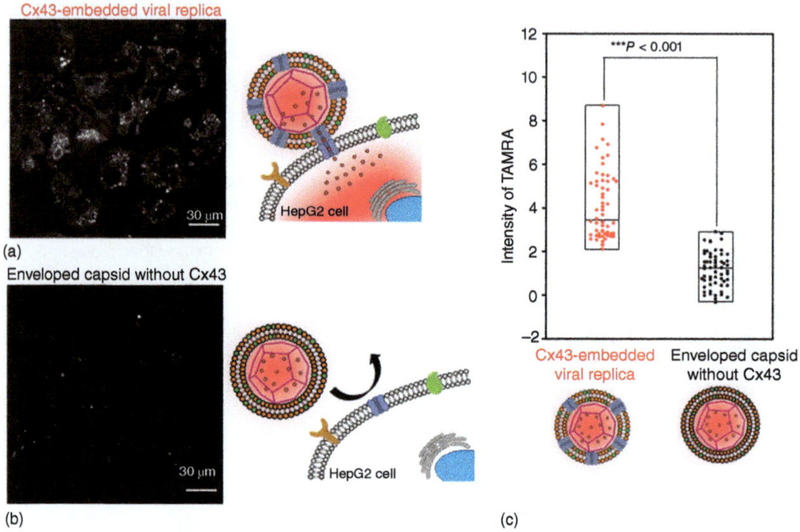

Figure 8.11 (a, b) Transport of TAMRA from Cx43-embedded viral replicas into HepG2 cells. CLSM images of HepG2 cells incubated with TAMRA-encapsulated Cx43-embedded viral replicas (a) and enveloped capsids without Cx43 (b). (c) Box plot of the fluorescence intensity distribution of TAMRA in HepG2 cells normalized outside of the cells ($N = 60$). Source: Reproduced with permission from ref. [47]/ Royal Society of Chemistry / CC BY 3.0.

8.5 Summary and Perspectives

In this chapter, the construction of artificial viral capsids that self-assembled from TBSV β-anulus peptides, the encapsulation of guest molecules, and decoration of their surface with functional proteins and lipid bilayers were summarized. The artificial viral capsids decorated with functional molecules are expected to be applied as nanoreactors, delivery materials for protein and nucleic acid medicines, basic materials for vaccines, and molecular robots. The advantage of these artificial viral capsids is that arbitrary functions can be provided by chemical design and synthesis.

In the near future, artificial virus systems endowed with functions such as selective internalization into cells, translocation to specific organelles, intracellular self-replication, selective genome packaging, budding from host cells, and blood–brain barrier penetration may be constructed using a chemical strategy. It will be possible to deepen our understanding of the actual virus infection mechanism and dynamics by analyzing intracellular dynamics using an artificial virus capsid as a model tool. The display of membrane proteins that are involved in viral infections, such as the spike protein of SARS-CoV-2 and the hemagglutinin of influenza virus, can be used to mimic natural viruses. These viral replicas could induce the activation of the immune system and the neutralizing antibody response without potential cytotoxicity. Constructing novel biomolecular systems using chemistry-based virus mimetic materials may contribute to next-generation pharmaceutical technologies.

References

1 Khudyakov, Y. and Pumpens, P. (ed.) (2016). *Viral Nanotechnology*. CRC Press.
2 Bronstein, L.M. (2011). Virus-based nanoparticles with inorganic cargo: what does the future hold? *Small* 7 (12): 1609–1618.
3 Van Rijn, P. and Schirhagl, R. (2016). Viruses, artificial viruses and virus-based structures for biomedical applications. *Adv. Healthc. Mater.* 5 (12): 1386–1400.
4 Wen, A.M. and Steinmetz, N.F. (2016). Design of virus-based nanomaterials for medicine, biotechnology, and energy. *Chem. Soc. Rev.* 45 (15): 4074–4126.
5 Lee, K.L., Twyman, R.M., Fiering, S., and Steinmetz, N.F. (2016). Virus-based nanoparticles as platform technologies for modern vaccines. *WIREs Nanomed. Nanobiotechnol.* **8** (4): 554–578.
6 Chen, M.Y., Butler, S.S., Chen, W., and Suh, J. (2019). Physical, chemical, and synthetic virology: Reprogramming viruses as controllable nanodevices. *WIREs Nanomed. Nanobiotechnol.* 11 (3): e1545.
7 Matsuura, K. (2017). Construction of functional biomaterials by biomolecular self-assembly. *Bull. Chem. Soc. Jpn.* 90 (8): 873–884.
8 Inaba, H. and Matsuura, K. (2019). Peptide nanomaterials designed from natural supramolecular systems. *Chem. Rec.* 19 (5): 843–858.
9 Matsuurua, K. (2014). Rational design of self-assembled proteins and peptides for nano- and micro-sized architectures. *RSC Adv.* 4 (6): 2942–2953.
10 Luo, Q., Hou, C., Bai, Y. et al. (2016). Protein assembly: versatile approaches to construct highly ordered nanostructures. *Chem. Rev.* 116 (22): 13571–13632.
11 Kuan, S.L., Bergamini, F.R.G., and Weil, T. (2018). Functional protein nanostructures: a chemical toolbox. *Chem. Soc. Rev.* 47 (24): 9069–9105.
12 Ramakers, B., Hest, J., and Löwik, D. (2014). Molecular tools for the construction of peptide-based materials. *Chem. Soc. Rev.* 43 (8): 2743–2756.
13 Lou, S., Wang, X., Yu, Z., and Shi, L. (2019). Peptide tectonics: encoded structural complementarity dictates programmable self-assembly. *Adv. Sci.* 6 (13): 1802043.
14 Matsuura, K. (2018). Synthetic approaches to construct viral capsid-like spherical nanomaterials. *Chem. Commun.* 54 (65): 8944–8959.
15 Matsuura, K., Watanabe, K., Matsuzaki, T. et al. (2010). Self-assembled synthetic viral capsids from a 24-mer viral peptide fragment. *Angew. Chem. Int. Ed.* 49 (50): 9662–9665.
16 Kobayashi, R., Inaba, H., and Matsuura, K. (2021). Fluorescence correlation spectroscopy analysis of effect of molecular crowding on self-assembly of β-annulus peptide into artificial viral capsid. *Int. J. Mol. Sci.* 22 (9): 4754.
17 Matsuura, K., Watanabe, K., Matsushita, Y., and Kimizuka, N. (2013). Guest-binding behavior of peptide nanocapsules self-assembled from viral peptide fragments. *Polym. J.* 45 (5): 529–534.
18 Fujita, S. and Matsuura, K. (2016). Encapsulation of CdTe quantum dots into synthetic viral capsids. *Chem. Lett.* 45 (8): 922–924.
19 Matsuura, K., Nakamura, T., Watanabe, K. et al. (2016). Self-assembly of Ni-NTA-modified β-annulus peptides into artificial viral capsids and encapsulation of His-tagged proteins. *Org. Biomol. Chem.* 14 (33): 7869–7874.

20 Fujita, S. and Matsuura, K. (2014). Inclusion of zinc oxide nanoparticles into virus-like peptide nanocapsules self-assembled from viral β-annulus peptide. *Nanomaterials* 4 (3): 778–791.

21 Nakamura, Y., Inaba, H., and Matsuura, K. (2019). Construction of artificial viral capsids encapsulating short DNAs via disulfide bonds and controlled release of DNAs by reduction. *Chem. Lett.* 48 (6): 544–546.

22 Nakamura, Y., Sato, Y., Inaba, H. et al. (2020). Encapsulation of mRNA into artificial viral capsids via hybridization of a β-annulus-dT_{20} conjugate and the poly(A) tail of mRNA. *Appl. Sci.* 10 (22): 8004.

23 Matsuura, K. (2020). Dressing up artificial viral capsids self-assembled from C-terminal-modified β-annulus peptides. *Polym. J.* 52 (9): 1035–1041.

24 Matsuura, K., Ueno, G., and Fujita, S. (2015). Self-assembled artificial viral capsid decorated with gold nanoparticles. *Polym. J.* 47 (2): 146–151.

25 Fujita, S. and Matsuura, K. (2017). Self-assembled artificial viral capsids bearing coiled-coils at the surface. *Org. Biomol. Chem.* 15 (23): 5070–5077.

26 Nakamura, Y., Yamada, S., Nishikawa, S., and Matsuura, K. (2017). DNA-modified artificial viral capsids self-assembled from DNA-conjugated β-annulus peptide. *J. Pept. Sci.* 23 (7–8): 636–643.

27 Sakamoto, K., Furukawa, H., Arafiles, J.V.V. et al. (2022). Artificial nanocage formed via self-assembly of β-annulus peptide for delivering biofunctional proteins into cell interiors. *Bioconjug. Chem.* 33 (2): 311–320.

28 Matsuura, K. and Honjo, T. (2019). artificial viral capsid dressed up with human serum albumin. *Bioconjug. Chem.* 30 (6): 1636–1641.

29 Matsuura, K., Shiomi, Y., Mizuta, T., and Inaba, H. (2020). Horseradish peroxidase-decorated artificial viral capsid constructed from β-annulus peptide via interaction between His-Tag and Ni-NTA. *Processes* 8 (11): 1455.

30 Raines, R. (1998). Ribonuclease A. *Chem. Rev.* 98 (3): 1045–1066.

31 Ardelt, W., Ardelt, B., and Darzynkiewicz, Z. (2009). Ribonucleases as potential modalities in anticancer therapy. *Eur. J. Pharmacol.* 625 (1): 181–189.

32 Makarov, A. and Ilinskaya, O. (2003). Cytotoxic ribonucleases: molecular weapons and their targets. *FEBS Lett.* 540 (1–3): 15–20.

33 Spalletti-Cernia, D., Sorrentino, R., Gaetano, S. et al. (2004). Highly selective toxic and proapoptotic effects of two dimeric ribonucleases on thyroid cancer cells compared to the effects of doxorubicin. *Br. J. Cancer* 90 (1): 270–277.

34 Wen, J., Kim, B., James, T., and Chan, W. (2008). Nanoparticle-mediated cellular response is size-dependent. *Nat. Nanotechnol.* 3 (3): 145–150.

35 Osaki, F., Kanamori, T., Sando, S. et al. (2004). A quantum dot conjugated sugar ball and its cellular uptake. On the size effects of endocytosis in the subviral region. *J. Am. Chem. Soc.* 126 (21): 6520–6521.

36 Aubin, M.-E., Morales, D.G., and Hamad-Schifferli, K. (2005). Labeling ribonuclease S with a 3 nm Au nanoparticle by two-step assembly. *Nano Lett.* 5 (3): 519–522.

37 Bastings, M.M.C., de Greef, T.F.A., van Dongen, J.L.J. et al. (2010). Macrocyclization of enzyme-based supramolecular polymers. *Chem. Sci.* 1 (1): 79–88.

38 Hamachi, I., Watanabe, J., Eboshi, R. et al. (2000). Incorporation of artificial receptors into a protein/peptide surface: a strategy for on/off type of switching of semisynthetic enzymes. *Pept. Sci.* 55 (6): 459–468.

39 Matsuura, K., Ota, J., Fujita, S. et al. (2020). Construction of ribonuclease-decorated artificial virus-like capsid by peptide self-assembly. *J. Org. Chem.* 85 (3): 1668–1673.

40 Liang, Y., Furukawa, H., Sakamoto, K. et al. (2022). Anticancer activity of reconstituted ribonuclease S-decorated artificial viral capsid. *ChemBioChem* 23 (15): e202200220.

41 Singh, R., Al-Jamal, K.T., Lacerda, L., and Kostarelos, K. (2008). Nanoengineering artificial lipid envelopes around adenovirus by self-assembly. *ACS Nano* 2 (5): 1040–1050.

42 Perrault, S.D. and Shih, W.M. (2014). Virus-inspired membrane encapsulation of DNA nanostructures to achieve in vivo stability. *ACS Nano* 8 (5): 5132–5140.

43 Furukawa, H., Inaba, H., Inoue, F. et al. (2020). Enveloped artificial viral capsids self-assembled from anionic β-annulus peptide and cationic lipid bilayer. *Chem. Commun.* 56 (52): 7092–7095.

44 Nomura, S.M., Kondoh, S., Asayama, W. et al. (2008). Direct preparation of giant proteo-liposomes by in vitro membrane protein synthesis. *J. Biotechnol.* 133 (2): 190–195.

45 Ohtsuka, T., Neki, S., Kanai, T. et al. (2011). Synthesis and in situ insertion of a site-specific fluorescently labeled membrane protein into cell-sized liposomes. *Anal. Biochem.* 418 (1): 97–101.

46 Kaneda, M., Nomura, S.M., Ichinose, S. et al. (2009). Direct formation of proteo-liposomes by in vitro synthesis and cellular cytosolic delivery with connexin-expressing liposomes. *Biomaterials* 30 (23–24): 3971–3977.

47 Furukawa, H., Inaba, H., Sasaki, Y. et al. (2022). Embedding a membrane protein into an enveloped artificial viral replica. *RSC Chem. Biol.* 3 (2): 231–241.

9

Peptide Assemblies on Surfaces: A Study Using Scanning Tunneling Microscopy

Lanlan Yu[1], Yanlian Yang[2], Chen Wang[2], and Chenxuan Wang[1]

[1] *School of Basic Medicine Peking Union Medical College, State Key Laboratory of Medical Molecular Biology, Department of Biophysics and Structural Biology, Institute of Basic Medical Sciences Chinese Academy of Medical Sciences, Beijing 100005, P. R. China*
[2] *CAS Key Laboratory of Biological Effects of Nanomaterials and Nanosafety, CAS Key Laboratory of Standardization and Measurement for Nanotechnology, Laboratory of Theoretical and Computational Nanoscience, CAS Center for Excellence in Nanoscience, National Center for Nanoscience and Technology, Beijing 100190, P. R. China*

9.1 Introduction

Peptide and protein self-assemblies have drawn great scientific interest due to their important implications in the physiological and pathological process *in vivo*. For instance, the abnormal amyloid protein misfolding into β-pleated sheet conformation and deposition in tissues and organs causes the occurrence of amyloidosis, including mad cow disease, Parkinson's disease, and heart failure associated with cardiac transthyretin amyloidosis [1, 2]. Common to different types of amyloidal proteins is that the aggregation behaviors are typical multistage β-sheet assembly processes [1]. A profound understanding of the structural conversion from native protein to pathological protein aggregates is pivotal for revealing the molecular mechanism underlying the occurrence and development of diseases and achieving a structure-based drug discovery against abnormal protein aggregation *in vivo*. In addition to its close relevance with diseases, peptide assembly has emerged as one of the most innovative biomaterials showing valuable applications in the field of biomedicine and biotechnology. The structural study manifesting the sequence-to-structure relationship for peptide assembly is essential for the rational design and optimization of peptide-assembled biomaterials.

Thanks to the advances made in the structural characterization methodology, we have learned a great deal about the structure of peptide assembly in the past decades. Researchers can analyze secondary structures of peptides using circular dichroism (CD) spectroscopy by measuring their specific absorption of polarized light, whereas

Peptide Self-Assembly and Engineering: Fundamentals, Structures, and Applications, First Edition.
Edited by Xuehai Yan.

that of using Fourier transformed infrared (FTIR) by measuring the specific IR absorption corresponding to the structural transitions of polypeptide backbone between a ground vibrational state and an excited vibrational state. X-ray crystallography determines the spatial arrangement of atoms for assemblies at atomic resolution by measuring the diffraction of X-rays on the crystals formed by assemblies. However, due to the difficulty of peptide assembly in crystallization, a limited number of peptide assembly structures have been atomically resolved by X-ray crystallography. As an alternative route to gain the three-dimensional structure of peptide assembly, nuclear magnetic resonance spectroscopy (NMR) measures the spatial constraints between two nuclei with unpaired spins and reconstructs the fine three-dimensional structure of assembly via chemical shift assignment, geometric restraint assignment, and structural calculation algorithms. Cryo-electron microscopy (cryo-EM) resolves the fine structure of macromolecular assembly by obtaining the three-dimensional (3D) electron density maps which are reconstructed from a set of two-dimensional (2D) cryo-EM images of individual assemblies. It should be noted that high-resolution structural analysis is still challenging due to the polymorphism and polydispersity of peptide assembly structures. Most structural analysis techniques inevitably implicate the average/extraction of the predominant/periodic portions of structural signals, masking the peptide intrinsic structures, especially for the minorly populated metastable states. Thus, a novel structural characterization technology for studying peptide assembly structure is still needed.

Benefiting from the pioneering efforts of generations of researchers, major technological breakthroughs have been achieved in the observation of individual molecules using scanning tunneling microscopy (STM). The single-molecule imaging based on STM is dependent on an electron tunneling process [3]. When an STM probe is brought close to the surface of a substrate with a separation distance of less than 1 nm, a bias voltage (V) applied between the probe and substrate allows electrons to penetrate through the classically impenetrable energy barrier separating them (Figure 9.1). The tunneling current (I) reflects the local density of states of the surface. The STM probe is extremely sharp with a probe apex radius at the atomic scale, and thus STM can distinguish the structural features on the surface with an

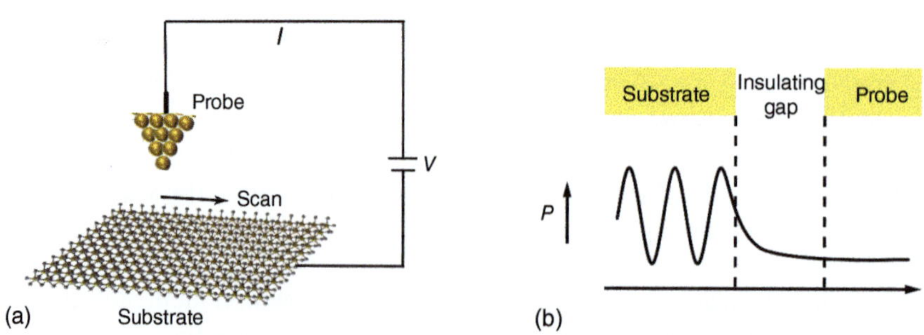

Figure 9.1 (a) Schematic representation of STM. An STM probe is brought close to the surface of a substrate. A bias voltage applied between the probe and substrate results in a tunneling current. (b) The decay behavior of the electron transmission probability (P) across the classically impenetrable energy barrier. Source: Lanlan Yu and Chenxuan Wang.

atomic resolution. The capability of precise control of the STM probe position above the surface helps us to monitor the structural changes as the probe moves across the surface. Using STM to image the absorbed peptide assembly present on a substrate surface, the local electron density of the state of peptide molecules can be detected. With the aid of STM, it becomes feasible to directly observe the structures of peptide assembly on the surface and resolve the polydisperse structures of peptide assembly at a single-molecule level. STM can be used in various environments, including ultrahigh vacuum (UHV), ambient conditions, and solid/liquid interfaces [3]. The high-resolution spatial structure analysis achieved by STM improves our understanding of the molecular mechanism governing assembly and the correlation between peptide assembly structure and physiological function, providing a theoretical basis for the treatment of relevant diseases and the design of targeted drugs.

9.2 Structural Characterization of Individual Peptides

The submolecular resolution of STM promises its application in the visualization and structural determination of single molecules in their surface-bounded states. The molecular front orbitals of adsorbates, including the highest occupied molecular orbitals (HOMO) and the lowest unoccupied molecular orbitals (LUMO), are the dominant contributors to the topography of individual molecules in STM images [3]. Extensive theoretical studies have been implicated in establishing the correlation between the experimentally observed structures and the intrinsic molecular orbitals. The adsorbate–substrate interactions, regulating the adsorption orientations of observed molecules on a surface, also influence the observed images. The same molecules might adopt distinct binding configurations on different substrate surfaces, displaying variable adsorption geometries in STM images. The capability of STM in the single-molecule recognition of polypeptides has been demonstrated in different systems, including linear oligopeptides, cyclic peptides, and proteins. Herein, we introduce the progress in the field of single-molecule recognition by STM as follows.

9.2.1 Isolated Peptide Molecules Immobilized on Surfaces

A peptide is made by the polymerization of amino acids with a general molecular weight of less than 5 kDa. As the simplest form of the peptide, dipeptides have been used as a model system for manifesting the capability of STM in the structural analysis with individual adsorbed peptides on a surface. Dipeptide L-Phe-L-Phe contains two phenylalanine residues with an *L* configuration (Figure 9.2a). By depositing dipeptides on a Cu(110) surface in a UHV chamber (base pressure $\approx 2\times10^{-10}$ mbar) at room temperature, L-Phe-L-Phe displays a dumbbell-like structure in the STM images in which two bright protrusions correspond to the π-electron-rich phenyl rings of Phe side chains and the dumbbell bar-like structure corresponds to the peptide bond linking two consecutive amino-acid residues (Figure 9.2b and c) [4]. The axis passing through the center of the two phenyl rings of single L-Phe-L-Phe is rotated 34° clockwise from the $[1\bar{1}0]$ substrate direction. By contrast, the STM images

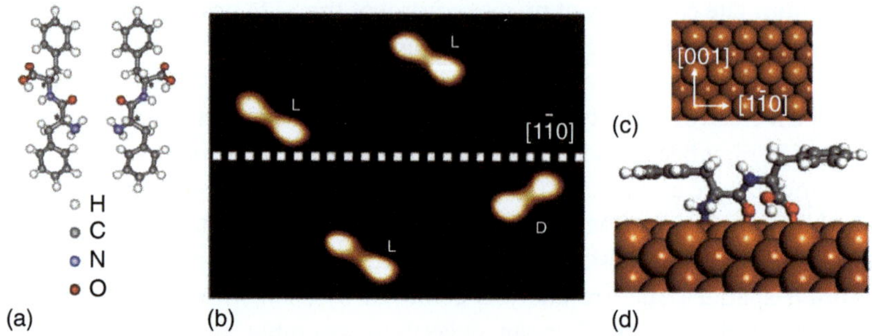

Figure 9.2 (a) Chemical structures of L-Phe-L-Phe (left) and D-Phe-D-Phe (right). (b) STM image of individual dipeptides on the Cu(110) surface. Tunneling conditions: $I = 0.9$ nA and $V = -0.9$ V. (c) Schematic model of the atomic arrangement of a Cu(110) surface. (d) Minimum-energy structure of an L-Phe-L-Phe on a Cu(110) surface obtained in DFT calculation. Source: Reproduced with permission from Ref. [4]; © 2023 Wiley-VCH Verlag GmbH & Co. KGaA, Weinheim.

of the enantiomer of L-Phe-L-Phe, i.e. D-Phe-D-Phe, display an enantiomorphous structure, where the molecular axis of D-Phe-D-Phe is rotated 34° counterclockwise from the [1$\bar{1}$0] substrate direction (Figure 9.2b). To provide insights into the adsorption structures of dipeptides, theoretical simulations were used to analyze the topography of a single dipeptide in STM images. As revealed by a density functional theory (DFT) simulation, three atoms, including the nitrogen atom of the amino group, one oxygen atom of the carboxyl group, and the oxygen atom of the carbonyl group, are key for bounding dipeptide on the Cu(110) surface (Figure 9.2d). The L-Phe-L-Phe/surface interaction energy was calculated to be 1.6 eV. We comment that not only limited to Cu surfaces, other conductive substrates with atomically flat surfaces can also be used in the UHV-STM experiments.

STM-based single-molecule recognitions have been conducted for several peptides/proteins with complex structures. Valinomycin is a cyclic peptide with three tetrapeptide repeats (L-Val-D-Hyv-D-Val-L-Lac), acting as a transporter of K^+ ions across cell membranes (Figure 9.3a). In the UHV-STM study, valinomycin was vapor-deposited onto a Cu(111) surface at a cryogenic temperature of 78 K [5]. Cooling samples to 78 K using liquid nitrogen or 4 K using liquid helium stabilizes the immobilization of monodisperse molecules on surfaces by reducing the diffusion mobility of adsorbed molecules on surfaces, yielding high-quality STM images. Two adsorption configurations have been identified in the STM images of valinomycin monomers on the Cu(111) surface (Figure 9.3b and c). The first configuration, denoted as M_L, presented as a cloverleaf-shaped structure with a central protrusion, whereas the second configuration, denoted as M_R, presented as a cloverleaf-shaped structure with a central cavity. M_L and M_R show a threefold symmetry in the STM images and are attributed to the different adsorption orientations of the planar structure of valinomycin upon binding on the surface. The pursuit of the pathways of molecular adsorption reactions has provided ground for single-molecule recognition experiments. For elucidating the role of hydration in affecting the biomolecular structural organization, waters were dosed onto the Cu(111) surface to form

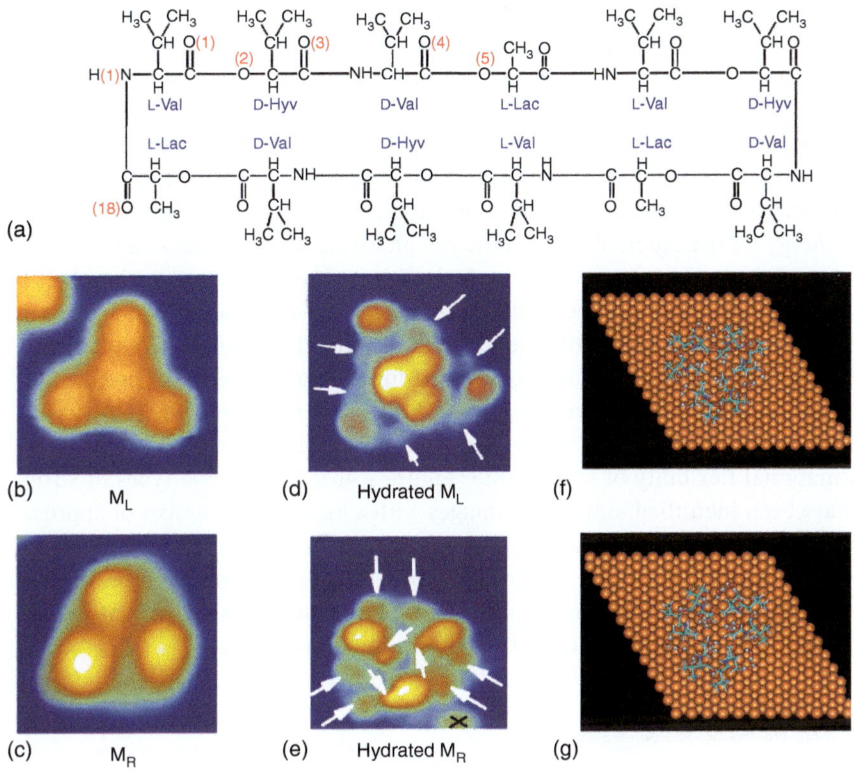

Figure 9.3 (a) Chemical structure of valinomycin. (b and c) STM images of valinomycin monomer in M_L configuration (b) and M_R configuration (c) on the Cu(111) surface at 78 K. (d and e) STM images of hydrated valinomycin in M_L configuration (d) and M_R configuration (e) on the Cu(111) surface at 78 K. The distinct differences between valinomycin and hydrated valinomycin are marked by the white arrows. Tunneling conditions: $V = -2$ V and $I = 50$ pA (b and c); $V = -2$ V and $I = 30$ pA (d and e). (f and g) Calculated structures of hydrated valinomycin in M_L configuration (f) and M_R configuration (g) on the Cu(111) surface in DFT simulation. Source: Reproduced with permission from Ref. [5]; © 2023 Springer Nature Limited.

hydrogen bonds with immobilized valinomycin. As shown in the STM images, hydration structurally converts the central protrusion of M_L to three smaller protrusions (Figure 9.3d). Differently, the hydrated M_R shows six additional protrusions at the edge of the monomer and an additional three protrusions at the center (Figure 9.3e). DFT calculations reveal that the distinct hydrated structures of valinomycin primarily originate from the change in the local electron density at the binding sites induced by the formation of hydrogen bonds between water and valinomycin (Figure 9.3f and g).

The UHV-STM-based single-molecule imaging with polypeptides and proteins is challenging due to the experimental difficulty in dosing macromolecules in a vacuum chamber without breaking macromolecular structures. Conventional deposition methods for the UHV-STM sample preparation, such as vaporization from a heated crucible, usually result in fragmentation due to the internal energy uptake occurring when macromolecules transfer from the gas phase to a substrate surface. To address the challenge, soft-landing electrospray ion beam deposition has been developed for the gentle deposition of macromolecules on a surface in a vacuum, used in conjunction with STM instruments [6]. Soft-landing refers to the deposition process in which the hyperthermal polyatomic species do not lose their structural integrity and produce intact deposition on surfaces. Generally, the landing species at low energy on the order of 10 eV in the gas phase can dissipate its kinetic energy and retain the intact structure on a surface without breaking covalent bonds [7]. Kern et al. designed soft-landing instruments equipped with a variable-temperature STM for recording the adsorption geometries of landed biomolecules. A peptide, bradykinin, composed of nine amino-acid residues (Figure 9.4a), was deposited on the Cu(110) surface for STM imaging at 300 K [6]. The single-molecule images of individual bradykinin on the Cu(110) surface show structural variability due to the conformational flexibility of the peptide (Figure 9.4b). Twenty-two types of structures have been identified in the STM images with a long molecular axis of approximately 3.0 nm and a short molecular axis of 1.5 nm. The adsorbate–substrate interactions play an important role in regulating the adsorption structures of peptides. Changing the substrate from Cu(110) to Cu(100) switches the assembled status of immobilized bradykinin on surfaces from monomer to dimer.

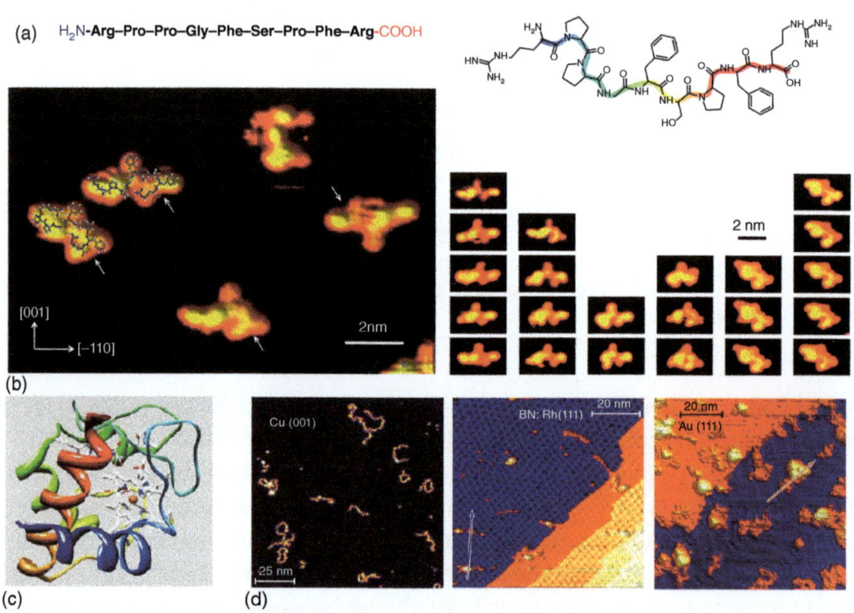

Figure 9.4 (a) Amino-acid sequence and structure of bradykinin. (b) STM image of bradykinin on the Cu(110) surface (left) and 22 types of structures identified in the STM images (right). Tunneling conditions: $V = -2$ V and $I = 50$ pA. (c) Structure of cytochrome c. (d) STM images of cytochrome c on the Cu(001), BN:Rh(111), and Au(111) surfaces from left to right. Source: Reproduced with permission from Ref. [6] (a and b) and Ref. [8] (c and d); © 2023 American Chemical Society.

The soft-landing electrospray deposition technology in conjunction with the UHV-STM has been leveraged to study the adsorption structures of proteins on surfaces [8]. The fine structures of cytochrome c (104 amino-acid residues, 12 kDa) in the surface-bounded state have been examined by STM on the Cu(001), Au(111), and monatomic insulating boron-nitride (BN) nanomesh layer on Rh(111) surfaces, respectively (Figure 9.4c and d). On the Cu(001) and BN:Rh(111) surfaces that are active to covalently interact with adsorbates, cytochrome c appears as low-height, string-like structures corresponding to the proteins in an unfolded state. On the Au(111) surface that is nearly inert to interact with adsorbates, cytochrome c appears as compact patches corresponding to the proteins in the folded state.

Taking advantage of soft-landing technology to shuttle macromolecules to an ultrahigh vacuum condition, the structural analysis of single glycans, including linear oligosaccharides and branched oligosaccharides, was achieved by applying the electrospray ion beam deposition in tandem arrangement with STM imaging [9–11]. The structural characterization of glycans is challenging for conventionally biophysical technologies due to the backbone flexibility and branched primary structures. This work shows the merits of the STM approach toward a profound understanding of glycan structures. It should be noted that, besides the fragmentation of macromolecules during sample deposition, the high electric field between the STM probe and sample also creates problems in the reproducibility of STM observations with macromolecules due to the bending and diffusion of macromolecules on the surface. To address this problem, Tanaka and Kawai reported the oblique pulse-injection method for the sample preparation, extending and fixing the adsorption structures of macromolecules, such as the long-chain single-stranded phage DNA molecules, on a Cu(111) surface and achieving the reproducible STM imaging and spectroscopy with macromolecules [12].

9.2.2 Single Peptides Immobilized by Molecular Matrix

For decreasing the diffusion mobility of molecules on surfaces to achieve the submolecular resolution with an STM, trapping the adsorbed molecules by assembling with adsorbate species is an alternative strategy instead of cooling-down temperature. An important example demonstrating the assembly-facilitated immobilization effects is the alkane-assisted adsorption and assembly of phthalocyanines and porphyrins [13]. The intermolecular interactions between adsorbed molecules remarkably stabilize the adsorption of molecules on surfaces, especially in the formation of multifarious frameworks [14–16]. The self-assembly of adsorbed molecules on a surface is a process in which adsorbed molecules adopt a defined arrangement with guidance from the interactions between adsorbed molecules and the interactions between adsorbed molecules and the surface. The self-assembled structures reflect the synergetic or competitive effects among different types of interactions, including covalent bonds, coordination bonds, and hydrogen bonds [17–19]. When the interactions between the surface and adsorbed molecules are dominant, the self-assembling structures are mainly determined by the surface properties [20]. When the interactions between the adsorbed molecules are stronger than the molecule–surface interactions, the self-assembly of molecules shows weak dependence on the topology and properties of a surface [21, 22]. Since the intermolecular interactions are sensitive to the separation between adsorbed molecules, the coverage of molecules on the surface also affects the assembling structures formed by adsorbates [23, 24].

The assembly-facilitated immobilization effects can be applied to analyze the single peptide structures within an assembly by STM. For example, the single-molecule observation of cyclic peptides is feasible by immobilizing cyclic peptides into the cavities of a molecular matrix. As illustrated in Figure 9.5a, a derivative of trimesic acid (StOF-$COOH_3$) assembles into a hexagonal, porous network with nanometer-sized cavities on the highly oriented pyrolytic graphite (HOPG) surface at room temperature [25]. The parameters of the unit cell for the porous network were measured to be $a = 4.7 \pm 0.2$ nm, $b = 4.7 \pm 0.2$ nm, and $\alpha = 60 \pm 1.5°$, suggesting the size of the individual inner cavity to be less than 4.7 nm. By adding the cyclic peptide, cyclosporine A (CsA), onto the StOF-$COOH_3$ self-assembled surface, an irregular polygonal structure was identified in the cavity, corresponding to the adsorption of cyclic peptide entrapped by the cavity of the hexagonal network (Figure 9.5b). It should be noted that single-molecule imaging with CsA is not feasible for the ambient-STM in the absence of a molecular template. The driving forces underlying the immobilization of CsA by molecular matrix are associated with the hydrophobic interactions between the hydrophobic side chains of CsA and StOF-$COOH_3$. The STM image contrast of individual CsA might reflect the topographic variation in peptide side chains. CsA can bind with Mg^{2+} in solution and form a 1:1 complex. The molecular recognition process of Mg^{2+} by CsA can be monitored by STM at a single molecular level (Figure 9.5c and d). Upon the introduction of Mg^{2+} into the CsA-/StOF-$COOH_3$ supramolecular network, the apparent diameter of the irregular polygon increases from 1.7 ± 0.3 nm (empty CsA) to 2.0 ± 0.4 nm (CsA–Mg^{2+} complex). Interestingly, two configurations of the CsA–Mg^{2+} complex were observed in the STM images, with the top-up configuration with a center-filled feature and the bottom-up configuration with a center-unfilled feature (Figure 9.5e and f). The relative energy difference between the two configurations is related to the ratio of their probabilities by the probability theory. Using a molecular template, the single-molecule imaging with other cyclic peptide structures, such as valinomycin and its structural conversion induced by the coordination with K^+, was also achieved by an STM working at a liquid/solid interface at room temperature [26].

9.3 Structural Characterization of Peptide Assemblies

The structural information of peptide assemblies provides insights into the folding and association mechanisms at a molecular level. STM has made great progress in the structural characterization of peptide assemblies. As demonstrated by the UHV-STM imaging at 40 K, ordered assembly structures are formed by angiotensin I (At-I, 10 amino-acid residues in sequence) on the Au(111) surface (Figure 9.6) [27]. Specifically, two At-I molecules are aligned antiparallelly and assemble into a dimeric building block with a length of 2.7 nm and a width of 2.3 nm by the van der Waals

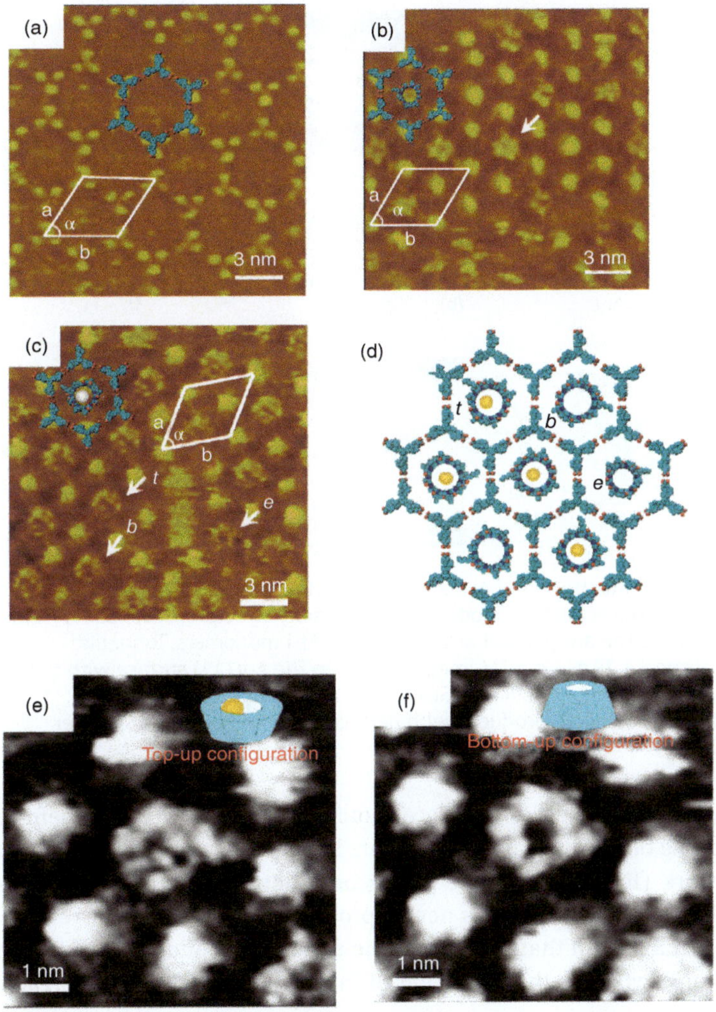

Figure 9.5 (a) STM image of StOF-$COOH_3$. Tunneling conditions: $V = 912.8$ mV and $I = 48.8$ pA. (b) STM image of CsA entrapped by the StOF-$COOH_3$ network. The white arrow points to the CsA. Tunneling conditions: $V = 949.7$ mV and $I = 48.8$ pA. (c) STM image of Mg^{2+}–CsA complex in the network of StOF-$COOH_3$. The white arrows *e*, *b*, and *t* point to the CsA molecules without Mg^{2+} and CsA–Mg^{2+} complexes with unfilled and filled centers, respectively. (d) Structural model of Mg^{2+}–CsA complex in the StOF-$COOH_3$ network, where *e*, *b*, and *t* stand for the CsA molecules without Mg^{2+} and CsA–Mg^{2+} complexes with unfilled and filled centers, respectively. (e and f) STM images of Mg^{2+}–CsA complex in the top-up configuration (e) and bottom-up configuration (f). Schematic model (upper right) illustrating the corresponding top-up and bottom-up configurations. Tunneling conditions: $V = 912.8$ mV and $I = 60.0$ pA. Source: Reproduced with permission from Ref. [25]; © 2023 American Chemical Society.

and Coulomb interactions between side chains. Different At-I dimers further associate linearly in a head-to-tail manner to form higher-ordered, assembled structures (Figure 9.6b). The peptide angiotensin II (At-II, 8 amino-acid residues) was designed to differ from the sequence of At-I with a deletion of two residues at the C-terminus

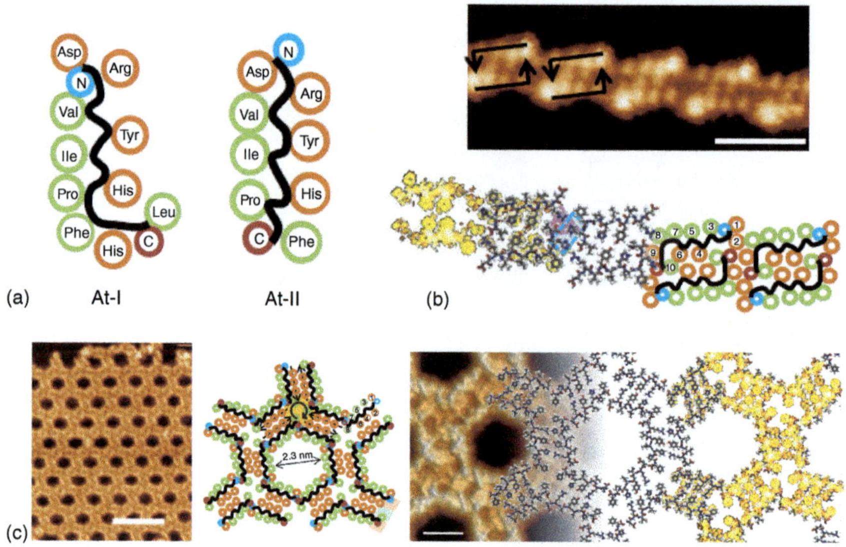

Figure 9.6 (a) Schematic illustration of At-I and At-II structures. (b) STM image of At-I on the Au(111) surface with the schematic model of At-I assembly structure. Scale bar is 4 nm. The black arrows indicate the antiparallel arrangement of At-I monomers. Tunneling conditions: V = 1.5 V and I = 29 pA. (c) STM image of At-II on the Au(111) surface with the schematic model of At-II assembly structure. Scale bars, 15 nm. Tunneling conditions: V = 1.3 V and I = 35.5 pA. Source: Reproduced with permission from Ref. [27] / Springer Nature / CC BY 4.0.

of At-I (Figure 9.6a). As a result of sequence truncation, At-II molecules assemble into dimeric building blocks but further aggregate into a long-range ordered, chiral honeycomb network (Figure 9.6c). The honeycomb networks represent cavities with an inner diameter of 2.3 nm. This new approach to 2D self-assembly on surfaces grants structural insight that will benefit the use of oligo- and polypeptides as promising building blocks for the formation of large, multifunctional bioorganic surfaces.

The single-molecule insights provided by STM greatly deepen our understanding of the evolutionary mechanisms underlying the biology-relevant structures in the complex hierarchical process. As revealed by the UHV-STM imaging on the surface of Cu(111) at 78 K, different landing surfaces of valinomycin deposition result in two binding configurations, right-handed chiral M_R and left-handed chiral M_L [28]. Strikingly, the chirality of valinomycin monomers can be amplified via a racemic tetramerization of monomers. By organizing one left-handed chiral M_L in the middle and three right-handed chiral M_R as three arms, the tetramer, i.e. enantiomeric mixture, is right-handed chiral in geometric topology (Figure 9.7a). The chirality of tetramers is further translated to a highly ordered assembly structure, where multiple right-handed chiral tetramers associate into a hex network with a right-handed chirality (Figure 9.7b). The parameters of the unit cell for the porous network were measured to be $a = 6.0 \pm 0.1$ nm, $b = 6.0 \pm 0.1$ nm, and $\alpha = 60°$. The single-molecule imaging with the self-assembly of valinomycin provides a novel evolutionary

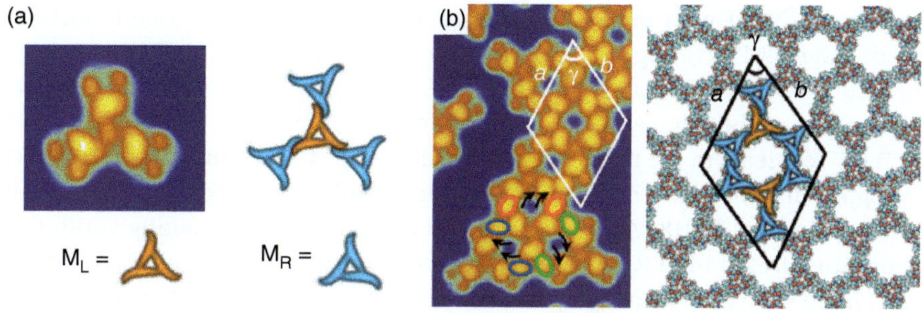

Figure 9.7 (a) STM image and schematic model of a tetramer formed by racemic valinomycin on the Cu(111) surface at 78 K. (b) STM image of the hex network formed by valinomycin on the Cu(111) surface at 78 K (left) and the schematic model of valinomycin assembly (right). The directional binding of valinomycin tetramer building blocks is marked by three groups of oval circles with red, blue, and green colors and black arrows. Tunneling conditions: $V = -2$ V and $I = 0.03$ nA. Source: Reproduced with permission from Ref. [28] / Springer Nature / CC BY 4.0.

mechanism that the homochirality can arise from the intermolecular recognition between distinct chiral monomers. The initial imbalance in tetramerization of heterochiral monomers, i.e. 3 M_R versus 1 M_L, results in a symmetry break in the homochiral structures of supramolecular networks.

The ultrahigh resolution of STM imaging with peptide assemblies exhibits a valuable application in the structural analysis with the disease-associated peptide assemblies. For instance, human islet amyloid polypeptide (hIAPP), a major component derived from pancreatic amyloid deposits, is associated with the occurrence of type 2 diabetes (T2D). The understanding of the hIAPP misfolding and aggregation structures is crucial for revealing the pathological mechanism of disease and inspiring the target drug discovery. Common to the aggregation structures of amyloidal proteins, the cross-β structure is the major supramolecular organization form identified for amyloid fibrils. In particular, successive protein molecules with a β-sheet conformation are stacked in parallel, forming a protofilament with a diameter of 25–60 Å [29]. In the protofilament, as evidenced by X-ray diffraction, each protein molecule is perpendicular to the direction of protofilament extension and ordered with a period of 4.7–4.8 Å [29, 30]. The protofilament is the assembly unit of a fibril where 2–6 protofilaments are intertwined with each other along the long-axis direction, forming a mature fibril with a diameter of 60–120 Å [29, 31]. High-resolution spatial analysis with amyloidal aggregation structure is still challenging due to the difficult crystallinity and polymorphism of amyloid aggregates. The advances achieved in the experimental and theoretical tools for structural study provide an opportunity for understanding the supramolecular organization of amyloid fibrils. For hIAPP, cryo-EM and scanning transmission electron microscopy (STEM) studies with the fibrils structures formed by hIAPP assembly found that mature hIAPP fibrils contain three protofilaments with a cross-sectional diameter of 4.5–5.5 nm [32]. The mass-per-length (MPL) of the fibril is 8.1 kDa/nm, and thus, the distance between each two polypeptide molecules arranged along the fibril axis is 0.47 nm, which is consistent with the β-sheet structural feature. A β-serpentine model for depicting the hIAPP folding structure was proposed by the Cryo-EM/STEM study in conjunction with theoretical simulation. This model suggests that hIAPP formed supramolecular fibrils in parallel β-sheet conformation in which the 12–17, 22–27, and 31–37 segments in the polypeptide formed three β-strand domains, making the polypeptide display a serpentine "S" shape. The polypeptide backbones form turns at the positions of His18, Asn21, Ser28, and Thr30. To experimentally verify the folding structures of hIAPP, an STM investigation was conducted at ambient conditions (Figure 9.8a) [33]. By imaging the assembly structures of hIAPP on the surface of HOPG under ambient conditions, it was found that hIAPP formed a bright, linear stripe-like structure, and the distance between the adjacent polypeptide backbones was measured to be 0.45 nm, indicating the formation of the peptide into β-sheet conformation on surfaces. In the high-resolution STM topography of hIAPP, the folding mode of the polypeptide main chain was resolved to be a parallel β-folded structure. The length distribution of β-strands was determined to be 3.4 ± 0.5 nm, equivalent to the length of 11 amino-acid residues in a β-strand. Not only limited to hIAPP, but the single-molecule imaging based on the ambient-STM can also be applied for the structural analysis with other amyloidal proteins, such as Aβ42 [34]. Aβ42 is a proteolytic product derived from the amyloid precursor protein (APP). As revealed by the ambient-STM images, Aβ42 molecules appear as bright lamellar structures and are arranged in parallel with each other (Figure 9.8b). The peptide lamellae are stacked perpendicular to the long-axis direction of the stripe. The separation between two neighboring peptide chains was measured to be 0.45 ± 0.05 nm, indicating the hydrogen bonding of β-strands into a β-sheet.

STM-based single-molecule imaging not only complements the structural findings obtained from conventional characterization techniques but also provides direct imaging evidence to facilitate the understanding of the molecular mechanism of peptide assemblies. Based on these advantages, STM can be further exploited to unveil the modulation effects on peptide assemblies (primary sequence, post-translational modification, chirality, etc.) and the mechanism of peptide–small molecule co-assemblies.

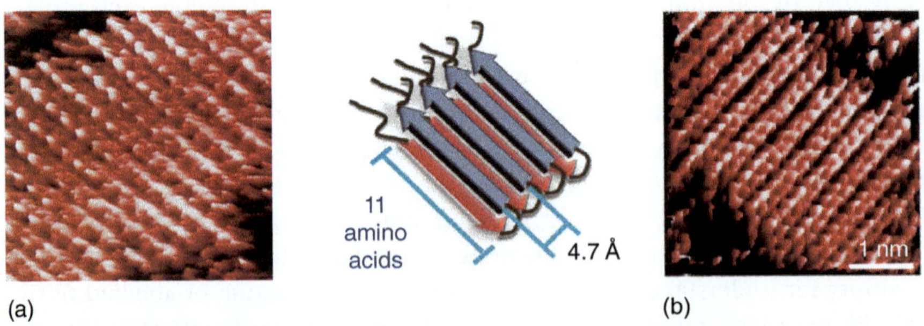

Figure 9.8 (a) STM image of hIAPP on the HOPG surface (left) and the schematic model of the β-sheet assembly formed by hIAPP (right). Tunneling conditions: $V = -675.4$ mV and $I = 537.1$ pA. (b) STM image of Aβ42 on the HOPG surface. Tunneling conditions: $V = -418.5$ mV and $I = 919.2$ pA. Source: Reproduced with permission from Ref. [33] (a) and Ref. [34] (b); © 2023 Elsevier Inc.

9.4 Modulation Effects on Peptide Assemblies

For peptide assemblies, the chemical nature of components and their combination is key to the peptide–peptide and peptide–substrate interactions. Mutations in peptide sequence (including single-site mutations, multiple-site mutations, truncation, and scramble), post-translational modification (PTM, such as phosphorylation and glycosylation), and chirality alteration significantly affect the peptide–peptide binding affinities and peptide assembly characteristics. Herein, the structural insights into peptide assemblies modulated by mutations, PTMs, and chirality are reviewed.

9.4.1 Peptide Assemblies Modulated by Mutations

Mutations in peptide sequences can influence peptide biological functions, especially for disease-related peptides. Mutation effects on these peptides have drawn considerable interest due to their relevance to currently intractable diseases, such as degenerative diseases and some types of cancers. For example, hIAPP is a major component in pancreatic amyloid deposits, found in patients with T2D [35]. The substitution of S20G, associated with a familial early-onset form of T2D, reinforces more aggregation propensity and increases the risk of disease [36, 37]. TAR DNA-binding protein 43 (TDP-43) is associated with amyotrophic lateral sclerosis (ALS) and frontotemporal lobar degeneration (FTLD) [38]. Over 30 mutations in the TDP-43 protein have been found among ALS patients [39, 40]. Detailed structural analysis is needed to unveil the mechanism of mutation regulating peptide aggregation and pathogenicity.

Herein, IAPP is selected to introduce the STM progress made toward understanding mutation-induced structural regulation. The assembly structures of hIAPP mutants probed by the ambient-STM are presented in Figure 9.9, including the assemblies formed by the 8–37 segment of hIAPP (hIAPP$_{8-37}$), rat IAPP$_{8-37}$ (rIAPP$_{8-37}$), and the single-point mutant rIAPP$_{8-37}$ R18H, respectively, on the HOPG surfaces (Figure 9.9a) [41]. To uncover the key sites implicated in the formation of β-hairpin folding structures of IAPP analogs, a small molecule 4,4′-bipyridine (4Bpy) was introduced to co-assemble with these peptides (Figure 9.9a). In the STM images, the linear arrays with bright protrusions are attributed to the 4Bpy molecules for the high local electron density of the conjugated π-electrons (Figure 9.9b). The lamellar structures with slightly low contrast are ascribed to the peptide β-strands involved in the assembly (Figure 9.9b). 4Bpy forms an O—H···N hydrogen bond with peptide between the carboxylate group at the C-terminus of the peptide and the nitrogen atom of 4Bpy, labeling the spatial position of the peptide C-terminus. The C-terminal β-motif sequence involved in the surface-bound peptide assembly can be deduced from the β-strand length measurement with the STM images (Figure 9.9c). The 4Bpy-labeling strategy can be applied to determine the key folding sites for different β-sheet assemblies consisting of multiple β-motifs. In particular, the folding sites for the hIAPP$_{8-37}$ β-hairpin C-terminal motifs were determined to be Phe23 and Gly24 (Figure 9.9d). The folding site of rIAPP$_{8-37}$ C-terminal motifs was estimated to be Leu23 (Figure 9.9e–g). By contrast, a single-site mutation

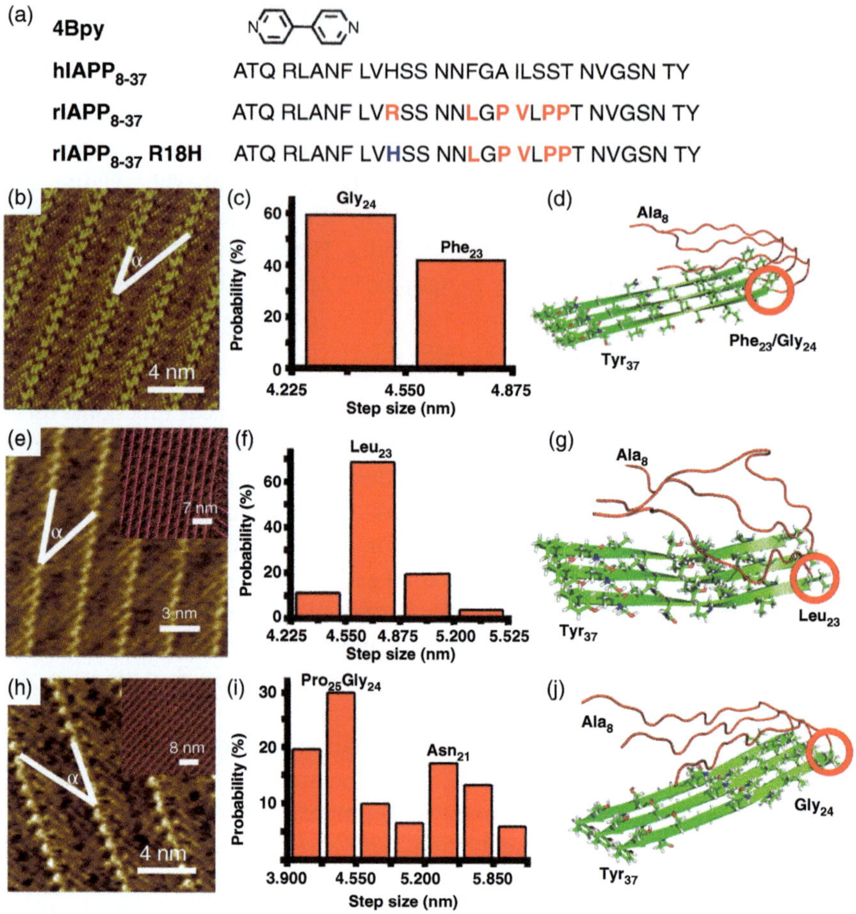

Figure 9.9 (a) Chemical structure of 4Bpy and the sequences of IAPP peptides. (b, e, h) High-resolution STM images of $hIAPP_{8-37}$ (b, α = 34 ± 2°, I = 314.3 pA, and V = 605.2 mV), $rIAPP_{8-37}$ (e, α = 36 ± 2°, I = 550.1 pA, and V = 322.8 mV), and $rIAPP_{8-37}$ R18H (h, α = 34 ± 2°, I = 612.9 pA, and V = 300.4 mV) co-assembled with 4Bpy molecules on the HOPG surfaces at ambient conditions. Insets in (e and h) show the corresponding large-scale STM images. (c, f, i) Length distribution histograms of $hIAPP_{8-37}$ (c), $rIAPP_{8-37}$ (f), and $rIAPP_{8-37}$ R18H (i) β-motifs measured from STM images. (d, g, j) Proposed structural models for the folding sites in $hIAPP_{8-37}$ (d), $rIAPP_{8-37}$ (g), and $rIAPP_{8-37}$ R18H (j) β-structure motifs, respectively. Source: Reproduced with permission from Ref. [41]; © 2023 Proceedings of the National Academy of Sciences of the United States of America.

R18H converts the folding sites of $rIAPP_{8-37}$ R18H to Gly24 (Figure 9.9h–j). Not only limited to the key folding sites but also the dispersity of C-terminal β-motif length shows a correlation with the amino-acid variations in peptide sequences. Relative to the length distribution of $hIAPP_{8-37}$ C-terminal β-motifs, i.e. 4.22–4.88 nm (Figure 9.9c), the length distribution of $rIAPP_{8-37}$ C-terminal β-motifs is broader, ranging from 4.22 to 5.52 nm (Figure 9.9f). A substitution of Arg18 by His with $rIAPP_{8-37}$ further increases the dispersity of β-motif length, where the length of $rIAPP_{8-37}$ R18H C-terminal β-motif ranges from 3.90 to 6.20 nm (Figure 9.9i).

It is necessary to validate the structural information obtained from the STM-based single-molecule experiments by other spectroscopic and microscopic technologies to confirm the structural details provided by STM. The suggested hIAPP β-structures based on the STM results are in qualitative agreement with the models proposed by solid-state NMR and the X-ray crystallography results for hIAPP segments ($hIAPP_{21-27}$ and $hIAPP_{28-33}$), where the core structures are formed by two β-strands (residues 8–17 and residues 28–37) joined by a loop (residues 18–27) with a pronounced bend in the peptide strand facilitated by Gly24. The results in the STM study suggesting the key site at Phe23 or Gly24 for hIAPP are in close agreement with the previous structural models.

To decipher mutation-induced regulation, homo-oligo(amino acid)s with length mutations were designed to decode the modulation effects of amino-acid sequences on homo-oligo(amino acid) assemblies. For example, the fine structures of the PolyQ7/4Bpy and PolyQ8/4Bpy co-assemblies on the HOPG surfaces were investigated by the ambient-STM to reveal the odd–even sequence effect on surface-mediated homo-oligo(amino acid) assemblies (Figure 9.10a and b) [42]. PolyQ7/4Bpy co-assembly displays a single predominant apparent β-strand length (4.55 nm) ascribed to 14 residues involved with two PolyQ7 β-strands, reflecting one predominant homo-oligo(amino acid) conformation in assembly structures. However, three major apparent β-strand lengths, corresponding to 14, 15, and 16 residues, can be identified in the PolyQ8/4Bpy co-assembly, reflecting the coexistence of three conformations, including 7+7, 7+8, and 8+8, in their assembly structures (Figure 9.10c). The difference in assembling characteristics suggests the asymmetric

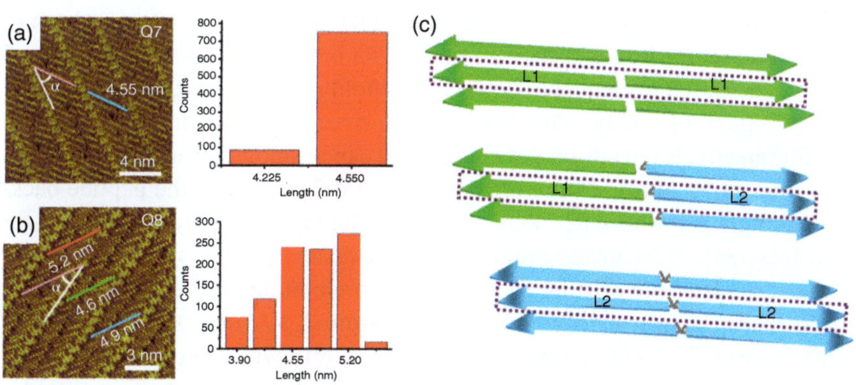

Figure 9.10 (a) STM images of PolyQ7/4Bpy co-assembly and the length distribution histograms of β-strands involved in the assembly. The angle α between peptide molecular axes (pink line) and the 4Bpy string directions (white line) was measured to be $35 \pm 3°$. Tunneling conditions: I = 399.3 pA and V = 600.2 mV. (b) STM images of PolyQ8/4Bpy co-assembly and the length distribution histograms of β-strands involved in the assembly. The angle α was measured to be $31 \pm 2°$. Tunneling conditions: I = 299.3 pA and V = 600.2 mV. (c) Three types of β-strand arrangements for PolyQ7/Q8 molecules. The lengths of peptide dimers of L1 + L1, L1 + L2, and L2 + L2 values correspond to the measured peptide lengths of 5.2, 4.9, and 4.6 nm in the STM images, respectively. Source: Reproduced with permission from Ref. [42]; © 2023 SIOC, CAS, Shanghai, & WILEY-VCH Verlag GmbH & Co. KGaA, Weinheim.

adsorption effect of homo-oligo(amino acid)s relating to the odd–even number of homo-oligo(amino acid) residues.

To further unveil sequence effects on peptide assemblies, block peptides with two homogeneous residue motifs were designed and co-assembled with 4Bpy to determine the stabilization effects of amino-acid side chains on surface-mediated peptide assemblies. The synthetic binary peptides, including F5H5, H5F5, F5A5, A5F5, F5N5, N5F5, H5D5, D5H5, H5A5, A5H5, N5D5, D5N5, N5A5, and A5N5, and their assembly structures on the HOPG surfaces were determined by STM at ambient conditions [43]. For instance, the histogram of the β-strand length distribution of H5D5 presents that the most probable length is between 4.22 and 4.55 nm, suggesting two peptides with one or two His residues and five Asp residues per peptide to form β-strands. For the reverse isomer of H5D5, i.e. D5H5, the β-strand length distribution ranges from 3.57 to 5.53 nm, suggesting two peptides with one Asp residue and five His residues per peptide to form β-strands. This result hints that His shows a higher β-sheet assembly propensity relative to Asp on the HOPG surfaces. Using the same method, the assembly propensity of the amino-acid residues in surface-mediated peptide assemblies can be ranked: Phe > Asn > Ala > His > Asp. The insights into the assembly propensity of residue side chains are informative for understanding the sequence dependence of aggregation propensity of surface-bound peptides.

Beyond the homogeneous residue motifs, more complex mutations with heterogeneous residue building blocks were investigated by STM imaging under ambient conditions. These building blocks endow peptide assemblies with more functionality. STM-based single-molecule imaging helps to reveal the assembly structures of functional peptides with different building blocks to facilitate the design and optimization of function [44]. For instance, the TTR1 peptide, derived from human transthyretin, is fibril-forming. Based on the TTR1 scaffold, TTR1–RGD functional peptide was designed to conjugate the TTR1 scaffold motif with an RGD motif by a GG linker (Figure 9.11a) [45]. The RGD motif, derived from the extracellular matrix protein fibronectin, is the minimal sequence that can promote cell adhesion [46, 47]. The GG linker was added to promote the structural flexibility of the peptide backbone, flanking a Ser-terminated RGD motif to enhance bioactivity and affinity for integrin receptors. STM investigations on TTR1-related peptide assemblies show that the β-strands of the pristine TTR1 peptides preferentially extend from Tyr1 to Tyr10 with the length distribution centered around 3.25 nm (Figure 9.11b–d), whereas those of TTR1–RGD extend from Tyr1 to Gly13 with the prolonged length distribution centered around 4.25 nm (Figure 9.11e–g). This observation provides site-specific evidence that the TTR1 scaffold with GG linker is fully included in the TTR1–RGD peptide assembly. The cell-adhesive RGD motif adopts a flexible conformation, facilitating the achievement of its function for promoting cell adhesion. The observation of the GG linker fully incorporated in the assembly β-strand inspires that the sequence can be further optimized by enhancing the flexibility of linker motifs, such as elongating the linker motif. The number of bright protrusions along one peptide strand in the STM image is consistent with the calculated number of residues obtained by quantitative length measurements (Figure 9.11e), which verifies the reliability of this statistical method and the capability of STM to determine

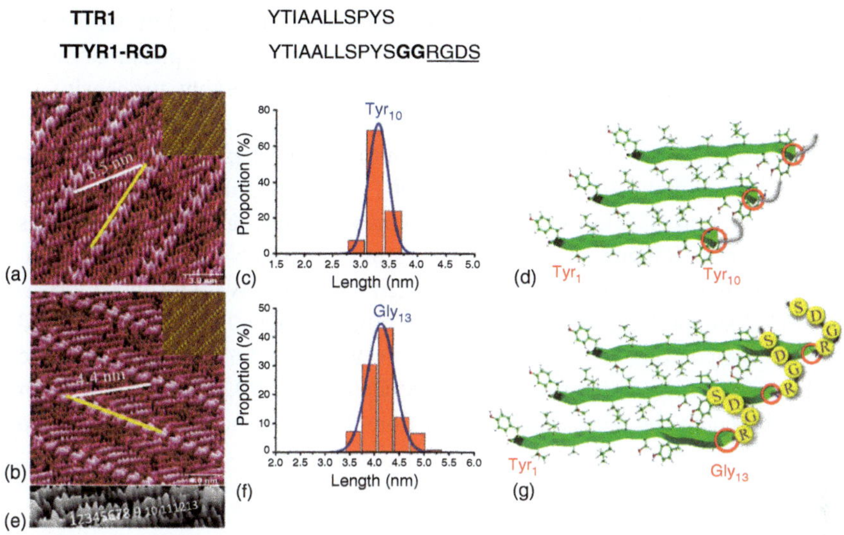

Figure 9.11 (a) TTR1-related peptide sequences. (b and e) High-resolution STM images of TTR1/4Bpy and TTR1–RGD/4Bpy co-assemblies, respectively. Tunneling conditions: I = 350.0 pA and V = 600.0 mV (b); I = 347.9 pA and V = 600.0 mV (e). Inset images in (b and e) are large-scale STM images with the size of 15 nm × 15 nm. White lines denote the peptide molecular axes, and yellow lines show the 4Bpy strings. The number of protrusions counted from the image (bottom) is 13. (c and f) Statistical histograms of the length distribution of the β-strands corresponding to (b) and (e). Both statistical histograms were fitted by Gaussian distribution (blue lines), and the most probable residues are also indicated in blue. (d and g) Proposed structural models of TTR1 (d) and TTR1–RGD (g), respectively. The thick green ribbons show the ordered β-strand motifs, and the flexible gray strings depict the flexible segments indiscernible by STM. Source: Reproduced with permission from Ref. [44]; © 2023 Springer Nature Switzerland AG.

the number of assembly-involved residues. In addition, the forward and reverse peptide sequences, despite the difference in chemistry, present similar assembling characteristics, illustrating the robust nature of peptide assemblies and providing more opportunities for diversity in the design of the peptide backbones. This site-specific observation deepens our understanding of peptide assemblies and would contribute to the rational design and optimization of sequence-programmed peptide architectures.

9.4.2 Peptide Assemblies Modulated by Post-Translational Modifications

PTM refers to the covalent modification of proteins after protein biosynthesis, generally taking place in the endoplasmic reticulum and the Golgi apparatus [48, 49]. PTMs include but are not limited to phosphorylation, glycosylation, acetylation, and lipidation and play an essential role in protein folding, cell signaling, cell adhesion, immunity response, etc. [48, 49]. PTMs introducing moieties (such as phosphoryl or saccharide groups) to peptide chains have a profound effect on peptide–peptide interactions, resulting in varied assembly characteristics.

The phosphorylation effects on the β-motif of amyloidogenic TDP-43 mutants have been explored by STM under ambient conditions [50]. An ALS-phosphorylated mutant, TDP-43 A315T(p), is reported to increase the assembly propensity of TDP-43 in solution and promote the formation of protein aggregates in cultured neurons relative to the wild type (WT) of TDP-43, resulting in accelerated neurodegeneration in transgenic animals [51]. The β-domains of TDP-43 mutant peptides, including A315T(p), A315T, and A315E, were observed to present the typical lamellar features on the HOPG surfaces (Figure 9.12a–e). The lengths of β-strands involved in the peptide assemblies can be ranked as A315E > A315T(p) > A315T = WT. The phosphorylation at the A315 site prolongs the β-motif length range from 2.60 to 4.55 nm centered at 3.57 nm (corresponding to 11 residues) (Figure 9.12c) to 3.57–5.52 nm, with the most probable β-domain consisting of 16 residues (Figure 9.12e). This elongation effect is attributed to the electrostatic interactions between the phosphoryl group

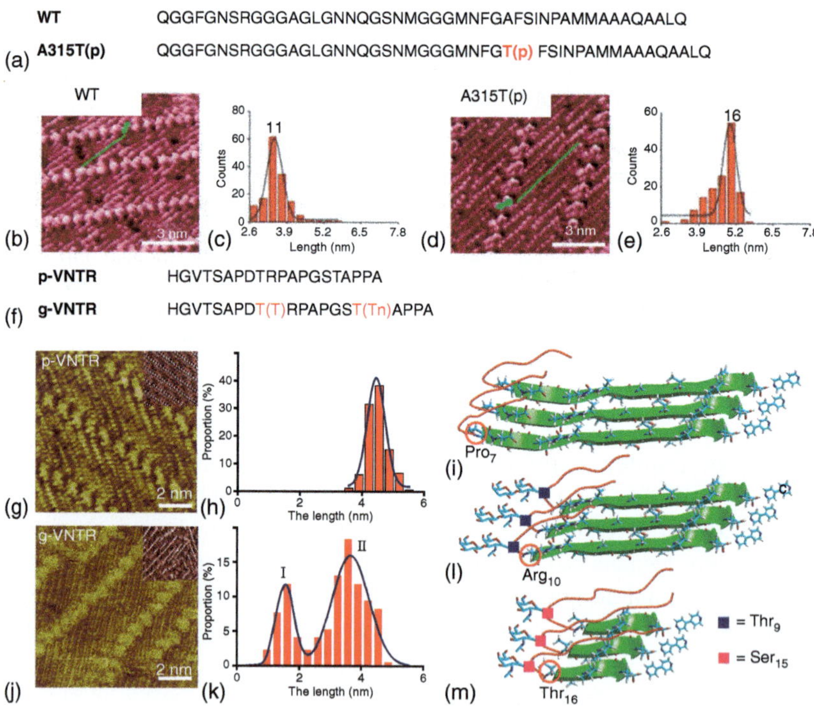

Figure 9.12 (a) WT and A315T(p) TDP-43 peptide sequences. (b and d) STM images of WT (b) and A315T(p) (d) TDP-43 peptides co-assembled with 4Bpy. (c and e) Length distributions of the β-strands involved in the assembly of the WT (c) and A315T(p) (e) TDP-43 peptides. (f) p-VNTR and g-VNTR peptide sequences. (g and j) STM images of p-VNTR/4Bpy and g-VNTR/4Bpy co-assemblies, respectively. The inset images are the corresponding large-scale STM images. (h and k) Length distributions of the β-strands involved in the assembly of p-VNTR and g-VNTR, respectively. (i) Structural model of the p-VNTR/4Bpy co-assembly. (l and m) Structural models of the g-VNTR/4Bpy co-assembly. Source: Reproduced with permission from Ref. [50] (a–e); © 2023 Elsevier B.V, and Ref. [52] (f–m); © 2023 American Chemical Society.

(negatively charged) and the R293 residue side chain (positively charged). The key folding site in the β-motif of A315T(p) peptide assembly is Phe316, proximal to the phosphorylated site Thr315. This work reveals a positive correlation between β-domain length at the solid–liquid interfaces and the aggregation propensity in solution and demonstrates the feasibility of understanding the aggregation behavior and its cytotoxicity by β-motif structural analysis at the single-molecule level.

Glycosylation also plays a significant role in biological events and influences the physicochemical properties of proteins. The glycosylation effect on the peptide assemblies on the HOPG surfaces was revealed with a glycosylated peptide derived from mucin 1 glycoprotein by the ambient-STM imaging. Glycoprotein mucin 1, containing a variable number of tandem repeats (VNTRs), is overexpressed in more than 90% of breast cancers, serving as a promising target for cancer immunotherapy [53]. The VNTR peptide, composed of 20 amino-acid residues, contains five potential *O*-glycosylation sites. The assembling characteristics of two peptides, pristine VNTR (p-VNTR) and the glycosylated VNTR (g-VNTR) with a Tn antigen (monosaccharide)-functionalized Ser15 and a T antigen (disaccharide)-modified Thr9, were determined to unravel the glycosylation effect on peptide assemblies (Figure 9.12f) [52]. In the STM images, the p-VNTR assembly presents highly homogeneous lamellar features separated by 4Bpy strings, and the histogram of the p-VNTR β-strand length distribution presents a sharp single peak with the most probable length around 4.55 nm assigned to 14 residues (Figure 9.12g–i). In contrast, the glycosylated peptide g-VNTR assembly shows apparent heterogeneity in lamellae width. The length of the β-motif participating in the assembly shows bimodal distribution centered at 1.62 and 3.57 nm, respectively (Figure 9.12j and k). This observation of shortened β-motif length verifies a glycosylation-induced destabilization effect on the glycopeptide assemblies. The length distribution of the assembly-involved β-motif of g-VNTR shows bimodal peaks, assigned to the key folding sites Thr16 and Arg10 (Figure 9.12l and m). Specifically, Thr16 and Arg10 are close to the glycosylated residues Ser15 and Thr9, resulting in the exposure of the saccharide moieties on the outside of the peptide β-sheet assembly. Based on the population distribution, disaccharides (T antigen) have a greater influence on the assembly stability than monosaccharides (Tn antigen), owing to the larger size of disaccharide moieties. The STM imaging approach demonstrates that glycosylation exerts a pronounced site-specific destabilization effect on peptide assemblies. These studies provide insights into the PTM effect on the local structural stability of peptide assemblies and benefit the rational interface design of biodevices and biomaterials associated with protein–surface interactions.

9.4.3 Peptide Assemblies Modulated by Chirality

Amino-acid chirality in peptide sequence has been recognized as an important factor in constructing peptide architectures via chirality-induced stereochemical effects. The titration of D-amino-acid residues into L-peptides can change their original folding structures [54–57]. Equimolar mixtures of L- and D-peptides would co-assemble into rippled β-sheet structures featuring alternating L- and D-sequences, rather than segregate into enantiomerically pure structures [58–60]. Heterochirality can even endow nonassembling peptides with self-organization capability [61]. Generally, chirality modulates the peptide folding structures, assembly structures, and the relevant bulk properties, such as mechanical rigidity, and proteolytic stability.

The chirality effect on the surface-mediated peptide assemblies on the HOPG surfaces has been studied using STM under ambient conditions [62]. The effects of chiral mutations at different positions and with various side-chain moieties on peptide assemblies were investigated with homo-oligo(amino acid)s, including Poly7Q, Poly7N, Poly7E, and Poly7D (Figure 9.13a). The assembly structures formed by

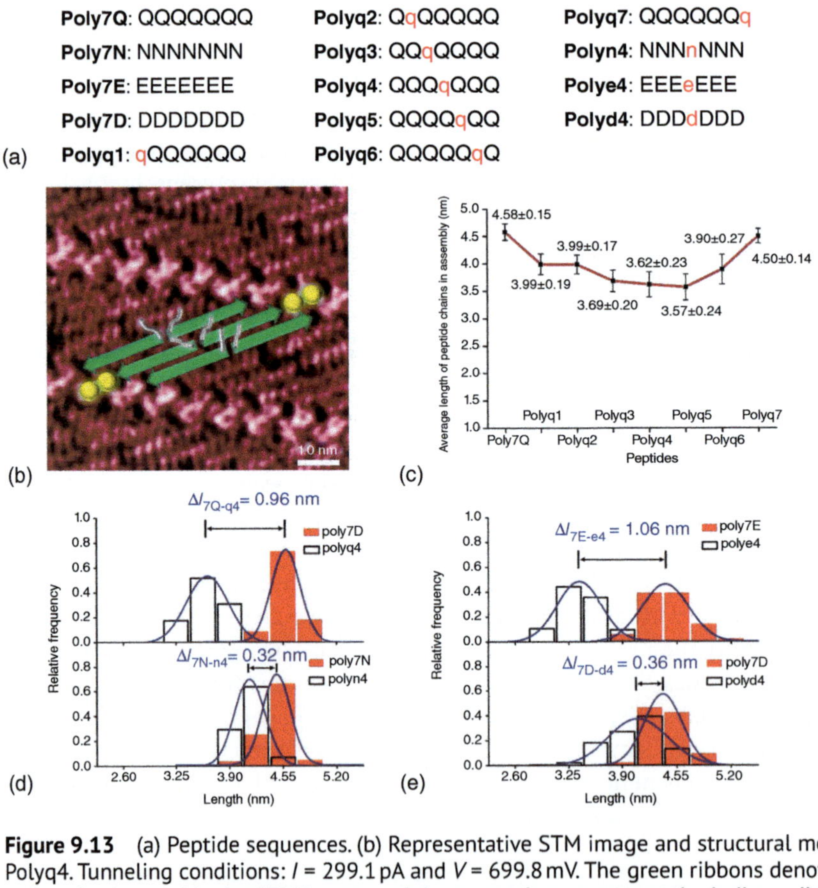

Figure 9.13 (a) Peptide sequences. (b) Representative STM image and structural model of Polyq4. Tunneling conditions: I = 299.1 pA and V = 699.8 mV. The green ribbons denote the β-strands observed in the STM image, and the gray strings represent the indiscernible residues in STM observations. The two yellow dots stand for the 4Bpy molecule. (c) Average lengths of peptide β-strands in the assemblies of Poly7Q and its seven variants. (d, e) Statistical histograms with Gaussian fitting of the length distributions of peptide β-strands in the assemblies formed by Poly7Q, Poly7N, Poly7E, Poly7D, and their mutants, respectively. $\Delta l_{7X\text{-}x4}$ (X = Q, N, E, D; x = q, n, e, d) is the difference value in the average length of β-strands involved in assembly before and after chiral mutation. Source: Reproduced with permission from Ref. [62]; © 2023 American Chemical Society.

PolyQ7 and its seven mutants ranging from Polyq1 to Polyq7 with a D-Gln substitution at different positions show that the chirality switching of a single residue modulates the length of β-strands formed by homo-oligo(amino acid)s (Figures 9.13a–c). The dependence of β-strand average lengths on the position of chiral substitution displays a groove-like feature, where Polyq3, Polyq4, and Polyq5 are located at the bottom with a decreased β-strand length of around 3.62 nm compared to Poly7Q (4.58 nm) (Figure 9.13c). The difference in the average length of β-strands formed by Poly7Q assembly versus Polyq4 assembly ($\Delta l_{7Q\text{-}q4} = 0.96$ nm) is larger than that of β-strands formed by Poly7N assembly versus Polyn4 assembly ($\Delta l_{7N\text{-}n4} = 0.32$ nm) (Figure 9.13d). The stereochemical influence becomes more pronounced in the Poly7E assembly ($\Delta l_{7E\text{-}e4} = 1.06$ nm) with bulky side chains than in the Poly7D assembly ($\Delta l_{7D\text{-}d4} = 0.36$ nm) (Figure 9.13e). Considering the one methylene difference in the pairs of Gln versus Asn and Glu versus Asp, it can be reasonably speculated that the disturbance effect of chirality is positively correlated with the steric hindrance of residue side chains. This work sheds light on the molecular insights of chirality-induced stereochemical interactions in peptide assemblies.

The effects of block chirality alterations on peptide assemblies were unveiled via the comparison between the block heterochiral peptide F5f5 (FFFFFfffff) assembly and the homochiral L-peptide F10 (FFFFFFFFFF) assembly (Figure 9.14a and b) [63]. Both peptides assemble into lamella-like structures with different widths, indicating the coexisting antiparallel β-sheet and parallel β-sheet conformations in their

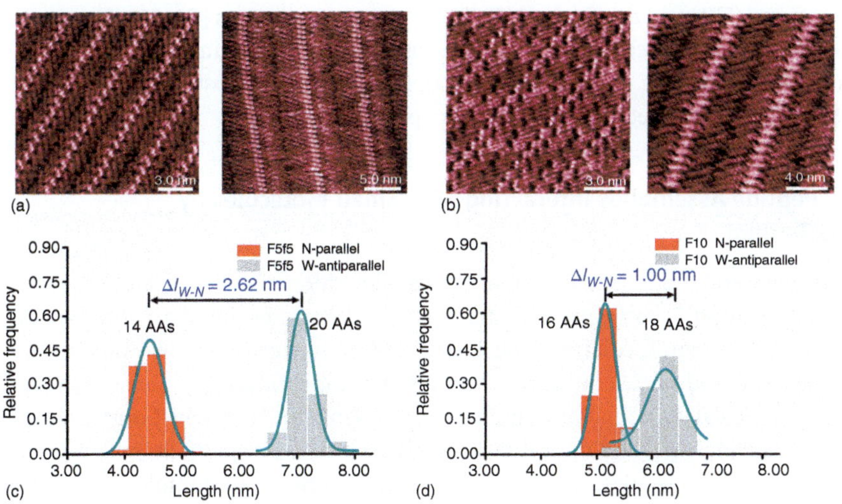

Figure 9.14 (a) Two types of co-assembly structures formed by F5f5/4Bpy on the HOPG surfaces under ambient conditions. Tunneling conditions: $I = 296.0$ pA and $V = 699.8$ mV. (b) Two types of co-assembly structures formed by F10/4Bpy on the HOPG surfaces. Tunneling conditions: $I = 296.0$ pA and $V = 699.8$ mV. (c, d) The length distributions of peptide β-strands involved in the F5f5/4Bpy parallel and antiparallel β-sheet structures (c) and F10/4Bpy parallel and antiparallel β-sheet structures (d). Δl_{W-N} is the difference value in the average lengths of β-strands involved in assembly between antiparallel and parallel β-sheet structures. Source: Reproduced with permission from Ref. [63]; © 2023 American Chemical Society.

assembly structures. For the F5f5 assembly, the antiparallel β-sheets involve more residues with an average assembly length of 7.09 nm than the parallel β-sheets with an average β-strand length of 4.47 nm (Figure 9.14c). By contrast, the F10 assembly presents the antiparallel β-sheet with a β-strand length of 6.15 nm and the parallel β-sheets with a β-strand length of 5.15 nm (Figure 9.14d). This observation hints that the staggered arrangement of heterochiral side chains across neighboring β-strands facilitates the elongation of F5f5 β-strands in antiparallel β-sheet conformation. These observations contribute to the single-molecular insights of stereochemical influences on directing peptide assemblies and the rational design of block heterochirality patterns for structural regulations.

9.5 Peptide–Organic Molecule Co-Assemblies

Introducing small molecules into peptide assemblies enriches the interaction categories in the co-assembly systems, thereby developing complementary venues for the construction of peptide–organic architectures and providing complementary strategies for probing the mechanism of peptide assemblies. Additionally, many drugs and dyes can interact with peptides as potential candidates for diagnosis and therapy. Several small molecules have been reported to co-assemble with peptides by noncovalent interactions, such as copper/zinc ion chelators, surfactants, well-known bioactive molecules (e.g. apomorphine, rifamycin, curcumin, and porphyrins), and dyes (e.g. Congo red (CR) and thioflavin T (ThT)), and their derivatives [64–66]. The structural insights into the mechanism of peptide assemblies interacting with small molecules contribute to the construction of peptide–organic nanostructures and the rational design of the specific drug/dye for peptide assemblies.

9.5.1 Peptide Assemblies Interacting with Small Molecules by Peptide Termini

The organic-peptide co-assemblies broaden the fabrication of novel 2D and 3D nanostructures for multifarious applications, such as templates for network growth, biosensors, and organic catalyst. The UHV-STM characterization of L-Phe-L-Phe molecules deposited on Cu substrates reveals that dipeptide molecules only form typically short isolated chains with a high density of kinks and prefer self-organization as one-dimensional (1D) chains. By co-depositing L-Phe-L-Phe peptides with terephthalic acid (TPA) molecules, 2D ordering and extending of supramolecular structures are constructed on the Cu(110) and Cu(100) surfaces [67]. L-Phe-L-Phe/TPA molecules assemble into a hybrid superlattice structure on a Cu(110) surface (Figure 9.15a and b). The parameters of a unit cell were measured to be $a = 2.41$ nm, $b = 0.77$ nm, and $\alpha = 87.9°$. These peptide–organic highly ordered nanostructures extend over an area of up to 150×150 nm^2 and keep stable up to 540 K. Besides promoting the formation of large domains, the TPA linker also improves the regularity of the L-Phe-L-Phe chains by reducing structural defects. In addition, L-Phe-L-Phe and TPA molecules organize with a self-selected stoichiometry of 1:1 for L-Phe-L-Phe

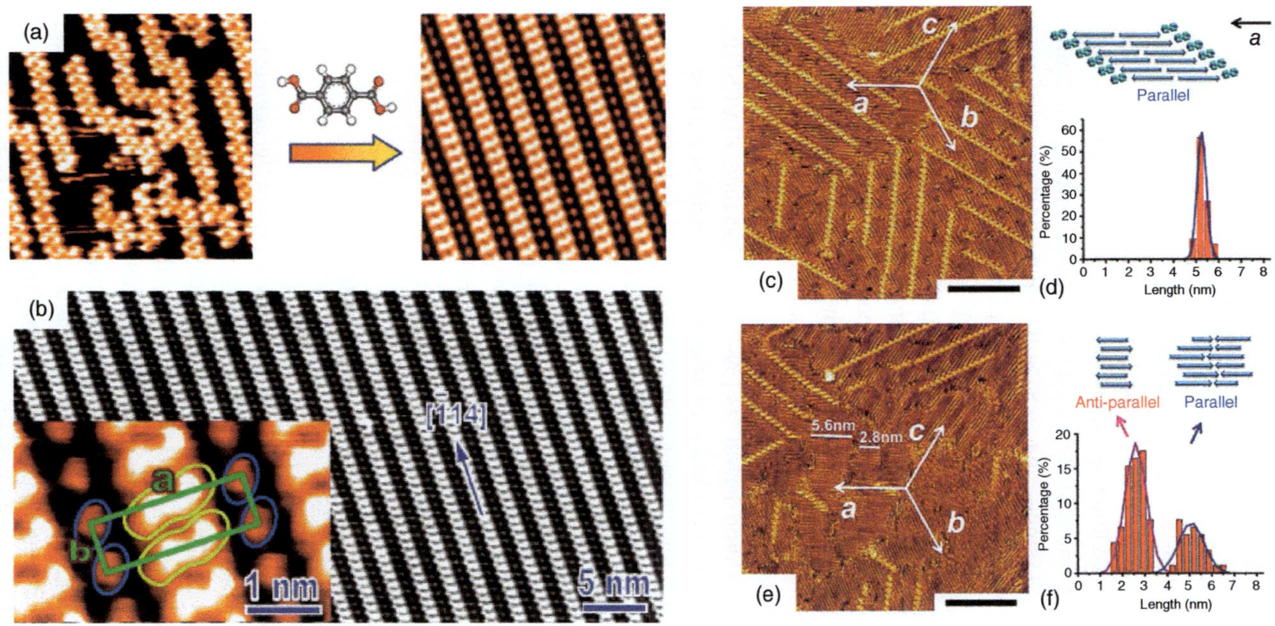

Figure 9.15 (a) L-Phe-L-Phe molecules form typically short isolated chains with a high density of kinks and prefer self-organization as 1D chains. Co-deposition of L-Phe-L-Phe peptides with TPA molecules builds 2D supramolecular structures on a Cu(110) surface. (b) STM images of 2D extended L-Phe-L-Phe/TPA hybrid motif on a Cu(110) surface. The quadrangle in the inset shows one unit of the molecular superlattice. The ellipse and dumbbell indicate TPA and L-Phe-L-Phe molecules, respectively. Tunneling conditions: I = 0.5 nA and V = −2.0 V for the large-scale image and I = 0.5 nA and V = −1.0 V for the zero-in image. (c and e) STM images of the Q11/4Bpy co-assembly on the HOPG surface (c) and the pristine Q11 assembly after removing the 4Bpy molecules from the central region (e). Arrows in the image present the three directions of peptide molecules. The STM tip scans in a-direction. Tunneling conditions for imaging: I = 300 pA and V = 600 mV. Scale bar = 10 nm. Tunneling conditions for manipulation: I = 600 pA and V = 10 mV. Scan rate = 15 Hz and scan area = 25 × 25 nm^2. (d and f) Length histograms and schematic models of peptide co-assemblies in a-direction with an ordered parallel β-sheet structure (d) and pristine peptide assemblies in antiparallel and parallel β-sheet structures (f). Source: Reproduced with permission from Ref. [67] (a and b); © 2023 American Chemical Society, and Ref. [68] (c–f); © 2023 John Wiley & Sons, Inc.

interacting with TPA and retain even for an excess of the initial L-Phe-L-Phe component on the Cu(110) and Cu(100) surfaces.

By forming hydrogen bonds with peptides, pyridine molecules promote the ordering of lamellar nanostructures of peptide assemblies, which was verified by the regularity decrease with the removal of pyridine molecules. By applying a bias voltage of 600 mV and tunneling current of 300 pA, the STM images of amyloidal Q11 peptides co-assembled with 4Bpy molecules present ordered lamellar structures on the HOPG surface with a relatively homogeneous length distribution of 5.3 ± 0.2 nm (Figure 9.15c and d) [68]. Upon a low-bias-voltage scan ($V = 10$ mV and $I = 600$ pA) in the center of the scan area to manipulate the Q11/4Bpy co-assemblies and remove 4Bpy molecules from the co-assemblies by an STM tip, pristine Q11 are agitated to assemble subsequently *in situ* (Figure 9.15e). Compared to the Q11/4Bpy co-assembled structures, the *in situ* Q11 assemblies present relatively disordered structures. The length of assembly-involved β-strands changes into a broad and bimodal distribution centered at 2.75 and 5.25 nm, corresponding to the antiparallel β-sheet conformation and parallel β-sheet conformation, respectively (Figure 9.15f). This STM experiment demonstrates that the co-assembly with pyridine molecules enhances the homogeneity and regularity of peptide assemblies.

Distinct supramolecular structures have been achieved by the co-assembly of peptides with different types of pyridine derivatives [69–71]. For instance, KLVFF/4Bpy co-assembly presents a parallel β-sheet-like conformation with sandwich-like striped structures. KLVFF/4′-chloro-2,2′:6′,2″-terpyridine (Cl-Ter) co-assembly forms a new closely packed ladder-like structure with an antiparallel β-sheet conformation [70]. By changing the molar ratio of KLVFF and Cl-Ter to 5:1, 1:1, and 1:5, KLVFF/Cl-Ter co-assemblies show stoichiometry-dependent structures, which can be attributed to the alterations of peptide secondary structures and pyridine configurations (Figure 9.16) [71]. All these works demonstrate that pyridine molecules can co-assemble with peptides to generate ordered domains and benefit to broaden the fabrication of novel 2D and 3D nanostructures.

9.5.2 Peptide Assemblies Interacting with Small Molecules by Peptide Side Chains

Small molecules have shown potency as molecular probes to exploit the aggregation process of amyloidal peptides and proteins [72, 73]. These small molecules can be used as compounds to affect the fibrotic process of peptides, making them potential candidates for diagnosis and therapy [74–76]. Deciphering the mechanism underlying small molecules targeting and binding peptide assemblies is essential for optimizing the molecular structures of the lead compounds for regulating the bioactivity of peptide assembly. Herein, the STM efforts for elucidating the interactions between small molecules and peptide assemblies are briefly reviewed.

STM can achieve real-space imaging of adsorption sites and configurations of small molecules atop their peptides/proteins target. For example, ThT is a probe molecule which is used to detect the assembly kinetics of amyloidal proteins [77–80]. STM studies have shown that ThT molecules bind to the prion peptide segment

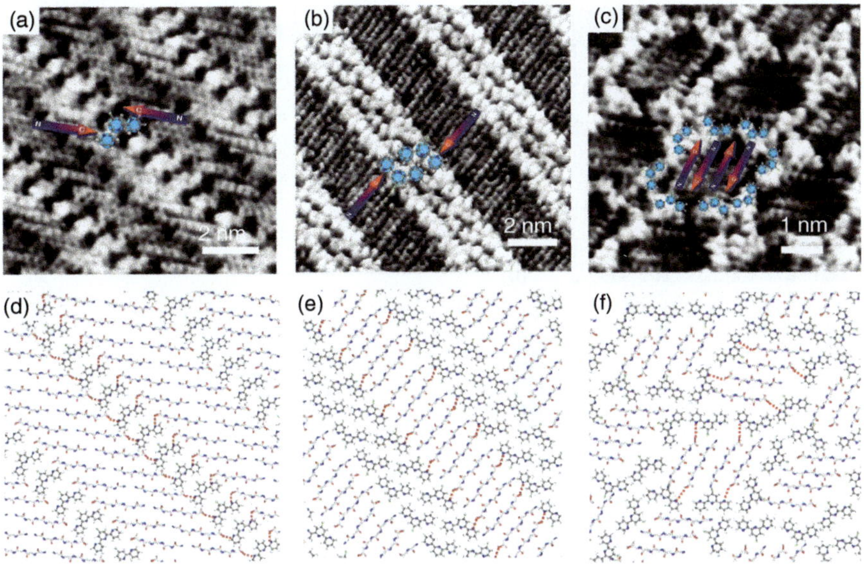

Figure 9.16 (a–c) STM images of KLVFF/Cl-Ter co-assemblies at different molar ratios: 5:1 (a), 1:1 (b), and 1:5 (c) on the HOPG surfaces under ambient conditions. Corresponding tunneling conditions: (a) I = 321.4 pA and V = 699.8 mV; (b) I = 433.3 pA and V = 630.2 mV; (c) I = 532.3 pA and V = 689.0 mV. (d–f) Structural models of KLVFF/Cl-Ter co-assemblies at different molar ratios: 5:1 (d), 1:1 (e), and 1:5 (f) on the HOPG surfaces. The O—H···N hydrogen bonds between the carboxyl groups of KLVFF peptides and the nitrogen atoms of Cl-Ter modulators are illustrated with red dotted lines. Source: Reproduced with permission from Ref. [71]; © 2023 American Chemical Society.

(GNNQQNY) assemblies in four different modes [81]. Another study demonstrates that ThT molecules adopt four oligomerization states to interact with its amyloidal protein receptor, the β-sheet assembly of $hIAPP_{8-37}$ [82]. $hIAPP_{8-37}$ peptides self-assemble into a 2D β-sheet monolayer on a freshly cleaved HOPG surface for ThT adsorption (Figure 9.17a). Based on the theoretical size of a single ThT molecule (approximately 1.3 nm × 0.4 nm), four types of ThT oligomers on $hIAPP_{8-37}$ assemblies were identified as dimers in two packing configurations ("1 × 2" and "2 × 1"), tetramers ("2 × 2"), and hexamers ("2 × 3") (Figure 9.17b). Two ThT molecules that organize with a side-by-side configuration (approximately 1.3 nm × 0.9 nm) atop the surface of $hIAPP_{8-37}$ assemblies are attributed to be a "1 × 2" mode. The dimerization in a head/tail configuration with the size of ca. 2.3 nm × 0.5 nm is assigned to be a "2 × 1" mode. Using the same analysis method, the other two oligomer types "2 × 2" and "2 × 3" were identified from STM images. The oligomerization states of ThT clusters are exclusively even numbers (2, 4, and 6) instead of odd numbers. The relative energy difference between two states can be estimated from the ratio of their probabilities based on the Maxwell–Boltzmann function (Figure 9.17c–e). The oligomerization state of the bounded ThTs is determined by adhesive interactions between ThT and the protein substrate and cohesive interactions among ThT molecules (Figure 9.17f and g). The work of the cohesive interaction between two head/tail ThTs is calculated to be $6.4\,k_BT$, and that between two side-by-side ThTs is

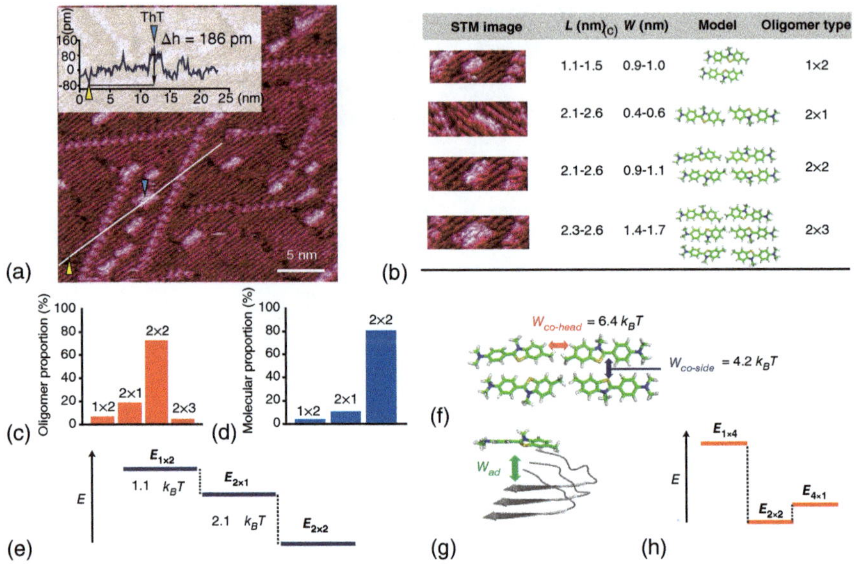

Figure 9.17 (a) A representative STM image of ThT adsorption at hIAPP$_{8-37}$/4Bpy surfaces. Tunneling conditions: I = 299.1 pA and V = 519.4 mV. Inset in (a) represents the cross-sectional profile corresponding to the white line. The blue and yellow triangles label the peak and the valley on the white line. (b) The measured and simulated oligomerization states of ThT on hIAPP$_{8-37}$/4Bpy co-assembly surfaces. (c and d) The number distribution of different types of ThT oligomers determined by the number of oligomers (c) and the number of constituent ThT units (d). (e) The relative energy difference among 1 × 2, 2 × 1, and 2 × 2 oligomerization states. (f and g) Schematic illustration of cohesive interactions within a 2 × 2 ThT tetramer (f) and adhesive interactions between a ThT monomer and hIAPP$_{8-37}$ β-sheet assembly (g). (h) A schematic illustration of the relative energy difference among the 2 × 2, 1 × 4, and 4 × 1 states. Source: Reproduced with permission from Ref. [82]; © 2023 American Association for the Advancement of Science.

4.2 k_BT (Figure 9.17f). The ThT oligomerization state exhibits a selection as the lack of observation of some types of oligomerization states, such as the "1×4" or "4×1" oligomers, verified by the calculated relative energy difference (Figure 9.17h). This single-molecular study provides valuable insights into the design of drug compounds using the selective oligomerization of molecular probes to recognize protein self-assemblies.

STM imaging can also detect the relative binding affinity of small molecules interacting with peptide assemblies. For instance, the labeling molecule, sodium copper phthalocyanine ($PcCu(SO_3Na)_4$), binding to polypeptides with specific amino-acid residues on the HOPG surfaces was investigated using the ambient-STM [83]. $PcCu(SO_3Na)_4$ is a dye molecule, used as a photodynamic therapy drug. The adsorption and distribution of $PcCu(SO_3Na)_4$ on four homo-oligo(amino acid)s, including PolyF8, PolyH8, PolyY8, and PolyQ7, were investigated (Figure 9.18a–e). By assuming the binding affinity of $PcCu(SO_3Na)_4$ to 4Bpy to be identical in these peptide–organic co-assembled systems, the relative binding affinities of $PcCu(SO_3Na)_4$ for the four peptide assemblies can be deduced as PolyH8:PolyQ7:PolyY8:PolyF8 = 3.32:1.29:0.79:0.34 (Figure 9.18f). The high affinity of $PcCu(SO_3Na)_4$ for PolyH8 is

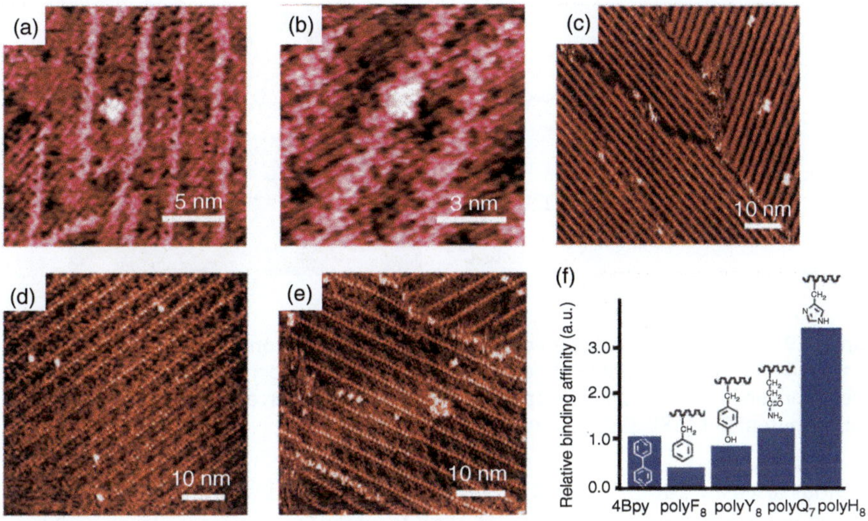

Figure 9.18 (a and b) STM images of $PcCu(SO_3Na)_4$ adsorption atop PolyQ7 (a) and 4Bpy (b), respectively. Tunneling conditions: I = 598.2 pA and V = 298.8 mV. (c–e) STM images of $PcCu(SO_3Na)_4$ adsorption on PolyF8-4Bpy (c), PolyH8-4Bpy (d), and PolyY8-4Bpy (e) co-assemblies, respectively. Tunneling conditions: (c) I = 344.5 pA and V = 620.0 mV; (d) I = 298.2 pA and V = 565.7 mV; (e) I = 612.2 pA and V = 311.0 mV. (f) The relative binding affinities of $PcCu(SO_3Na)_4$ on PolyF8, PolyQ7, PolyH8, and PolyY8 based on the statistical results. Source: Reproduced with permission from Ref. [83]; © 2023 The Royal Society of Chemistry.

attributed to the electrostatic interactions between the positively charged imidazole moiety of His and the negatively charged sulfonic acid group of $PcCu(SO_3Na)_4$. The thermodynamic and dynamic parameters related to peptide assembly–drug complex can reveal the binding affinities and enrich the understanding of interactions between molecules and peptide assemblies. The STM-based binding studies of dyes or drug molecules on homo-oligo(amino acid)s to explore molecular interactions at the molecular level help gain insight into the binding molecules on protein motifs or amino-acid residues.

9.5.3 Assemblies of Peptide–Organic Conjugates

The incorporation of functional moieties into biologically relevant systems offers the potential to construct exquisite biomaterials for a variety of applications. For example, the peptide–drug conjugates, which integrate bioactive peptides and small molecule drugs to achieve increased therapeutic outcomes, leverage the inherent and unique pharmacological abilities of the peptide and the small molecules [84]. Benefiting from the development of linker technology and conjugation chemistries, diverse conjugates, such as the growth hormone-inhibiting hormone (GHIH)-177Lu and luteinizing hormone-releasing hormone (LHRH)-doxorubicin, have been achieved and applied as the therapeutic interventions against endocrine diseases,

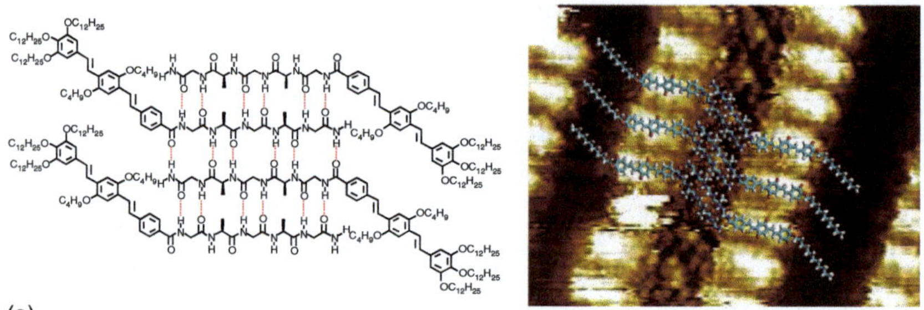

(a) (b)

Figure 9.19 (a) Molecular structure of an antiparallel β-sheet formed by OPV-peptide assembly. (b) An STM image of OPV-peptide assembly with a superimposed molecular model. Tunneling conditions: I = 28.0 pA and V = −860.0 mV. Source: Reproduced with permission from Ref. [88]; © 2023 American Chemical Society.

infectious diseases, autoimmune diseases, etc. [84–87]. By designing the noncovalent interactions encoded by the structure of peptide–organic conjugates, the assembly of conjugates can lead to the emergence of ordering on length scales. De Feyter, Meijer, Schenning, and coworkers reported the aggregation of oligo(*p*-phenylenevinylene)-peptide (OPV-peptide) conjugates into cylindrical micelles in an aqueous solution [88]. As revealed by STM imaging, the hydrogen bonds between neighboring peptide segments drive the self-association of peptides into antiparallel β-sheets at the 1-octanoic acid/HOPG interface (Figure 9.19a). The formation of β-sheets by peptide segments induced the spatial organization of OPV moieties into π-electron-rich supramolecular chains on the HOPG surface (Figure 9.19a and b) [88]. Highly ordered arrays of π-electrons have the potential to lead to emerging applications, particularly promising ones for the construction of effective charge-transporting materials. Taken together, the large diversity of self-assembled peptide–organic conjugates constitutes an interesting category that can be regarded as practical and promising building blocks for functional, hierarchical architectures.

9.6 Conclusions and Perspectives

This chapter has reviewed the progress in obtaining fine structural information for monodispersed peptides and peptide assemblies using the single-molecular STM technique. The modulation effects on assembly structures by mutations, post-translational modifications, and chirality alterations are discussed, and the interactions between small molecules and peptide assemblies are also briefly summarized. These garnered structural insights into the sequence-dependent assembly features could contribute to the rational design of peptide-based architectures. Further endeavors should be made toward deciphering the effects of environmental conditions including buffers, temperature, and substrates on peptide assemblies. *In situ* monitoring of the aggregation process by advanced technologies with both spatial- and time-resolved capabilities would be highly desirable to investigate the structural details of peptide assemblies.

Overall, decoding the mechanism of peptide assembly and aggregation pathway is of great significance to deepen the understanding of the assembly process and facilitate the rational design and optimization of peptide-based applications. Although current state-of-the-art techniques for the structural study of peptide assemblies, including cryo-EM, X-ray crystallography, and NMR spectroscopy, have revealed the precise three-dimensional arrangement of atoms within an assembly, the STM-based single-molecule experiments provide a unique and irreplaceable view for understanding the interpeptide recognition and association on a surface. For example, by contrast to the conventional opinion that peptide building blocks adopt a single conformation, the STM-based fundamental studies draw attention to the structural heterogeneity of an assembled peptide, where peptides in distinct conformations were observed to coexist within a supramolecular structure, such as the structures of hIAPP and bradykinin established based on the STM results [6, 41]. This observation highlights the contribution of peptide metastable conformations to the interpeptide association, inspiring future efforts to investigate the cooperativity between different, coexisting peptide conformations to regulate the structures and properties of an assembly.

Acknowledgments

This work was supported by the National Natural Science Foundation of China (31901007, 32201243), Beijing Natural Science Foundation (2222039), Haihe Laboratory of Cell Ecosystem Innovation Fund (HH22KYZX0011), the CAMS Innovation Fund for Medical Sciences (CIFMS 2021-I2M-1-022), and State Key Laboratory Special Fund (2060204).

Abbreviations

1D	One-dimensional
2D	Two-dimensional
3D	Three-dimensional
4Bpy	4,4′-Bipyridine
AD	Alzheimer's disease
AFM	Atomic force microscopy
ALS	Amyotrophic lateral sclerosis
APP	Amyloid precursor protein
At-I	Angiotensin I
At-II	Angiotensin II
Aβ	β-Amyloid
BN	Boron nitride
CD	Circular dichroism
Cryo-EM	Cryo-electron microscopy
CsA	Cyclosporine A

DFT	Density functional theory
DPE	1,2-Di(4-pyridyl)-ethylene
FTIR	Fourier transform infrared spectroscopy
FTLD	Frontotemporal lobar degeneration
GHIH	Growth hormone-inhibiting hormone
g-VNTR	Glycosylated variable number of tandem repeat
hIAPP	Human islet amyloid polypeptide
HOMO	Highest occupied molecular orbital
HOPG	Highly oriented pyrolytic graphite
IAPP	Islet amyloid polypeptide
LHRH	Luteinizing hormone-releasing hormone
LUMO	Lowest unoccupied molecular orbital
MD	Molecular dynamics
MPL	Mass-per-length
NMR	Nuclear magnetic resonance
OPV-peptide	Oligo(*p*-phenylenevinylene)-peptide
$PcCu(SO_3Na)_4$	Copper phthalocyanine tetrasulfonate sodium
PTM	Post-translational modification
p-VNTR	Pristine variable number of tandem repeat
rIAPP	Rat islet amyloid polypeptide
STEM	Scanning transmission electron microscopy
STM	Scanning tunneling microscopy
T2D	Type 2 diabetes
TDP-43	TAR DNA-binding protein 43
ThT	Thioflavin T
TSE	Transmissible spongiform encephalopathy
UHV	Ultrahigh vacuum
VNTRs	Variable number of tandem repeats
WT	Wild type

References

1 Wang, C.X., Yang, A.H., Li, X. et al. (2012). Observation of molecular inhibition and binding structures of amyloid peptides. *Nanoscale* 4 (6): 1895–1909.

2 He, S., He, X., Liu, L. et al. (2021). The structural understanding of transthyretin misfolding and the inspired drug approaches for the treatment of heart failure associated with transthyretin amyloidosis. *Front. Pharmacol.* 12: 628184–628192.

3 Wang, C. and Bai, C.L. (2006). *Single Molecule Chemistry and Physics – An Introduction*. Germany: Springer.

4 Lingenfelder, M., Tomba, G., Costantini, G. et al. (2007). Tracking the chiral recognition of adsorbed dipeptides at the single-molecule level. *Angew. Chem.-Int. Ed.* 46 (24): 4492–4495.

5 Chen, Y.M., Deng, K., Qiu, X.H., and Wang, C. (2013). Visualizing cyclic peptide hydration at the single-molecule level. *Sci. Rep.* 3: 2461–2467.
6 Rauschenbach, S., Rinke, G., Gutzler, R. et al. (2017). Two-dimensional folding of polypeptides into molecular nanostructures at surfaces. *ACS Nano* 11 (3): 2420–2427.
7 Cyriac, J., Pradeep, T., Kang, H. et al. (2012). Low-energy ionic collisions at molecular solids. *Chem. Rev.* 112 (10): 5356–5411.
8 Deng, Z., Thontasen, N., Malinowski, N. et al. (2012). A close look at proteins: submolecular resolution of two- and three-dimensionally folded cytochrome C at surfaces. *Nano Lett.* 12 (5): 2452–2458.
9 Wu, X., Delbianco, M., Anggara, K. et al. (2020). Imaging single glycans. *Nature* 582 (7812): 375–378.
10 Anggara, K., Zhu, Y., Delbianco, M. et al. (2020). Exploring the molecular conformation space by soft molecule-surface collision. *J. Am. Chem. Soc.* 142 (51): 21420–21427.
11 Anggara, K., Zhu, Y., Fittolani, G. et al. (2021). Identifying the origin of local flexibility in a carbohydrate polymer. *Proc. Natl. Acad. Sci. U. S. A.* 118 (23): e2102168118.
12 Tanaka, H. and Kawai, T. (2009). Partial sequencing of a single DNA molecule with a scanning tunnelling microscope. *Nat. Nanotechnol.* 4 (8): 518–522.
13 Qiu, X.H., Wang, C., Zeng, Q.D. et al. (2000). Alkane-assisted adsorption and assembly of phthalocyanines and porphyrins. *J. Am. Chem. Soc.* 122 (23): 5550–5556.
14 Chang, S., Liu, R., Wang, L. et al. (2016). Formation of ordered coronene clusters in template utilizing the structural transformation of hexaphenylbenzene derivative networks on graphite surface. *ACS Nano* 10 (1): 342–348.
15 Geng, Y., Xu, J., Xue, J. et al. (2015). Study of the edge-on self-assembly of axially substituted silicon(IV) phthalocyanine derivatives in a template on the HOPG surface. *Langmuir* 31 (49): 13394–13401.
16 Xu, J., Xiao, X., Deng, K., and Zeng, Q. (2016). Transformation of self-assembly of a TTF derivative at the 1-phenyloctane/HOPG interface studied by STM-from a nanoporous network to a linear structure. *Nanoscale* 8 (3): 1652–1657.
17 Shi, D.X., Ji, W., Lin, X. et al. (2006). Role of lateral alkyl chains in modulation of molecular structures on metal surfaces. *Phys. Rev. Lett.* 96 (22): 226101–226104.
18 Yang, B., Wang, Y., Cun, H. et al. (2010). Direct observation of enantiospecific substitution in a two-dimensional chiral phase transition. *J. Am. Chem. Soc.* 132 (30): 10440–10444.
19 Lu, J., Bao, D.L., Dong, H. et al. (2017). Construction of two-dimensional chiral networks through atomic bromine on surfaces. *J. Phys. Chem. Lett.* 8 (2): 326–331.
20 Ren, J., Bao, D.-L., Dong, L. et al. (2017). Lattice-directed construction of metal–organic molecular wires of pentacene on the Au(110) surface. *J. Phys. Chem. C* 121 (39): 21650–21657.
21 Barth, J.V., Weckesser, J., Trimarchi, G. et al. (2002). Stereochemical effects in supramolecular self-assembly at surfaces: 1-D versus 2-D enantiomorphic ordering for PVBA and PEBA on Ag(111). *J. Am. Chem. Soc.* 124 (27): 7991–8000.

22 Marschall, M., Reichert, J., Weber-Bargioni, A. et al. (2010). Random two-dimensional string networks based on divergent coordination assembly. *Nat. Chem.* 2 (2): 131–137.

23 Cun, H., Wang, Y., Du, S. et al. (2012). Tuning structural and mechanical properties of two-dimensional molecular crystals: the roles of carbon side chains. *Nano Lett.* 12 (3): 1229–1234.

24 Wu, R., Yan, L., Zhang, Y. et al. (2015). Self-assembled patterns and Young's modulus of single-layer naphthalocyanine molecules on Ag(111). *J. Phys. Chem. C* 119 (15): 8208–8212.

25 Wang, Y.B., Niu, L., Li, Y.B. et al. (2010). Single molecule studies of cyclic peptides using molecular matrix at liquid/solid interface by scanning tunneling microscopy. *Langmuir* 26 (21): 16305–16311.

26 Li, Y., Liu, C., Xie, Y. et al. (2013). Single-molecule observation of the K+-induced switching of valinomycin within a template network. *Chem. Commun.* 49 (79): 9021–9023.

27 Abb, S., Harnau, L., Gutzler, R. et al. (2016). Two-dimensional honeycomb network through sequence-controlled self-assembly of oligopeptides. *Nat. Commun.* 7: 10335–10341.

28 Chen, Y., Deng, K., Lei, S. et al. (2018). Single-molecule insights into surface-mediated homochirality in hierarchical peptide assembly. *Nat. Commun.* 9 (1): 2711–2718.

29 Fandrich, M., Schmidt, M., and Grigorieff, N. (2011). Recent progress in understanding Alzheimer's beta-amyloid structures. *Trends Biochem. Sci.* 36 (6): 338–345.

30 Eisenberg, D. and Jucker, M. (2012). The amyloid state of proteins in human diseases. *Cell* 148 (6): 1188–1203.

31 Goldsbury, C., Baxa, U., Simon, M.N. et al. (2011). Amyloid structure and assembly: insights from scanning transmission electron microscopy. *J. Struct. Biol.* 173 (1): 1–13.

32 Kajava, A.V., Aebi, U., and Steven, A.C. (2005). The parallel superpleated beta-structure as a model for amyloid fibrils of human amylin. *J. Mol. Biol.* 348 (2): 247–252.

33 Mao, X., Ma, X., Liu, L. et al. (2009). Structural characteristics of the beta-sheet-like human and rat islet amyloid polypeptides as determined by scanning tunneling microscopy. *J. Struct. Biol.* 167 (3): 209–215.

34 Ma, X., Liu, L., Mao, X. et al. (2009). Amyloid beta (1–42) folding multiplicity and single-molecule binding behavior studied with STM. *J. Mol. Biol.* 388 (4): 894–901.

35 Lorenzo, A., Razzaboni, B., Weir, G.C., and Yankner, B.A. (1994). Pancreatic-islet cell toxicity of amylin associated with type-2 diabetes-mellitus. *Nature* 368 (6473): 756–760.

36 Sakagashira, S., Hiddinga, H.J., Tateishi, K. et al. (2000). S20G mutant amylin exhibits increased in vitro amyloidogenicity and increased intracellular cytotoxicity compared to wild-type amylin. *Am. J. Pathol.* 157 (6): 2101–2109.

37 Sakagashira, S., Sanke, T., Hanabusa, T. et al. (1996). Missense mutation of amylin gene (S20G) in Japanese NIDDM patients. *Diabetes* 45 (9): 1279–1281.

38 Chen-Plotkin, A.S., Lee, V.M.Y., and Trojanowski, J.Q. (2010). TAR DNA-binding protein 43 in neurodegenerative disease. *Nat. Rev. Neurol.* 6 (4): 211–220.

39 Neumann, M., Sampathu, D.M., Kwong, L.K. et al. (2006). Ubiquitinated TDP-43 in frontotemporal lobar degeneration and amyotrophic lateral sclerosis. *Science* 314 (5796): 130–133.

40 Lagier-Tourenne, C., Polymenidou, M., and Cleveland, D.W. (2010). TDP-43 and FUS/TLS: emerging roles in RNA processing and neurodegeneration. *Hum. Mol. Genet.* 19: R46–R64.

41 Mao, X.B., Wang, C.X., Wu, X.K. et al. (2011). Beta structure motifs of islet amyloid polypeptides identified through surface-mediated assemblies. *Proc. Natl. Acad. Sci. U. S. A.* 108 (49): 19605–19610.

42 Guo, Y., Wang, C., Hou, J. et al. (2012). Odd-even sequence effect of surface-mediated peptide assemblies observed by scanning tunneling microscopy. *Chin. J. Chem.* 30 (9): 1987–1991.

43 Guo, Y., Hou, J., Zhang, X. et al. (2017). Stabilization effect of amino acid side chains in peptide assemblies on graphite studied by scanning tunneling microscopy. *ChemPhysChem* 18 (8): 926–934.

44 Yu, L., Zheng, Y., Xu, J. et al. (2018). Site-specific determination of TTR-related functional peptides by using scanning tunneling microscopy. *Nano Res.* 11 (1): 577–585.

45 Gras, S.L., Tickler, A.K., Squires, A.M. et al. (2008). Functionalised amyloid fibrils for roles in cell adhesion. *Biomaterials* 29 (11): 1553–1562.

46 Pierschbacher, M.D. and Ruoslahti, E. (1984). Variants of the cell recognition site of fibronectin that retain attachment-promoting activity. *Proc. Natl. Acad. Sci. U. S. A.-Biol. Sci.* 81 (19): 5985–5988.

47 Hautanen, A., Gailit, J., Mann, D.M., and Ruoslahti, E. (1989). Effects of modifications of the RGD sequence and its context on recognition by the fibronectin receptor. *J. Biol. Chem.* 264 (3): 1437–1442.

48 Muller, M.M. (2018). Post-translational modifications of protein backbones: unique functions, mechanisms, and challenges. *Biochemistry* 57 (2): 177–185.

49 Bah, A. and Forman-Kay, J.D. (2016). Modulation of intrinsically disordered protein function by post-translational modifications. *J. Biol. Chem.* 291 (13): 6696–6705.

50 Xu, M., Zhu, L., Liu, J. et al. (2013). Characterization of beta-domains in C-terminal fragments of TDP-43 by scanning tunneling microscopy. *J. Struct. Biol.* 181 (1): 11–16.

51 Guo, W., Chen, Y., Zhou, X. et al. (2011). An ALS-associated mutation affecting TDP-43 enhances protein aggregation, fibril formation and neurotoxicity. *Nat. Struct. Mol. Biol.* 18 (7): 822–830.

52 Yu, L.L., Sun, Z.Y., Yu, Y. et al. (2016). Molecular evidence of glycosylation effect on the peptide assemblies identified with scanning tunneling microscopy. *J. Phys. Chem. C* 120 (12): 6577–6582.

53 Singh, R. and Bandyopadhyay, D. (2007). MUC1: a target molecule for cancer therapy. *Cancer Biol. Therapy* 6 (4): 481–486.

54 Fujii, N., Fujii, N., Kida, M., and Kinouchi, T. (2010). Influence of Lbeta-, Dalpha- and Dbeta-Asp isomers of the Asp-76 residue on the properties of alphaA-crystallin 70–88 peptide. *Amino Acids* 39 (5): 1393–1399.

55 Shah, N.K., Brodsky, B., Kirkpatrick, A., and Ramshaw, J.A. (1999). Structural consequences of D-amino acids in collagen triple-helical peptides. *Biopolymers* 49 (4): 297–302.

56 Clarke, D.E., Wu, G., Wu, C., and Scherman, O.A. (2021). Host-guest induced peptide folding with sequence-specific structural chirality. *J. Am. Chem. Soc.* 143 (17): 6323–6327.

57 De Rosa, L., Diana, D., Capasso, D. et al. (2022). Switching the N-capping region from all-L to all-D Amino acids in a VEGF mimetic helical peptide. *Molecules* 27 (20): 6982–6994.

58 Nagy-Smith, K., Beltramo, P.J., Moore, E. et al. (2017). Molecular, local, and network-level basis for the enhanced stiffness of hydrogel networks formed from coassembled racemic peptides: predictions from Pauling and Corey. *ACS Central Sci.* 3 (6): 586–597.

59 Pauling, L. and Corey, R.B. (1953). Two rippled-sheet configurations of polypeptide chains, and a note about the pleated sheets. *Proc. Natl. Acad. Sci. U. S. A.* 39 (4): 253–256.

60 Swanekamp, R.J., Dimaio, J.T., Bowerman, C.J., and Nilsson, B.L. (2012). Coassembly of enantiomeric amphipathic peptides into amyloid-inspired rippled beta-sheet fibrils. *J. Am. Chem. Soc.* 134 (12): 5556–5559.

61 Marchesan, S., Easton, C.D., Kushkaki, F. et al. (2012). Tripeptide self-assembled hydrogels: unexpected twists of chirality. *Chem. Commun.* 48 (16): 2195–2197.

62 Zheng, Y., Yu, L., Zou, Y. et al. (2019). Steric dependence of chirality effect in surface-mediated peptide assemblies identified with scanning tunneling microscopy. *Nano Lett.* 19 (8): 5403–5409.

63 Zheng, Y., Luo, W., Yu, L. et al. (2022). Heterochirality-mediated cross-strand nested hydrophobic interaction effects manifested in surface-bound peptide assembly structures. *J. Phys. Chem. B* 126 (3): 723–733.

64 Nie, Q., Du, X.-G., and Geng, M.-Y. (2011). Small molecule inhibitors of amyloid beta peptide aggregation as a potential therapeutic strategy for Alzheimer's disease. *Acta Pharmacol. Sin.* 32 (5): 545–551.

65 Porat, Y., Abramowitz, A., and Gazit, E. (2006). Inhibition of amyloid fibril formation by polyphenols: Structural similarity and aromatic interactions as a common inhibition mechanism. *Chem. Biol. Drug Des.* 67 (1): 27–37.

66 Stains, C.I., Mondal, K., and Ghosh, I. (2007). Molecules that target beta-amyloid. *ChemMedChem* 2 (12): 1674–1692.

67 Wang, Y., Lingenfelder, M., Classen, T. et al. (2007). Ordering of dipeptide chains on Cu surfaces through 2D cocrystallization. *J. Am. Chem. Soc.* 129 (51): 15742–15743.

68 Yu, Y., Yang, Y., and Wang, C. (2015). Identification of core segment of amyloidal peptide mediated by chaperone molecules by using scanning tunneling microscopy. *ChemPhysChem* 16 (14): 2995–2999.

69 Liu, L., Zhang, L., Niu, L. et al. (2011). Observation of reduced cytotoxicity of aggregated amyloidogenic peptides with chaperone-like molecules. *ACS Nano* 5 (7): 6001–6007.

70 Niu, L., Liu, L., Xu, M. et al. (2014). Transformation of beta-sheet structures of the amyloid peptide induced by molecular modulators. *Chem. Commun.* 50 (64): 8923–8926.

71 Niu, L., Liu, L., Xi, W.H. et al. (2016). Synergistic inhibitory effect of peptide-organic coassemblies on amyloid aggregation. *ACS Nano* 10 (4): 4143–4153.

72 Ivanova, M.I., Sawaya, M.R., Gingery, M. et al. (2004). An amyloid-forming segment of beta 2-microglobulin suggests a molecular model for the fibril. *Proc. Natl. Acad. Sci. U. S. A.* 101 (29): 10584–10589.

73 Fowler, D.M., Koulov, A.V., Balch, W.E., and Kelly, J.W. (2007). Functional amyloid – from bacteria to humans. *Trends Biochem. Sci.* 32 (5): 217–224.

74 Sato, T., Kienlen-Campard, P., Ahmed, M. et al. (2006). Inhibitors of amyloid toxicity based on beta-sheet packing of A beta 40 and A beta 42. *Biochemistry* 45 (17): 5503–5516.

75 Blanchard, B.J., Chen, A., Rozeboom, L.M. et al. (2004). Efficient reversal of Alzheimer's disease fibril formation and elimination of neurotoxicity by a small molecule. *Proc. Natl. Acad. Sci. U. S. A.* 101 (40): 14326–14332.

76 Heiser, V., Scherzinger, E., Boeddrich, A. et al. (2000). Inhibition of huntingtin fibrillogenesis by specific antibodies and small molecules: Implications for Huntington's disease therapy. *Proc. Natl. Acad. Sci. U. S. A.* 97 (12): 6739–6744.

77 Millen, R.K., Buhimschi, S.C., Zhao, G. et al. (2018). Serum and urine Thioflavin-T–enhanced fluorescence in severe preeclampsia. *Hypertension* 71 (6): 1185–1192.

78 Groenning, M., Norrman, M., Flink, J.M. et al. (2007). Binding mode of Thioflavin T in insulin amyloid fibrils. *J. Struct. Biol.* 159 (3): 483–497.

79 Biancalana, M. and Koide, S. (2010). Molecular mechanism of Thioflavin-T binding to amyloid fibrils. *Biochim. Biophys. Acta* 1804 (7): 1405–1412.

80 Groenning, M. (2010). Binding mode of Thioflavin T and other molecular probes in the context of amyloid fibrils-current status. *J. Chem. Biol.* 3 (1): 1–18.

81 Mao, X.B., Guo, Y.Y., Wang, C.X. et al. (2011). Binding modes of thioflavin T molecules to prion peptide assemblies identified by using scanning tunneling microscopy. *ACS Chem. Neurosci.* 2 (6): 281–287.

82 Yu, L., Zhang, W., Luo, W. et al. (2020). Molecular recognition of human islet amyloid polypeptide assembly by selective oligomerization of thioflavin T. *Science. Advances* 6 (32): eabc1449.

83 Wang, C.X., Mao, X.B., Yang, A.H. et al. (2011). Determination of relative binding affinities of labeling molecules with amino acids by using scanning tunneling microscopy. *Chem. Commun.* 47 (38): 10638–10640.

84 He, R., Finan, B., Mayer, J.P., and Dimarchi, R.D. (2019). Peptide conjugates with small molecules designed to enhance efficacy and safety. *Molecules* 24 (10): 1855–1888.

85 Strosberg, J., El-Haddad, G., Wolin, E. et al. (2017). Phase 3 trial of (177)Lu-dotatate for midgut neuroendocrine tumors. *N. Engl. J. Med.* 376 (2): 125–135.

86 Emons, G., Gorchev, G., Harter, P. et al. (2014). Efficacy and safety of AEZS-108 (LHRH agonist linked to doxorubicin) in women with advanced or recurrent endometrial cancer expressing LHRH receptors: a multicenter phase 2 trial (AGO-GYN5). *Int. J. Gynecol. Cancer* 24 (2): 260–265.

87 Emons, G., Gorchev, G., Sehouli, J. et al. (2014). Efficacy and safety of AEZS-108 (INN: zoptarelin doxorubicin acetate) an LHRH agonist linked to doxorubicin in women with platinum refractory or resistant ovarian cancer expressing LHRH receptors: a multicenter phase II trial of the ago-study group (AGO GYN 5). *Gynecol. Oncol.* 133 (3): 427–432.

88 Matmour, R., De Cat, I., George, S.J. et al. (2008). Oligo(p-phenylenevinylene)-peptide conjugates: synthesis and self-assembly in solution and at the solid-liquid interface. *J. Am. Chem. Soc.* 130 (44): 14576–14583.

10

Molecular Assembly of Cyclic Dipeptide: Structure to Function

Debasis Ghosh and Thimmaiah Govindaraju

Bioorganic Chemistry Laboratory, New Chemistry Unit and School of Advanced Materials (SAMat), Jawaharlal Nehru Centre for Advanced Scientific Research, Jakkur P.O, Bengaluru, Karnataka, India

10.1 Introduction

Molecular architectonics is the custom design and engineering of molecular assembly of functional building blocks to produce advanced molecular and material architectures with novel functional properties and applications [1–5]. The molecular building blocks are designed by integrating the tailorable functional core with suitable assembly-directing auxiliary units [6, 7]. Nature has developed many elegant and sophisticated molecular and material architectures of DNA, proteins, carbohydrates, and lipids, which serve as a guiding principle in the scheme of molecular architectonics [8, 9]. CDPs are the simplest class of cyclic peptides found use in a range of materials and biomedical applications [10–15]. CDPs comprise conformationally constrained six-membered cyclo-lactam (2,5-diketopiperazine: DKP) ring with two amide groups at opposite positions in the ring (Figure 10.1) [16]. In contrast to linear dipeptides, structurally robust CDPs exhibit unique properties such as strong intermolecular interactions through multiple hydrogen bonds as well as various noncovalent interactions among the substituent functionalities [17]. Numerous natural compounds containing CDP scaffolds have been extracted, and their characteristics and biosynthesis have been studied [11, 18]. The natural origin, biological relevance, and drug-like capabilities of CDP scaffolds have promoted their use in biological applications that range from drug targets to drug delivery tools [10, 18–22]. CDPs have been identified as a promising chemical in drug discovery and delivery research due to their structural rigidity and multiple hydrogen-bonding abilities of cyclolactam ring, and scaffold diversity through the substituents, biocompatibility, and drug-like characteristics combined with improved proteolytic stability, and high-affinity binding with biological targets (biological receptors) [11, 18, 20].

Peptide Self-Assembly and Engineering: Fundamentals, Structures, and Applications, First Edition.
Edited by Xuehai Yan.

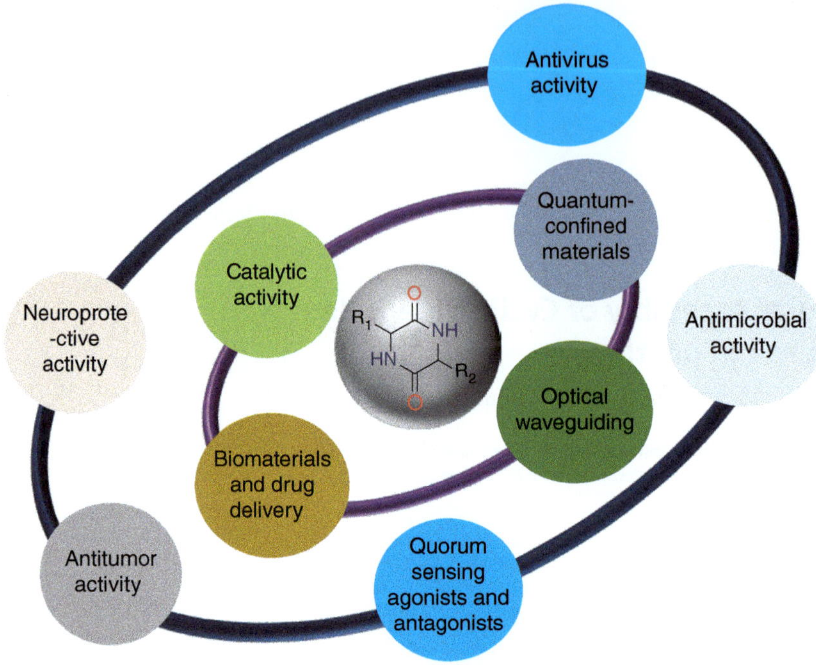

Figure 10.1 Schematic illustration of cyclic dipeptide (CDP) scaffold (R1 and R2 = amino-acid side chain) and its molecular assembly to produce functional architectures with material and biomedical applications.

The rationally designed CDPs have been developed as low-molecular-weight gelators, AIEgenic systems, antioxidant CDPs, high-strength organic materials, and quantum-confined materials [14, 23, 24]. Furthermore, they have been used as a preferred scaffold in producing complex natural products [11, 25]. Overall, the structural and functional diversity of the CDP scaffolds exemplifies their significance and utility across the domains of chemistry, biology, and materials science. The diverse applications of CDP-derived molecular and material architectures are shown in Figure 10.1. Overall, the CDP scaffold is considered an exemplary modular building block in the scheme of molecular architectonics [3, 10, 26].

10.2 The Synthesis of CDPs

The design and synthesis of CDPs is crucial for the development of novel molecular and material architectures. There are two major routes for synthesizing CDPs: (i) solution and (ii) solid-phase synthesis involving various cyclization protocols [11]. The dipeptide ester cyclization has been the commonly used technique for the synthesis of CDPs. In recent years, solid-phase synthesis has gained prominence for the parallel synthesis of CDPs [27]. The biosynthetic pathways of CDPs are mostly divided into nonenzymatic and enzymatic pathways. The CDP biosynthesis is specifically mediated by two families of enzymes: i) nonribosomal peptide synthetases and ii) CDP synthetases [18].

10.3 Molecular Assembly of CDP

The CDP with multiple hydrogen bond donors and acceptors and stereochemistry at α-carbons play key role in intermolecular interaction and assembly to form molecular chains and layers (Figure 10.2a) [10, 13, 17]. The α-substituents of the constituent amino acids offer distinct and diverse noncovalent interactions such as π–π, hydrophobic, hydrophilic, and electrostatic interactions to support the hydrogen-bonding-guided molecular assembly of CDPs. The molecular assembly of CDP manifests into ordered nano-, macro-, and hierarchical architectures with interesting functional properties and applications [10–12, 26].

10.3.1 CDP Nanoarchitectures

The diverse assembly architectures are the result of distinct ordered intramolecular packing governed by various noncovalent forces among the CDP building blocks. The judiciously modulated intermolecular interactions and assembly conditions produce 0D, 1D, 2D, 3D, and hierarchical nanoarchitectures such as nanospheres, nanofibers, nanotubes, nanorods, nanowires, nanoribbons, and nanosheets [15–17, 30, 31]. In the following section, we discuss various factors that influence molecular assembly of CDPs to form well-defined nanoarchitectures.

10.3.1.1 Amino-acid Composition and Stereochemistry

The controlled molecular assembly of designer CDP building blocks can be achieved by modulating noncovalent interactions through the scheme of molecular architectonics [10]. The molecular assembly can be engineered by controlling intermolecular noncovalent interactions using suitable natural or unnatural amino-acid constituents of designer CDP building blocks. For instance, aromatic amino acids such as Phe, Tyr, and Trp impart aromatic π–π stacking and hydrophobic interactions.

Electrostatic forces were produced via charged amino acids such as Arg and Lys. Kleinsmann and Nachtsheim demonstrated improved gelation ability of Phe-containing CDPs [32]. The scanning electron microscopy (SEM) study showed nanofibers morphology for cyclo(L-Phe-Gly) and cyclo(L-Phe-L-Ser). Cyclo(L-Phe-L-Glu) showed clearly defined porosity networks, and cyclo(L-Phe-L-Cys) formed irregular nanosheets. Cyclo(L-Phe-L-Lys) and cyclo(L-Phe-L-His) are self-assembled into compact lamellar sheets with connected nanofibers.

The stereochemistry of CDP plays a decisive role in its molecular assembly process. For instance, cyclo(L-Tyr-D-Ala) produced peptide microtubes, while cyclo(L-Tyr-L-Ala) assembled into nanotubes and nanowires. Water molecules inside the crystal lattice of cyclo(L-Tyr-L-Ala) can be thermally removed without changing the crystal structure, but cyclo(L-Tyr-D-Ala) displayed reversible behavior. Thus, water molecules within the CDP assemblies can potentially determine the outcome of molecular assembly architectures [15]. We have described single-crystalline self-assembled structures of cyclo(L-Ala-L-Ala), cyclo(L-Ala-D-Ala), cyclo(L-Phg-L-Phg), cyclo (L-Phg-L-Phg), and (L-Phg-D-Phg) (Figure 10.2b) [14]. Cyclo-(L-Phg-L-Phg), cyclo (L-Ala-L-Ala), and cyclo(L-Ala-D-Ala) all showed thin single-crystalline sheets, but

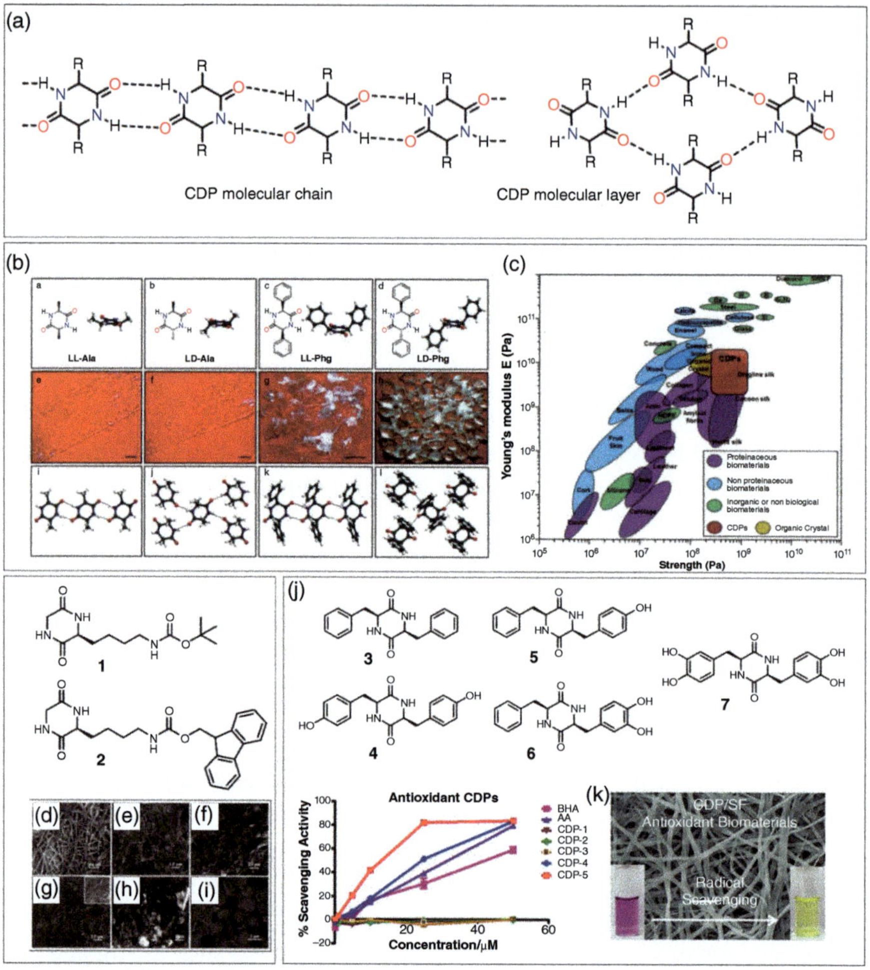

Figure 10.2 (a) CDP molecular self-assembly to construct molecular chains and layers via numerous hydrogen-bonding centers. (b, c) Self-assembled single-crystalline architectures of different CDPs (cyclo-L, L-Ala, L, D-Ala, L, L-Phe, and LD-Phg) and their elastic modulus plot. *Source:* (a–c) Ref. [14] / Springer Nature / CC BY 4.0. Young's modulus (*E*) vs strength plot showed very high mechanical strength of CDP material architectures in contrast to numerous materials including silk. The CDP gelators **1** and **2** and the ordered microporous hierarchical structure images (FESEM) of xerogels formed by **1** in (d) $CHCl_3$, (e) CCL_4, (f) hexane, and (g) toluene and that of **2** in (h) toluene and (i) $CHCl_3$. *Source:* Panels (d–i) are adapted with permission. Copyright 2012, Royal Society of Chemistry [28]. The antioxidant CDPs (**3**–**7**) were studied to evaluate their radical scavenging properties and a relative plot of antioxidant properties of all CDPs. (k) Antioxidant property of CDPs after addition of electrospun silk fibroin (SF). Native SF and CDP incorporated SF electrospun scaffolds as seen in SEM micrographs (left to right). The inset demonstrates how the SF and CDP-ability SFs to scavenge radicals differ by changing the coloration of DPPH. Panels (j and k) are adapted with permission. Wiley-VCH [29], Copyright 2017.

cyclo(L-Phg-D-Phg) was obtained as rhombohedral-shaped crystals. The stereochemistry (LL) was found to facilitate the formation of 1D molecular chains, while LD stereochemistry promoted 2D molecular layers which manifest into 2D architectures. The nanoindentation technique was employed to evaluate the mechanical properties of these CDP crystals (Figure 10.2c). Interestingly, CDPs with mixed stereochemistry (LD) at the side chain (α-substituent) exhibited better mechanical strength due to molecular layer arrangement compared to the same stereochemistry (LL or DD) that facilitated assembly through molecular chains. Cyclo(L-Phg-D-Phg) exhibited superior mechanical property ($E = 20.5$ GPa and $H = 1163.9$ MPa). The elastic modulus (E) is higher than that estimated for diphenylalanine-based nanotubes ($E = 19$ GPa) and twice that of spider dragline silk (~10 GPa). Furthermore, cyclo(L-Phg-D-Phg) shows high yield strength, σ_y (estimated using the relation $\sigma_y = \mathrm{H}/3$) of 388 MPa. This work demonstrated the possibility of designing bioinspired low-density organic materials with tunable mechanical characteristics by adapting the principles of molecular architectonics. CDP-based low-molecular-weight gelators (LMWGs) were developed with excellent mechanical strength, biocompatibility, and efficacy in biomaterial applications through a reductionistic approach. The cyclo(Gly-L-Lys) ε-amino derivatives containing tert-butyloxycarbonyl (**1**) and fluorenyl methoxycarbonyl (**2**) Lys-amino protection were designed and synthesized [20]. CDPs **1** and **2** were evaluated for their tendency to form hydrogel and organogels in both aqueous and organic solvents employing the sequential heating–cooling method. CDP **1** was found to form organogels in hexane, carbon tetrachloride, toluene, and chloroform, while CDP **2** formed both organogels and hydrogels (Figure 10.2d–i). Remarkably, organogels formed by CDP **2** with Fmoc group in chloroform exhibited substantial mechanical strength, as the microporous hierarchical architecture is supported by both intermolecular hydrogen-bonding and π–π aromatic interactions. Further, drug (curcumin) and dye (rhodamine B) molecule encapsulation within the gel network without altering the structural integrity of the gel was demonstrated, which suggests that these gels could be used for drug delivery applications. The antioxidant property of a set of CDPs with variable proteinogenic and nonproteinogenic aromatic or phenolic amino acids was assessed (Figure 10.2j) [29]. The comprehensive structure–activity study revealed that CDPs containing 3,4-dihydroxy-L-phenylalanine (DOPA, **7**) units exhibit excellent antioxidative properties compared to natural antioxidant (ascorbic acid) and well-known food additive (butylated hydroxytoluene: BHT). The cellular experiments have shown that the antioxidant CDPs are biocompatible and not cytotoxic to PC12 (adrenal phaeochromocytoma) cells. In addition, a biocompatible electrospun silk fibroin (SF) scaffold was prepared by incorporating antioxidant CDP **6** (CDP/SF composite, Figure 10.2k). CDP/SF composite showed good radical scavenging characteristics, suggesting the potential of derived scaffold in tissue engineering and therapeutic applications.

10.3.1.2 The Effect of Molecular Assembly Conditions

The molecular assembly of peptides is normally performed in aqueous solvents. Safiullina et al. reported the insight of the molecular assembly of cyclo(L-Leu-L-Leu) [33]. It forms different structures with changing solvent medium. For the preparation of gel, CDP was added to an organic solvent or a mixture of organic solvents with water. It was observed that excess water can form gels with linear (*n*-hexane and *n*-heptane) and cyclic alkanes (cyclohexane). However, in aromatic solvent (toluene), cyclo(L-Leu-L-Leu) forms organogels without the water addition and without the influence of any mechanistic and thermal stimuli. It was mentioned that instead of water, the hydrogen-bonding solvents such as alcohol can be used to initiate the gel formation. Solvent and CDP concentration both play crucial roles in the formation of assembly architectures. Cyclo(L-Ile-L-Ala) assembly was susceptible to many factors such as solvent and concentration of CDP compared to the assembly of its linear dipeptide (L-Ile-L-Ala). Cyclo(L-Ile-L-Ala) assembled into ordered 2D plates in hexafluoro-isopropanol (HFIP) and at high concentration assembled into larger plates. Upon changing solvent from HFIP to methanol, cyclo (L-Ile-L-Ala) formed fibrous architectures [23]. CDP chiral auxiliary-functionalized naphthalene diamides (NCDPs) and their solvent-dependent helical assembly and chiroptical switching (M- to P-type) have been confirmed through a series of concentration-, temperature-, and solvent-dependent circular dichroism spectroscopy (CD) studies [34]. NCDPs form M-helical assembly in HFIP which undergoes chiral inversion to form P-helical assembly architecture in DMSO/HFIP. These observations reveal that the molecular assembly of CDPs can be engineered to produce desired nanoarchitectures by judiciously choosing constituent amino acids with variable α-substituents and appropriate experimental conditions.

10.3.2 CDP Molecular Assembly for Various Applications

CDP derivatives have attracted researchers' interest because of their exceptional molecular assembly and gelation behavior [10]. CDP-derived gels are suitable for drug delivery applications because of their tunable, reversible, and flexible assembly and drug encapsulation properties. In addition, CDP-derived assembly architectures and gels exhibit excellent optical, mechanical, and electrical properties [24, 25].

10.3.2.1 CDP Amphiphiles for Drug Encapsulation and Delivery

CDP amphiphiles (CDPAs) are excellent candidates for the gelation of organic, aqueous, and ionic solvents. The CDP hydrogels have shown stability in a variety of environments owing to the robust molecular interactions within the derived architectures [35]. Introduction or modification of certain functional groups on CDP scaffolds can serve as a handle to modulate molecular assembly through controlled van der Waals forces, π–π interactions, and hydrophobic and hydrophilic interactions. For instance, CDPs with simple protecting groups such as t-Boc and Fmoc can facilitate the gelation ability [21, 28, 36]. The structural fine-tuning and controlled molecular assembly of CDPs impart dynamic and stimulus-responsive properties to the derived gels, which is useful for drug delivery applications [14].

In order to investigate the molecular gelation in organic and aqueous solvents, four CDP derivatives were designed and synthesized: cyclo(L-Phe-L-Asp(OtBu), cyclo(L-Leu-L-Glu), cyclo(L-Tyr-L-Glu (OtBu), and cyclo(L-Phe-L-Glu (OtBu) (Figure 10.3e) [21]. Cyclo(L-Phe-L-Glu (OtBu)) was found to be a supergelator

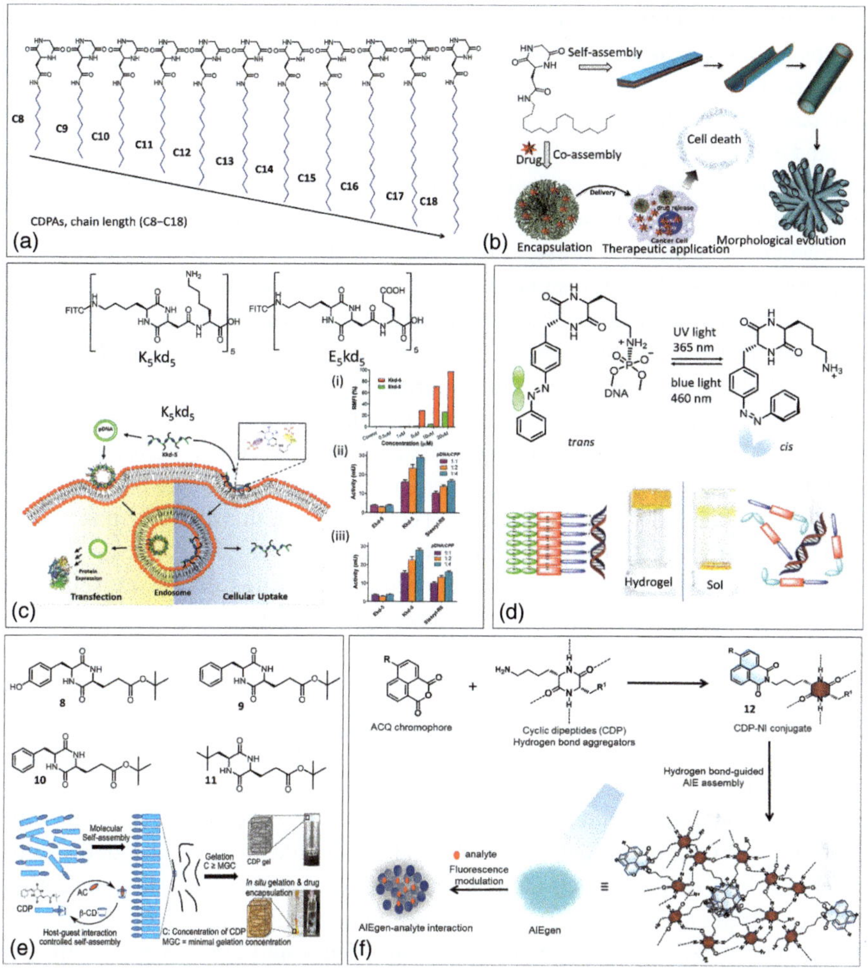

Figure 10.3 (a, b) Illustrative structures of CDPAs (**C8–C18**), and different architectures, and use of CDPA architecture as drug delivery vehicle for cancer cell death. *Source:* Panel (a) is adapted with permission. Copyright 2020, American Chemical Society [35]. (c) Chemical structures of CDP peptidomimetics K_5kd_5 and E_5kd_5. Schematic representation showing cellular uptake, DNA delivery, and transfection efficiency of K_5kd_5: (i) Cellular uptake study. (ii, iii) *In vitro* transfection efficiencies of pDNA (p-CMV-SPORT-β-gal) delivered with K_5kd_5, E_5kd_5, and Stearyl-R_8 in HeLa cells and HEK293 cells, determined using β-galactosidase activity against varying pDNA:CPP ratios (1:1, 1:2, and 1:4). *Source:* Panel (c i, ii, iii) is adapted with permission. Copyright 2017, Royal Society of Chemistry. (d) Schematic representation of the switching of PAP-DKP-Lys from self-assembled hydrogels to liquid upon irradiation of UV light. *Source:* Panel (d) Reproduced by permission of The Royal Society of Chemistry [20]. (e) CDP gelators **8–11** and schematic representation of self-assembly displaying the gel formation in the presence of guest molecules β-cyclodextrin. *Source:* Panel (e) is adapted with permission. Copyright 2017, American Chemical Society [21]. (f) The synthesis scheme of CDP-NMI conjugate **12** and AIEgenic assembly behavior, usage of the detection of NAC (picric acid) and phenolic drugs (doxorubicin and rifampicin) in aqueous media. ACQ is aggregation-caused quenching. *Source:* Panel (f) is adapted with permission. Copyright 2019, American Chemical Society [37].

with *in situ* hydrogelation ability. Such hydrogels with good cytocompatibility and drug encapsulation are potential candidates for drug delivery and controlled release [21, 38].

CDP-based molecular architectonics strategy was utilized to develop cytocompatibility and efficient drug delivery systems [35].

Several CDPAs (**C8–C18**) were designed and synthesized by conjugating cyclo(Gly-L-Asp) with alkyl chains of varying lengths (n = C8-C18) as shown in Figure 10.3a. The molecular assembly of these CDPAs was engineered to construct a range of nanoarchitectures such as nanospheres, nano-/microsheets, nanotubes, and flowers in organic or aqueous solvents. The formation of various molecular assembly architectures was affected by the hydrogen-bonding (CDP core) and hydrophobic (alkyl chains) interactions. The elasticity and hardness of the derived materials' architectures can be tuned by varying the length of alkyl chains. The co-assembly of **C15** with BrdU was demonstrated as a successful anticancer delivery system (Figure 10.3b). The cellular study of co-assembly architectures of **C15**-BrdU showed BrdU-triggered cellular cytotoxicity to HeLa cells. Further, this study demonstrated that cytocompatibility CDPAs are effective molecular building blocks for the construction of novel molecular and material architectures driven by self- and co-assembly with implications in drug delivery applications.

In another study, unnatural CDP amino-acid-incorporated peptidomimetics were designed and studied for cell-penetrating ability and DNA delivery applications [20]. CDP peptidomimetics (K_5kd_5 and E_5kd_5) were synthesized by incorporating unnatural amino acid kd (a lysine and aspartic acid-based CDP) at alternative positions in lysine (K) and glutamic acid (E) decapeptides, respectively (Figure 10.3c). The cationic charges from five K units and rigidity ascribed by kd units rendered K_5kd_5 with high serum stability, bioavailability, and excellent cellular uptake and delivery of DNA cargos into the HeLa cell compared to the known cationic cell-penetrating peptide (CPP) such as stearyl R_8. E_5kd_5 was found ineffective for DNA delivery applications due to its anionic nature. The detailed small-angle X-ray scattering (SAXS) and nuclear magnetic resonance (NMR) studies have shown that K_5kd_5 interacts with membrane lipid and DNA through specific hydrogen bonding of backbone CDP units (kd) as well as electrostatic interactions provided by the K units. Such noncovalent interactions of K_5kd_5 with DNA and lipid membrane mediated selective complexation and effective delivery of DNA through the membrane into cells in comparison to nonspecific interaction-driven delivery by cationic CPPs and polymers. The concentration-dependent cellular uptake study in HeLa cells revealed that 97% K_5kd_5 was taken up by cells at 20 μM, while E_5kd_5 uptake was insignificant (Figure 10.3c i). The delivery of DNA cargo (pCMV-SPORT-β-gal) in HeLa cells and HEK293 was validated by the β-galactosidase activity (Figure 10.3c ii, iii). This study demonstrated the significant role of CDP in DNA complexation, cellular delivery, and effective transfection. Overall, CDP-incorporated peptidomimetics can be utilized as molecular platforms for the delivery of therapeutics into cells.

CDPs coupled with azobenzene had the capacity to form hydrogenation by the assembly, and the hydrogels showed an ionic strength-dependent behavior [39].

Self-healing biomolecular hydrogels showed light-dependent releasing properties after entrapping DNA oligomers (Figure 10.3d). The hydrogels deteriorated and dissolved with a slight shaking of the vial, after 30 minutes of UV exposure. The light-stimulated photoisomerization disturbed the π-stacking between CDP molecules and made the fibers scatter, and finally, the gels transformed into liquid. Correspondingly, the anticancer drug doxorubicin can be entrapped inside the hydrogels and released in a light-dependent manner. Thus, CDPs assembly-based low-molecular-weight hydrogels show reversible and stimuli-responsive behavior in comparison to conventional polymer gels and are the potential for drug delivery.

Four CDPs, cyclo(L-Tyr-L-Glu (OtBu)) (**8**), cyclo(L-Phe-L-Glu (OtBu)) (**9**), cyclo(L-Phe-L-Asp(OtBu)) (**10**), and cyclo(L-Leu-L-Glu(OtBu)) (**11**), were designed, synthesized, and evaluated for gelation ability [21]. CDP **10** was found to be a supergelator to form cytocompatibility and thermostable ambidextrous gels with a minimum gelation concentration of 0.3 wt% (Figure 10.3e). The self-assembled 2D fibrous networks of **10** in the hydrogel become stable by intermolecular N–H---O hydrogen bonding, while t-Boc and phenyl group provided π–π stacking and hydrophobic interactions. Notably, CDP **10** produced an injectable hydrogel entrapped with curcumin under ambient conditions, which proved its potential use in drug delivery-related applications.

A set of CDP-tethered naphthalimides (CDP-NIs) was designed to extend CDP auxiliary-driven chromophore assembly architecture strategy [37]. The CDP units are linked to NI core to explore the aggregation-induced emission (AIE) characteristics of the latter. The fascinating intermolecular hydrogen-bonding property of the CDP auxiliary was expected to control the assembly of NI chromophore and transform it from aggregation-caused quenching (ACQ) to aggregation-induced emission (AIE) state in aqueous media. Remarkably, CDP-NI **12** showed maximum AIE quenching in the presence of picric acid, with a limit of detection (LOD) of 0.4 ppm (1.7 μM). The AIEgenic characteristics of CDP-NI **12** are employed for the detection of phenolic drugs in an aqueous medium. CDP-NI **12** detects doxorubicin and rifampicin with LOD in the range of 18 nM (9.7 ppb) and 202 nM (164 ppb), respectively. This study demonstrated the CDP-modulated AIEgenic characteristics of chromospheres, which inspire the development of smart sensor systems for a variety of bioactive analytes.

10.3.2.2 CDP-based Catalytic Architecture

The self-assembly of amyloid-forming short peptides and peptidomimetics generates assembly architectures in the presence of metal ions with catalytic activity in various chemical transformations. In a very recent study, Aβ peptide-based CDP peptidomimetics were judiciously designed and studied in their metal-mediated assembly to produce differential architectures with distinguishable catalytic activities [40]. Peptidomimetics (Akd^{NMC} and Akd^{NMC}Py) of Aβ14–23 incorporated with unnatural CDP amino acid cyclo(L-Lys–L-Asp) (kd) and pyrene (Py) units at specific positions are shown in (Figure 10.4a). Aβ14–23 (^{14}HQKLVFFAE^{23}D) has a self-recognition motif KLVFF and metal-binding affinity. The assembly of Akd^{NMC}Py

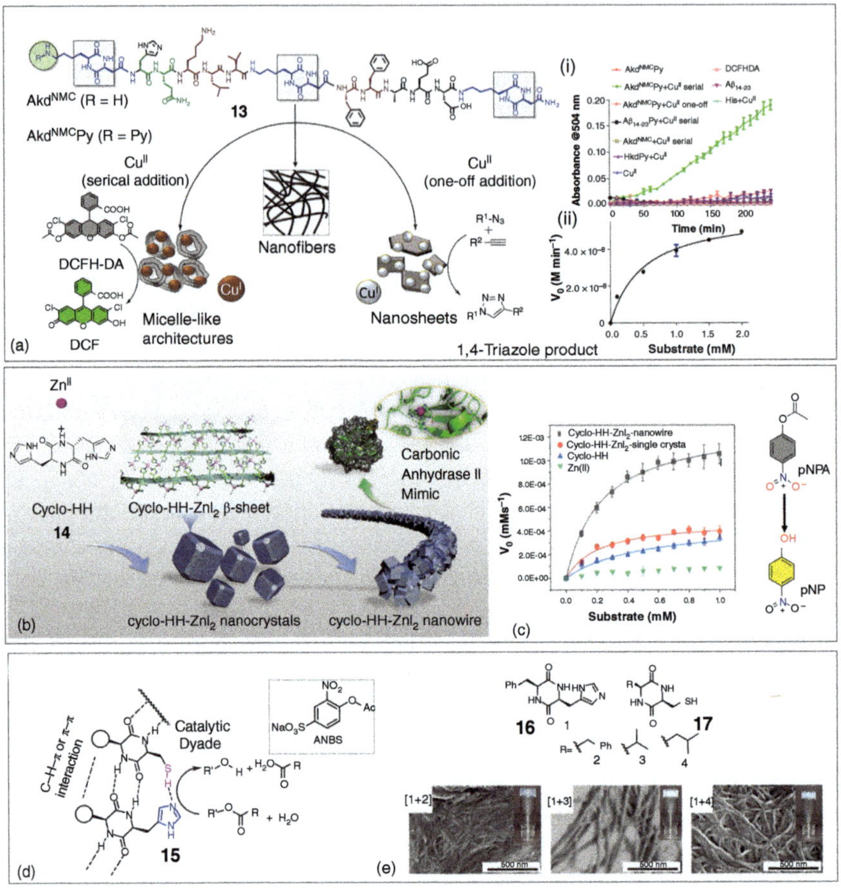

Figure 10.4 (a) Molecular assembly of Akd^{NMC}Py upon one-off and serial additions of Cu^{II} to produce distinct architectures that stabilize Cu^{II} and Cu^{I} states with catalytic activity in tandem oxidative hydrolysis of DCFDA (2′,7′-dichlorofluorescein diacetate) and the alkyne–azide click reaction, respectively. *Source:* Panels (a i, ii) are adapted with permission. Copyright 2022, Royal Society of Chemistry [40]. (b, c) Molecular self-assembly process of cyclo-HH and the kinetic assay of pNPA hydrolysis by cyclo-HH-ZnI_2 nanowires, cyclo-HH-ZnI_2 single crystal, only cyclo-HH, and Zn (II). *Source:* Panels (b, c) are adapted with permission. Copyright 2021, Wiley-VCH [41]. (d, e) Histidine- and cysteine-containing cyclic peptide-based sodium 4-acetoxy-3-nitrobenzene sulfonate (ANBS) hydrolysis. Nanofibrillar morphology of co-assemblies showed by TEM images. *Source:* Panels (d, e) reproduced from the Royal Society of Chemistry, Copyright 2020 [42].

(15 μM) was studied by the addition of Cu^{II} in two distinct ways. In one-off addition, the assembly of Akd^{NMC}Py is guided by the complexation and reduction of Cu^{II} to Cu^{I} to generate nanosheet (thermodynamically stable) architectures. Furthermore, serial addition headed to the precise complexation of Akd^{NMC}Py to Cu^{II} subsequent in step-by-step or cascade assembly processes, which forms metastable (kinetically controlled) micelle-like core–shell architectures. These meta-stable architectures ultimately altered into thermodynamically stable nanosheets upon complete

complexation and reduction of Cu^{II} to Cu^{I} state. This observation allowed us to exploit their functional activities toward Cu^{II} and Cu^{I}-catalyzed chemical transformations. The catalytic utility of micelle-like architectures of $Akd^{NMC}Py + Cu^{II}$ was investigated in a tandem reaction involving oxidative hydrolysis of 2′,7′-dichlorofluorescein diacetate (DCFH-DA). The transformation of DCFH-DA into DCF was observed with an initial rate of 8.7 nM min^{-1}, and the calculated K_M and catalytic efficiency (k_{cat}/K_M) were 0.5 mM and 120 M^{-1} min^{-1}, respectively (Figure 10.4a, i). The activity of nanosheets of $Akd^{NMC}Py$ in Cu^{I}-catalyzed azide-alkyne cycloaddition (CuAAC) reaction was evaluated. It was found that $Akd^{NMC}Py$-Cu^{I} catalyzes the CuAAC reaction faster than the standard $CuSO_4$ + sodium ascorbate catalyst system (Figure 10.4a, ii). This work demonstrated that a designer peptidomimetics and its differential assembly architectures can modulate the oxidation states of metal (Cu) and catalyze multiple chemical transformations.

Recently, another approach is reported on developing a catalytic system with small peptide sequences that can recreate the hydrolase activity. The histidine-containing CDP (cyclo-HH) gained access to a nanowire superstructure (cyclo-HH-ZnI_2 nanowire) in the presence of Zn (II) (Figure 10.4b) [41]. Self-assembled nanostructures with rich catalytically active sites displayed rapid hydrolytic activity up to 93.21 M^{-1} s^{-1} (Figure 10.4c). Additionally, the author also mentioned that this cyclo-HH-ZnI_2 catalyst was also found to be stable and reusable, which is of noteworthy in the field of small molecule-based chemical transformations.

CDP-guided catalytic assembly was investigated by Nachtsheim and coworkers [42]. The heterologous combination of His (1) and Cys (2, 3, and 4) containing cyclic peptides produced a self-assembling gel via C–H–π and noncovalent hydrogen-bond formation, which is supported by Hartree–Fock calculations and HRMAS NOE experiments (Figure 10.4d,e). Sodium 4-acetoxy-3-nitrobenzenesulfonate (ANBS) was used as a substrate to examine the esterase activity by coaggregates (1 + 2, 1 + 3, and 1 + 4) individually. The proximity of the dyad consisting of cysteine and histidine facilitated the catalytic activity of each coaggregate. Taking inspiration from the histidine-enriched catalytic surface and manifesting hydrolytic activity, Marchesan and colleagues reported the tetrapeptides D-Ser-D-His-L-Phe-L-Phe and D-Ser-D-His-L-Phe-NH_2 [43]. These studies have shown that CDP-based architectures can successfully catalyze different biochemical transformations.

10.3.2.3 CDP Assembly: Optical Waveguiding

Assembly architectures of CDP are demonstrated with novel physical characteristics including optical waveguiding effects and piezoelectric capabilities that were not present in the monomers [44–46]. CDPs are capable of delivering intriguing materials for optical waveguiding because of their superior optical characteristics and elongated morphology of the derived nanoarchitectures.

The molecular packings showed that the lattice of the cyclo(L-Phe-L-Phe) crystal was made of four cyclo(L-Phe-L-Phe) molecules. The π–π stacking interactions between cyclo(L-Phe-L-Phe) molecules play a crucial role during the formation of an extended chain in the direction of longitudinal axis. In cyclo(L-Phe-L-Phe) molecule, there is only one carbonyl group that participates in hydrogen-bonding interactions. The association of π-stacking and hydrogen-bonding interactions resulted in the crystalline structures as hierarchically oriented crystalline platelets (Figure 10.5a) [44, 48]. The cyclo(L-Phe-L-Phe) assembly architectures emitted blue photoluminescence upon excitation at 330–380 nm. It was stated that the platelets' bodies displayed somewhat weaker emission, while very bright photoluminescent spots emerged at both ends, which was a classic optical waveguiding phenomenon (Figure 10.5b). There were prominent photoluminescent red patches only at the ends of the platelets following incorporation of Nile Red into the crystalline structures and upon excited at 515–560 nm. The middle of the platelets displayed considerably less luminescence. A solvothermal treatment technique was subsequently created to speed up the crystallization and intramolecular cyclization of linear dipeptide. Cyclo(L-Phe-L-Phe) produced ultralong peptide crystalline nanobelts. Intriguingly, optical waveguides with regular straight and curved shapes were seen in the peptide crystal nanobelts, which have a thickness of hundreds of micrometers. After the incorporation of Nile Red in the nanobelts, it showed optical waveguide characteristics.

Subsequently, Tao et al. reported a series of CDPs based on Trp and investigated the physical characteristics of the assembled structures. Cyclo(Gly-L-Trp) is very much able to assemble into monoclinic crystals. The backbone of diketopiperazines of cyclo(Gly-L-Trp) was orientated antiparallelly by hydrogen bonding. The pyrrole rings of the indole molecules form herringbone type of aromatic interaction. Upon excitation between 300 and 400 nm, cyclo(Gly-L-Trp), **14** crystals exhibited waveguiding of blue light emission, but cyclo(Gly-L-Trp) monomer emission could be observed in the UV region [45]. Additionally, upon excitation, the assemblies displayed stronger emission at the terminals edge (Figure 10.5c), which suggested typical optical waveguiding capabilities. However, the cyclo(L-Phe-L-Trp), **15** backbone CDPs offered several hydrogen-bonding contacts in parallel structure. Additionally, the significant aromatic contacts in the cyclo(L-Phe-L-Trp) crystallographic structure between phenyl rings/phenyl and indole rings led to the creation of a hydrophobic zone. The substantial aromatic contacts and hydrogen bonding that occurred during molecular assembly rendered CDP-based materials with exceptional thermal stability and optical waveguiding abilities. Cyclo(L-Phe-L-Trp) crystals showed wide band visible photoluminescence with emission varying from 395 to 550 nm. Interestingly, the intensity of the emitted light can be tuned by modulating the excitation wavelength (Figure 10.5c, d) [49].

10.3.2.4 CDP-based Quantum-confined Materials

Small peptides have the ability to self-assemble into nanoscale quantum-confined structures through hydrogen-bonding and hydrophobic interactions. Numerous investigations have shown that short peptides containing aromatic amino acids such as Phe and Trp produce a variety of nanoarchitectures with interesting physiochemical features (**16–17**) [50–52]. The Trp-containing CDPs formed quantum-restricted microstructures through molecular assembly of building block cyclo(L-Trp-L-Trp), which dimerize into quantum dots. The π-stacking interactions between indole rings of Trp drive the dimerization process. The molecular

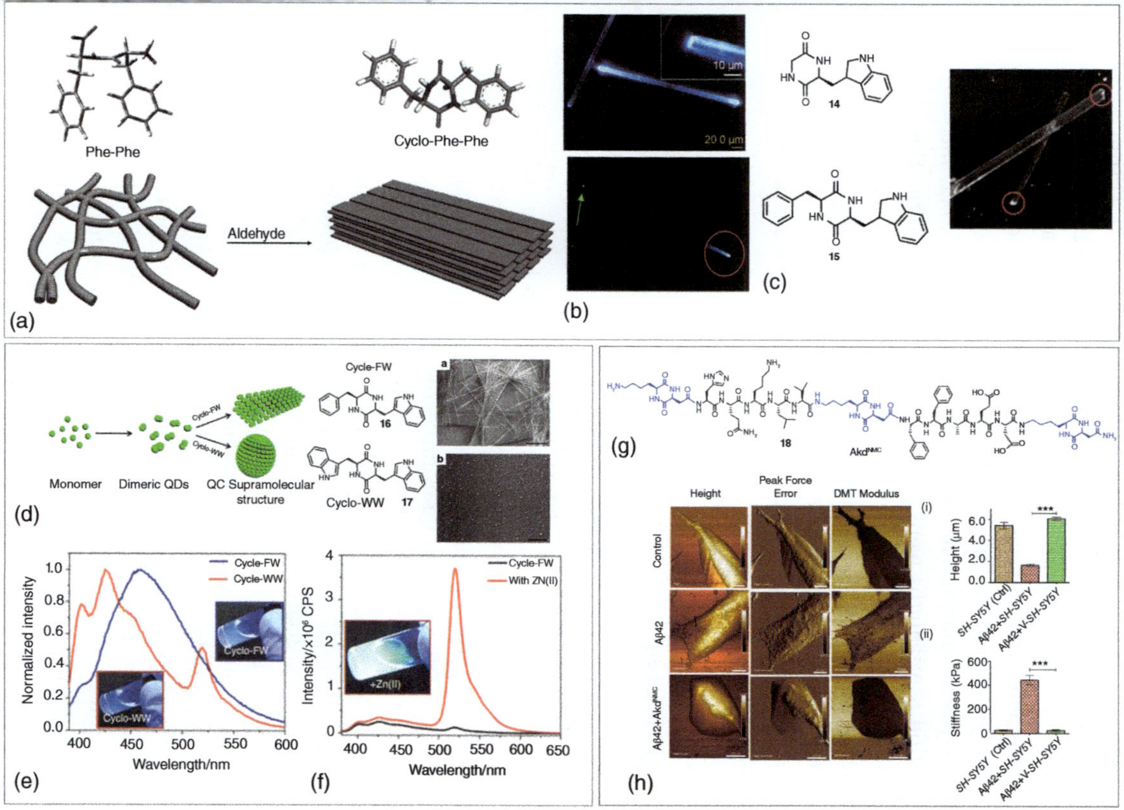

Figure 10.5 (a) Schematic presentation of the development of organized, lengthy, crystalline platelets established on self-assembling fibrous networks. The formation of cyclo(L-Phe-L-Phe) and phase transition from gels to crystals can both be facilitated by the incorporation of aldehyde, which can also produce the cyclization of diphenylalanine. (b) Self-assembled cyclo(L-Phe-L-Phe) crystalline platelets' ability to guide optical waves. A photoluminescence image of a single platelet shows greater blue emission at both ends than the platelet body (the top), upon excitation between 330 and 380 nm. The green arrow indicates the outcoupling of photoluminescence emission at the opposite end, and the red circle denotes the excitation area (the down). (c) The structure of cyclo(Gly-L-Trp) and fluorescent image of peptide crystals. The brighter light with optical waveguiding properties is represented by the red circle. *Source:* Panels (a–c) reproduced with permission Copyright 2011 John Wiley and Sons [44]. (d) Cyclic dipeptide (CDP) monomers self-assembly for quantum-confined architectures; scanning electron microscopy shows needle-like cyclo(L-Phe-L-Trp) crystals and cyclo(L-Trp-L-Trp) nanospheres; and their molecular structures. (e, f) The fluorescent emission of cyclo(L-Phe-L-Trp) and cyclo(L-Trp-L-Trp) self-assemblies and the fluorescent emission of cyclo(L-Trp-L-Trp) + $Zn^{(II)}$. *Source:* Panels (d–f) reproduced with permission Copyright 2019 Springer Nature [46]. (g–j) Chemical structure of CDP (kd)-inserted peptidomimetic and PF QNM-AFM images of SH-SY5Y cells in the height, peak force error, and DMT modulus modes under both normal and amyloid-induced stress conditions. (h) PBS, pH 7.4 (control), Aβ42, and Aβ42 + 18 were treated to the cells. (i, j) Stress fibers are depicted by arrows in the Aβ-42-treated cells. (b, c) From the PF QNM-AFM data, height and mechanical stiffness were calculated. Number of experiments = 3, mean SD, $^{*}P = 0.001$, number of experiments. *Source:* Panels (g–j) reproduced from the Royal Society of Chemistry, Copyright 2022 [47].

assemblies (dimeric architectures) displayed intrinsic photoluminescent property due to molecular restriction and significant quantum-confined effects. The photoluminescence can be modulated from visible to the near-infrared (NIR) range by varying solvent, metal ion coordination, stereochemistry, and amino-acid composition, which impact the molecular assembly process. Cyclo(L-Trp-L-Trp) showed fluorescence in the visible region with a smaller peak at 520 nm, when excited at 425 nm (Figure 10.5e). The emission intensity of cyclo(L-Trp-L-Trp) was significantly increased after coordination with Zn ions, showing a narrow peak at 520 nm (Figure 10.5f). The nanosphere structure and both visible and NIR fluorescence were displayed by cyclo(L-Trp-L-Trp) assemblies after changing from methanol to dimethyl sulfoxide (DMSO) as the solvent. The emission peaks were at 615, 712, and 780 nm when the excitation wavelengths were adjusted from 520 to 780 nm. The self-assembled nanostructure changed when other aromatic moieties, such as cyclo(L-Tyr-L-Tyr), cyclo(L-His-L-His), and distinct self-assembling morphologies were displayed with different photoluminescence. The constructed structures based on CDPs with broad-spectrum emission properties can be employed for the building of light-emitting diodes.

10.4 Biological Activities

10.4.1 Neuroprotective CDPs

Cyclo(His-Pro) is a known neuroprotective agent with blood–brain barrier (BBB) permeability [53]. Specifically, cyclo(His-Pro) has been shown to lower oxidative and endoplasmic reticulum (ER) stress and is an anti-inflammatory agent. Inspired by the BBB crossing ability of CDPs and their neuroprotective effects, the rationally designed CDP-based molecules were discovered as neuro therapeutics agents [18].

Tau is a microtubule-associated protein containing intrinsically disordered regions (IDRs) that aggregate and undergo LLPS (liquid–liquid phase separation) to produce pathogenic species. The accumulation of intracellular tau tangles (AD) causes the Alzheimer's disease [54–56]. Few chemical strategies are available to control protein LLPS, which could be used to treat neurodegenerative disorders. Recently, CDP-based small-molecule modulators (SMMs) consist of multiple chemical functionalities that offer various chemical interactions to modulate the chemical grammar of tau LLPS [19]. The lead CDP containing Trp and Zn(II)-binding moiety effectively inhibits and dissolves Zn-mediated tau LLPS condensates and tau aggregation. The approach of designing SMMs of LLPS provides a new platform with possible implications for the development of therapeutics for neurodegenerative diseases [19].

It was believed that the amalgamation of rigid, proteolytically stable CDP units into Aβ14–23 at definite positions would overcome the boundaries such as proteolytic instability, flexibility issues, and cytotoxicity, associated with peptide-based amyloid modulators [47, 57]. The excellent intermolecular hydrogen-bonding ability and bioactivity of CDP are expected to modulate amyloidogenic stress and associated adverse cellular mechanics through interactions with Aβ monomers or aggregation species. Aβ14–23 peptidomimetics incorporated with CDP units (kd) efficiently combat amyloid-caused membrane toxicity, excessive ROS production, and unfavorable cellular mechanics under stress conditions. It has been shown that Aβ14–23 peptidomimetics with three kd units at the N- and C-termini and middle (Akd^{NMC}, **18**) is a potent inhibitor of amyloid aggregation of Aβ14–23 as well as Aβ42 (Figure 10.5g). Incorporation of kd units at the middle and C-terminal proximal to the FF region of Aβ14–23 was found to show effective aggregation-inhibition ability. Meanwhile, a single kd unit at the N-terminus distant from the FF region turns out to be an aggregation promoter, due to its self-aggregation behavior. The mechanical properties of neuronal cells (SH-SY5Y cells) including height and DMT (Derjaguin–Muller–Toporov) modulus were measured using advanced PeakForce Quantitative NanoMechanics-AFM (PFQNM AFM) technique [52, 58]. The PFQNM AFM images of untreated cells (control) exhibited smooth topography with a height profile of $5.4 \pm 0.6\,\mu m$, and mechanical stiffness of $30 \pm 5\,kPa$, and stress fiber formation was not observed (Figure 10.5h–j). After treatment with Aβ42, the cells deformed (height: $1.6 \pm 0.2\,m$) with increased stiffness of $417 \pm 39\,kPa$, and formation of extensive stress fibers networks was observed (indicated by the arrows). The emergence of stress owing to Aβ-membrane interactions is indicated by the stress-fibers-containing deformed cellular characteristics with rigid mechanics (high stiffness). Remarkably, Aβ42 + Akd^{NMC}-treated cells displayed cytoskeletal organization (height $6.4 \pm 0.4\,\mu m$), stiffness $21 \pm 9\,kPa$, and cell shape comparable to the control cells. The PF QNM-AFM study showed that the neuronal cells were developed with healthy cell stiffness and other features in the presence of Akd^{NMC} under amyloid-induced stress conditions. The enzymatic stability, biocompatibility, and non-self-aggregating properties combined with the efficient modulation of Aβ aggregation and ROS generation thus demonstrated Akd^{NMC} (**18**) as a most probable candidate to ameliorate amyloid-induced toxicity and adverse cellular mechanics.

10.4.2 Antitumor activity

There has been a lot of interest in CDPs and their derivatives as potential antitumor agents [11]. The natural CDPs derived from fungus and bacteria have been tested for their cytotoxicity in various cancer cell lines, and some of them have proven to have antiproliferative effects. There are several CDPs reported, viz. cyclo(Gly-L-Val), cyclo(Gly-D-Val), cyclo(Gly-L-Leu), cyclo(Gly-L-Ile), cyclo(L-Phe-L-Cys), and cyclo(L-Tyr-L-Cys), to exhibit antitumor activity effect on MCF-7, Hela, and HT-29 cells. [59, 60] Among them, cyclo(L-Tyr-L-Cys) showed better antitumor activity compared to others. Notably, there are drug candidates based on CDP scaffolds already in the advanced stage of clinical trials or approved. Interestingly, phenylahistin-based drug plinabulin (NPI-2358) is undergoing phase-III trial for chemotherapy-induced neutropenia (CIN) and non-small cell lung carcinoma (NSCLC) [11, 61, 62].

10.4.3 Antivirus activity

It has been reported that DKP inhibits the viral hemorrhagic septicemia virus [63]. Spiro-DKP is an example of a CDP derivative that was used to construct the chemokine receptors for selective C–C chemokine receptor type 5 (CCR5) antagonists, which exhibit potential antiviral effects [64]. The 1,4,9-triazaspiro [5.5] undecane-2,5-dione template was used to build novel compounds due to the equivalence to spiro-piperidine advantaged structures in several G-protein-coupled receptor (GPCR) ligands. Some of the candidates showed strong antiviral activity *in vitro* against multidrug-resistant HIV strains [63]. Aplaviroc **6** stood out among them due to its strong affinity for CCR5, potent antagonistic action against a number of R5-HIV-1 isolates, and is in a clinical trial for AIDS treatment [65].

10.5 Conclusions

An overview of CDPs and their assembly material architectures with potential applications is presented. CDPs exhibit a variety of bioactivities, including neuroprotective, anticancer, antiviral, antibacterial, and antioxidant activities. CDPs are excellent modular building blocks for constructing functional molecular and material architectures through the scheme of molecular architectonics. CDPs assembly architectures have the potential to be used in biomaterials, therapeutics, drug delivery, catalysis, sensors, high-strength organic materials, quantum-confined devices, and optical waveguiding, among others. Thus, molecular architectonics of CDPs serve as a promising avenue for expanding their use in nanotechnology and biomedicine.

Acknowledgments

The authors thank JNCASR, CEFIPRA (grant: IFC/A/62T10-3/2020/963), DST, SERB (grant: CRG/2020/004594), New Delhi, Government of India, for financial support.

References

1 Avinash, M.B. and Govindaraju, T. (2018). Architectonics: design of molecular architecture for functional applications. *Acc. Chem. Res.* 51: 414–426.
2 Ghosh, D., Datta, L.P., and Govindaraju, T. (2020). Molecular architectonics of DNA for functional nanoarchitectures. *Beilstein J. Nanotechnol.* 11: 124–140.
3 Roy, B. and Govindaraju, T. (2019). Amino acids and peptides as functional components in arylenediimide-based molecular architectonics. *Bull. Chem. Soc. Jpn.* 92: 1883–1901.
4 Yu, G., Jie, K., and Huang, F. (2015). Supramolecular amphiphiles based on host–guest molecular recognition motifs. *Chem. Rev.* 115: 7240–7303.
5 Roy, B. and Govindaraju, T. (2022). Biomolecules-guided molecular architectonics to nanoarchitectonics. *Concepts Design Mater. Nanoarchitect.* 337–360.

6 Govindaraju, T. and Ariga, K. (2021). *Molecular Architectonics and Nanoarchitectonics (Nanostructure Science and Technology)*. Singapore: Springer Nature.

7 Konar, M., Roy, B., and Govindaraju, T. (2020). Molecular architectonics-guided fabrication of superhydrophobic and self-cleaning materials. *Adv. Mater. Interfaces* 7: 2000246.

8 Avinash, M. and Govindaraju, T. (2012). Amino acid derivatized arylenediimides: a versatile modular approach for functional molecular materials. *Adv. Mater.* 24: 3905–3922.

9 Datta, L.P., Ghosh, D., and Govindaraju, T. (2022). Functional Molecule-Templated DNA Molecular Architectonics. In: *Molecular Architectonics and Nanoarchitectonics*, 281–305. Springer.

10 Manchineella, S. and Govindaraju, T. (2017). Molecular self-assembly of cyclic dipeptide derivatives and their applications. *ChemPlusChem* 82: 88–106.

11 Balachandra, C., Padhi, D., and Govindaraju, T. (2021). Cyclic dipeptide: a privileged molecular scaffold to derive structural diversity and functional utility. *ChemMedChem* 16: 2558–2587.

12 Konar, M. and Govindaraju, T. (2022). Molecular architectonics. In: *Molecular Architectonics and Nanoarchitectonics*, 3–34. Springer.

13 Govindaraju, T., Pandeeswar, M., Jayaramulu, K. et al. (2011). Spontaneous self-assembly of designed cyclic dipeptide (Phg-Phg) into two-dimensional nano-and mesosheets. *Supramol. Chem.* 23: 487–492.

14 Avinash, M., Raut, D., Mishra, M.K. et al. (2015). Bioinspired reductionistic peptide engineering for exceptional mechanical properties. *Sci. Rep.* 5: 16070.

15 Zhao, K., Xing, R., and Yan, X. (2021). Cyclic dipeptides: biological activities and self-assembled materials. *J. Pept. Sci.* 113: e24202.

16 Corey, R.B. (1938). The crystal structure of diketopiperazine. *J. Am. Chem. Soc.* 60: 1598–1604.

17 Govindaraju, T. (2011). (2011) Spontaneous self-assembly of aromatic cyclic dipeptide into fibre bundles with high thermal stability and propensity for gelation. *Supramol. Chem.* 23: 759–767.

18 Borthwick, A.D. (2012). 2, 5-Diketopiperazines: synthesis, reactions, medicinal chemistry, and bioactive natural products. *Chem. Rev.* 112: 3641–3716.

19 Ramesh, M., Balachandra, C., Baruah, P., and Govindaraju, T. (2022). Cyclic dipeptide-based small molecules modulate zinc-mediated liquid-liquid phase separation of tau. *J. Pept. Sci.* 29 (5), e3465.

20 Madhu, C., Voshavar, C., Rajasekhar, K., and Govindaraju, T. (2017). Cyclic dipeptide based cell-penetrating peptidomimetics for effective DNA delivery. *Org. Biomol. Chem.* 15: 3170–3174.

21 Manchineella, S., Murugan, N.A., and Govindaraju, T. (2017). Cyclic dipeptide-based ambidextrous supergelators: minimalistic rational design, structure-gelation studies, and in situ hydrogelation. *Biomacromolecules* 18: 3581–3590.

22 Sánchez-Navarro, M. and Giralt, E. (2022). Peptide shuttles for blood–brain barrier drug delivery. *Pharmaceutics* 14: 1874.

23 Tao, K., Fan, Z., Sun, L. et al. (2018). Quantum confined peptide assemblies with tunable visible to near-infrared spectral range. *Nat. Commun.* 9: 3217.

24 Roy, B., Ghosh, D., and Govindaraju, T. (2021). Molecular-architectonics-guided dynamic assembly to generate fluorescent organic nanoclusters with implications for optical imaging. *ACS Appl. Nano Mater.* 4: 979–984.

25 Shende, V.V., Khatri, Y., Newmister, S.A. et al. (2020). Structure and function of NzeB, a versatile C–C and C–N bond-forming diketopiperazine dimerase. *J. Am. Chem. Soc.* 142: 17413–17424.

26 Moorthy, H., Datta, L.P., and Govindaraju, T. (2021). Molecular architectonics-guided design of biomaterials. *Chem. Asian J.* 16: 423–442.

27 Scarel, M. and Marchesan, S. (2021). Diketopiperazine gels: new horizons from the self-assembly of cyclic dipeptides. *Molecules* 26: 3376.

28 Manchineella, S. and Govindaraju, T. (2012). Hydrogen bond directed self-assembly of cyclic dipeptide derivatives: gelation and ordered hierarchical architectures. *RSC Adv.* 2: 5539–5542.

29 Manchineella, S., Voshavar, C., and Govindaraju, T. (2017). Radical-scavenging antioxidant cyclic dipeptides and silk fibroin biomaterials. *Eur. JOC* 2017 (30), 4363–4369.

30 Govindaraju, T. (2019). *Templated DNA Nanotechnology: Functional DNA Nanoarchitectonics*. CRC Press.

31 Roy, B., Ghosh, D., and Govindaraju, T. (2019). Functional molecule–templated DNA nanoarchitectures. In: *Templated DNA Nanotechnology*, 69–106. Jenny Stanford Publishing.

32 Kleinsmann, A.J. and Nachtsheim, B.J. (2013). Phenylalanine-containing cyclic dipeptides–the lowest molecular weight hydrogelators based on unmodified proteinogenic amino acids. *Chem. Commun.* 49: 7818–7820.

33 Safiullina, A.S., Ziganshina, S.A., Lyadov, N.M. et al. (2019). Role of water in the formation of unusual organogels with cyclo (leucyl–leucyl). *Soft Matter.* 15: 3595–3606.

34 Manchineella, S., Prathyusha, V., Priyakumar, U.D., and Govindaraju, T. (2013). Solvent-induced helical assembly and reversible chiroptical switching of chiral cyclic-dipeptide-functionalized naphthalenediimides. *Chem. A Eur. J.* 19: 16615–16624.

35 Pandurangan, K., Roy, B., Rajasekhar, K. et al. (2020). Molecular architectonics of cyclic dipeptide amphiphiles and their application in drug delivery. *ACS Appl. Bio Mater.* 3: 3413–3422.

36 Khazi, M.I., Balachandra, C., Shin, G. et al. (2020). Co-solvent polarity tuned thermochromic nanotubes of cyclic dipeptide–polydiacetylene supramolecular system. *RSC Adv.* 10: 35389–35396.

37 Balachandra, C. and Govindaraju, T. (2019). Cyclic dipeptide-guided aggregation-induced emission of naphthalimide and its application for the detection of phenolic drugs. *J. Org. Chem.* 85: 1525–1536.

38 Datta, L.P., Manchineella, S., and Govindaraju, T. (2020). Biomolecules-derived biomaterials. *Biomaterials* 230: 119633.

39 Pianowski, Z.L., Karcher, J., and Schneider, K. (2016). Photoresponsive self-healing supramolecular hydrogels for light-induced release of DNA and doxorubicin. *Chem. Commun.* 52: 3143–3146.

40 Ghosh, D., Konar, M., Mondal, T., and Govindaraju, T. (2022). Differential copper-guided architectures of amyloid β peptidomimetics modulate oxidation states and catalysis. *Nanoscale Adv.* 4: 2196–2200.
41 Chen, Y., Yang, Y., Orr, A.A. et al. (2021). Self-assembled peptide nano-superstructure towards enzyme mimicking hydrolysis. *Angew. Chem. Int. Ed.* 133: 17301–17307.
42 Kleinsmann, A.J. and Nachtsheim, B.J. (2020). A minimalistic hydrolase based on co-assembled cyclic dipeptides. *Org. Biomol. Chem.* 18: 102–107.
43 Kurbasic, M., Garcia, A.M., Viada, S., and Marchesan, S. (2022). Heterochiral tetrapeptide self-assembly into hydrogel biomaterials for hydrolase mimicry. *J. Pept. Sci.* 28: e3304.
44 Yan, X., Su, Y., Li, J. et al. (2011). Uniaxially oriented peptide crystals for active optical waveguiding. *Angew. Chem. Int. Ed.* 123: 11382–11387.
45 Tao, K., Hu, W., Xue, B. et al. (2019). Bioinspired stable and photoluminescent assemblies for power generation. *Adv. Mater.* 31: 1807481.
46 Tao, K., Xue, B., Li, Q. et al. (2019). Stable and optoelectronic dipeptide assemblies for power harvesting. *Mater. Today* 30: 10–16.
47 Konar, M., Ghosh, D., Samanta, S., and Govindaraju, T. (2022). Combating amyloid-induced cellular toxicity and stiffness by designer peptidomimetics. *RSC Chem. Biol.* 3: 220–226.
48 Yan, X., Zhu, P., and Li, J. (2010). Self-assembly and application of diphenylalanine-based nanostructures. *Chem. Soc. Rev.* 39: 1877–1890.
49 Yuan, C., Ji, W., Xing, R. et al. (2019). Hierarchically oriented organization in supramolecular peptide crystals. *Nat. Rev. Chem.* 3: 567–588.
50 Gan, Z., Wu, X., Zhu, X., and Shen, J. (2013). Light-induced ferroelectricity in bioinspired self-assembled diphenylalanine nanotubes/microtubes. *Angew. Chem. Int. Ed.* 52: 2055–2059.
51 Esin, A., Baturin, I., Nikitin, T. et al. (2016). Pyroelectric effect and polarization instability in self-assembled diphenylalanine microtubes. *Appl. Phys. Lett.* 109: 142902.
52 Nguyen, V., Zhu, R., Jenkins, K., and Yang, R. (2016). Self-assembly of diphenylalanine peptide with controlled polarization for power generation. *Nat. Commun.* 7: 1–6.
53 Bellezza, I., Peirce, M.J., and Minelli, A. (2014). Cyclic dipeptides: from bugs to brain. *Trends Mol. Med.* 20: 551–558.
54 Govindaraju, T. (2022) Alzheimer's Disease: Recent Findings in Pathophysiology, Diagnostic and Therapeutic Modalities. Royal Society of Chemistry.
55 Rajasekhar, K. and Govindaraju, T. (2018). Current progress, challenges and future prospects of diagnostic and therapeutic interventions in Alzheimer's disease. *RSC Adv.* 8: 23780–23804.
56 Rajasekhar, K., Chakrabarti, M., and Govindaraju, T. (2015). Function and toxicity of amyloid beta and recent therapeutic interventions targeting amyloid beta in Alzheimer's disease. *Chem. Commun.* 51: 13434–13450.
57 Ramesh, M. and Govindaraju, T. (2022). Multipronged diagnostic and therapeutic strategies for Alzheimer's disease. *Chem. Sci.* 13: 13657–13689.

58 Ramesh, M., Balachandra, C., Andhare, P., and Govindaraju, T. (2022). Rationally designed molecules synergistically modulate multifaceted Aβ toxicity, microglial activation, and neuroinflammation. *ACS Chem. Nerosci.* 13: 2209–2221.

59 Van der Merwe, E., Huang, D., Peterson, D. et al. (2008). The synthesis and anticancer activity of selected diketopiperazines. *Peptides* 29: 1305–1311.

60 Graz, C., Grant, G., Brauns, S. et al. (2000). Cyclic dipeptides in the induction of maturation for cancer therapy. *J. Pharm. Pharmacol.* 52: 75–82.

61 Mita, M.M., Spear, M.A., Yee, L.K. et al. (2010). Phase 1 first-in-human trial of the vascular disrupting agent plinabulin (NPI-2358) in patients with solid tumors or lymphomas. *Clin. Cancer Res.* 16: 5892–5899.

62 Singh, A.V., Bandi, M., Raje, N. et al. (2011). A novel vascular disrupting agent plinabulin triggers JNK-mediated apoptosis and inhibits angiogenesis in multiple myeloma cells. *Blood Am. J. Hematol.* 117: 5692–5700.

63 Mas, V., Falco, A., Brocal, I. et al. (2006). Identification of selective inhibitors of VHSV from biased combinatorial libraries of N, N′-disubstituted 2, 5-piperazinediones. *Antiviral Res.* 72: 107–115.

64 Maeda, K., Nakata, H., Koh, Y. et al. (2004). Spirodiketopiperazine-based CCR5 inhibitor which preserves CC-chemokine/CCR5 interactions and exerts potent activity against R5 human immunodeficiency virus type 1 in vitro. *Virol. J.* 78: 8654–8662.

65 Lemoine, C. and Wanner, J. (2010). Small molecule antagonists of the chemokine receptor CCR5. *Curr. Top. Med. Chem.* 10: 1299–1338.

11

Minimalistic Peptide Assemblies for Sustainable Optoelectronics

Yan Zhang[1,2], Jiahao Zhang[1,2], Junbai Li[3], Ehud Gazit[2,4], and Kai Tao[1,2]

[1] State Key Laboratory of Fluid Power and Mechatronic Systems, Key Laboratory of Advanced Manufacturing Technology of Zhejiang Province, School of Mechanical Engineering, Zhejiang University, Hangzhou, China
[2] Zhejiang-Israel Joint Laboratory of Self-Assembling Functional Materials, ZJU-Hangzhou Global Scientific and Technological Innovation Center, Zhejiang University, Hangzhou, China
[3] Beijing National Laboratory for Molecular Sciences, CAS Key Lab of Colloid, Interface and Chemical Thermodynamics Institute of Chemistry, Chinese Academy of Sciences, Beijing, China
[4] The Shmunis School of Biomedicine and Cancer Research George S. Wise Faculty of Life Sciences, Department of Materials Science and Engineering, Iby and Aladar Fleischman, Tel Aviv University, Israel

11.1 Introduction

The intricate quantum confinement (QC) regions within peptide nanostructures can induce hierarchical electron transmissions. These structures exhibit a unique type of intrinsic peptide fluorescence that is specific to structures containing a high proportion of β-sheets. This fluorescence emission occurs in the visible range and appears to be independent of the presence of aromatic residues within the polypeptide structures. This phenomenon has been observed in a range of protein aggregates and crystals [1–3], including those associated with protein misfolding diseases such as Alzheimer's disease, Parkinson's disease, and various types of amyloidosis [4, 5]. This discovery of intrinsic amyloid fluorescence has facilitated the monitoring of amyloid formation from disease-relevant polypeptides in a label-free manner with high specificity.

In addition to the photoluminescent property in the visible light region, the self-assembling peptide photoluminescence can even exhibit a bathochromic-shift extent up to the near-infrared region [6]. Similarly, minimalistic peptides display various excellent photoactive properties. The use of minimalistic peptide sequences reduces the complexity in structural design and synthesis costs, thereby making it easier to establish the relationship between structures and their functionalities with their self-assemblies composed of noncovalent interactions.

In this chapter, we focus on the photoactive properties of minimalistic peptide self-assembly systems and their diverse applications in technology. Firstly, we

Peptide Self-Assembly and Engineering: Fundamentals, Structures, and Applications, First Edition.
Edited by Xuehai Yan.

highlight the mechanisms of QC-based intrinsic photoluminescence of minimalistic peptide structures as well as the effects of various influencing factors. Next, we review the optical waveguide properties of minimalistic peptide structures, along with the methods used to obtain aligned minimalistic peptide assemblies with optical waveguide properties that bring new prospects for oriented optical devices.

11.2 Photoluminescent Properties

11.2.1 Quantum Confinement-induced Inherent Photoluminescence

QC is a significant quantum mechanical phenomenon that leads to substantial deviation in the electron energy spectrum, resulting in unique optical absorption, exciton luminescence, and laser effects at room temperature [7]. Self-assembled minimalistic peptide nanostructures exhibit inherent optical absorption and photoluminescence in both UV and visible regions [8, 9], which is related to the QC phenomena when the structural dimensions become comparable to the de-Broglie wavelength of the confined charges [10, 11]. The confinement can be divided into three orders: 0D-Quantum Dot (0D-QD), 1D-quantum wire, and 2D-Quantum Walls (2D-QW), among which the 2D-QW is the most common QC structure in peptide self-assemblies [12–14]. Figure 11.1a illustrates the differences between such QC structures and bulky materials, along with the density of electronic states (DOS) of each structure. As shown in the schematic image, the DOS of 0D-QD and 2D-QW are characterized by spike- and step-like behavior, respectively [9, 16, 17]. This striking alternation of the DOS in QC structures causes dramatic changes in electronic and optical properties, which are widely used in the field of light-emitting diodes and optical waveguiding devices [18].

The unique optical properties of the minimalistic peptides that self-assemble in a quantum limit region, mediated by electronic–vacuum interaction, are primarily due to the strong Coulomb interaction between the electron–hole pairs [9, 19]. A striking example of this phenomenon is observed in aromatic dipeptides, which have been found to self-assemble into supramolecular semiconductors [14, 20]. Density functional theory (DFT) calculations have been used to investigate the electron band structures of self-assembled diphenylalanine (FF), dityrosine (YY), and phenylalanine-tryptophan (FW) nanotubes [21]. The results show that FW and FF form a wide direct bandgap semiconductor, with electrons jumping from Z and Γ points, akin to gallium nitride. Conversely, YY forms an indirect bandgap semiconductor, akin to silicon carbide [21, 22]. The absorption spectrum of the aligned FF nanotubes exhibits two distinct steps located at 245–264 and 300–370 nm, in contrast to the narrow absorption peak of the FF monomers at 257 nm. This typical step-like optical absorption behavior suggests the formation of a crystal structure with two-dimensional quantum wells in the FF nanotubes [23, 24]. Furthermore, the same phenomenon was later observed in Fmoc-FF self-assembling hydrogels, providing further evidence that the peptide structures were organized into two-dimensional quantum wells in the gel state [25].

In comparison to the arrays, discrete FF nanotubes exhibit spike-like optical absorbance and photoluminescence characteristic of quantum dot (QD) structures [26]. Interestingly, when the solvent was changed from water to methanol, the nanotubes disassembled into single dimerized QDs, providing further evidence for the existence of the dimer as a single entity. This approach enables the preparation of FW into QD nanoparticles with different diameters (2.1 nm for cyclo-FW and 1.1–1.3 nm for FF QDs), allowing for the simple modification of peptide morphologies between superstructures and QDs. This feature is critical in the development of multifunctional photoluminescence devices [27, 28]. Cyclo-FW and cyclo-WW have the ability to combine and form dimeric QDs [15]. These QDs then act as building blocks, allowing the cyclodipeptides to self-assemble into larger quantum-confined nanostructures. As depicted in Figure 11.1b, fluorescent characterization of the self-assembled structures showed a redshift in the molecular excitation to 305 nm compared to 285 nm for the monomers, indicating successful self-assembly. Furthermore, UV–Vis absorption spectra showed a spike-like absorbance with three characteristic peaks at 273, 280, and 289 nm, demonstrating the formation of QD structures (Figure 11.1c). The diameter of the QDs was calculated to be around 2.24 nm, approximately twofold the dimension of the monomer. Therefore, it can be concluded that the cyclodipeptide monomers initially formed dimers that acted as QDs

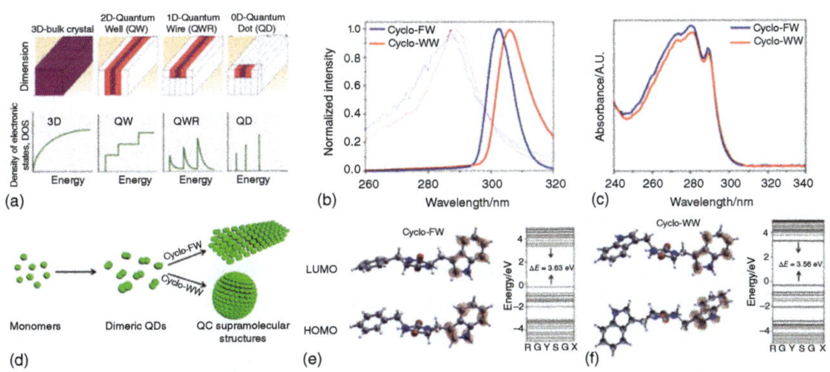

Figure 11.1 (a) Schematic illustration of different QC structures in comparison to the bulky crystal, along with the corresponding density of electronic states. Intrinsic optical properties of self-assembled aligned FF nanotubes. Source: From Rosenman et al. (2010) [9]/Figure 11.4/with permission of John Wiley and Sons. (b) Excitation spectra of cyclodipeptide monomers (thin light curves, 0.05 mM) and self-assemblies (thick dark curves, 5.0 mM) in MeOH. The excitation wavelengths redshifted from 285 to ~305 nm after self-assembly. (c) UV–Vis absorption spectra of the cyclodipeptides after self-assembly, showing spike-like absorptions at 273, 280, and 289 nm, characteristic of the formation of QD structures. (d) Schematic representing the process of cyclodipeptides self-assembly: the monomers form dimeric QDs, which serve as the building blocks to self-assemble into larger QC architectures. (e, f) Calculated molecular orbital amplitude plots and energy levels of the highest occupied and lowest unoccupied molecular orbitals of cyclo-FW and cyclo-WW, showing bandgaps of 3.63 and 3.56 eV, respectively. (e) cyclo-FW. (f) cyclo-WW. Source: Tao et al. [15]/Springer Nature/CC BY 4.0.

and served as the fundamental building blocks for further organization into supramolecular structures, as illustrated in Figure 11.1d. Theoretical calculations using DFT showed that the highest occupied and lowest unoccupied molecular orbitals of the dipeptides were concentrated mainly on the side-chain indole rings. As shown in Figure 11.1e, f, the bandgaps of cyclo-FW and cyclo-WW were calculated to be 3.63 and 3.56 eV, respectively, indicating their wide-gap semiconductive nature. This suggests that dimerization was primarily driven by the π–π interactions between the aromatic side chains, specifically the indole rings. The aromatic interactions resulted in through-space electron cloud conjugation from adjacent indole rings, thus restricting molecular movement and creating a molecular basis for QC regions. The QC effects and restricted molecular activity lead to the release of excitation energy solely as emitted light. These bioinspired nanostructures have inherent photoluminescence through various regulatory self-assembly processes, with an emission range adjusted from the visible region to the near infrared (NIR) region (420–820 nm). These nanostructures could serve as biocompatible labeling agents for *in vivo* imaging and as phosphors for engineering light-emitting diodes [15]. The crystals assembled by cyclo-FW exhibit wide-spectrum photoluminescence in the visible region (395–550 nm) and have intensity-controlled optical waveguiding [29]. A simple cyclodipeptide molecule, cyclo-glycine-tryptophan (cyclo-GW), was designed to form monoclinic crystals with thermal stability up to 370 °C and Young's modulus of 24.0 GPa, showing excellent mechanical strength [30]. The extensive and oriented hydrogen bond and aromatic interaction network could reduce the optical bandgap and promote aggregation-induced emission in the visible light range, optical waveguiding, and, especially, stable photoluminescence, rather than general aggregation-induced quenching [30].

11.2.2 Influence of Solvents

Water molecules play a critical role in the assembly and resulting photoluminescent properties of FF nanotubes [33]. The absence of water molecules can alter the bandgap of the nanotubes, which are formed through interpeptide head-to-tail bonds, peptide-water hydrogen bonds, and T-shaped aromatic layer interaction networks [34, 35]. Typical laminated FF nanotubes are prepared at a relative humidity of 33%, exhibiting a hexagonal head and overlapping hexagonal nanotubes at the tail (Figure 11.2a shows a schematic of the FF tubular structure) [31]. UV–visible–NIR diffuse reflectance spectra show that the nanotubes' bandgap increases with FF concentration, closely related to the highly ordered subnanometer crystalline structure of FF nanotubes with weak water bonding (Figure 11.2b) [31]. The UV photoluminescence peak shift is most likely affected by relative humidity and FF concentration. As the FF concentration increases to 160 mg mL^{-1}, the photoluminescence peak becomes closely related to the number of water molecules attached to the FF nanotubes, indicating that the bandgap increases with the water molecules. DFT calculations show that the valence band peak splits obviously with the addition of water molecules into the channel, leading to the shift and split of photoluminescence peak.

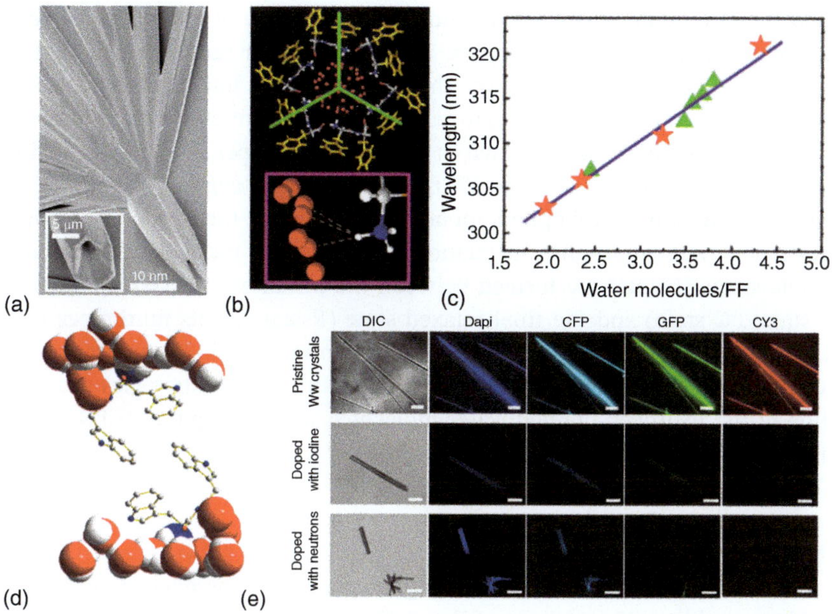

Figure 11.2 (a) Field-emission SEM image of the hierarchical FF nanotubes. The inset shows the enlarged head of the nanotube. (b) Schematic of the weak water molecule bonding (red balls) onto the FF molecule in the channel core of a hexagonal unit cell (inset) in the highly ordered FF subnanometer crystalline structure. (c) Linear dependence of the UV photoluminescence peak position versus the average number of water molecules per FF molecules. Source: From Wang et al. [31]/with permission of John Wiley and Sons. (d) Crystallographic structure showing the side-chain indole rings of W*w* surrounded by a hydrogen-bonded, motion-restricted medium composed of water molecules and polar atoms (oxygen and nitrogen) of peptide backbones. (e) Fluorescent microscopy images of pristine W*w* crystals (upper row; scale bar = 30 mm), doped with iodine (middle row; scale bar: 50 mm) and neutrons (lower row; scale bar = 200 mm), showing the attenuated fluorescence upon doping. DIC: differential interference contrast mode; DAPI: 40, 6-diamidino-2-phenylindole filter (Ex: 340–380 nm, Em: 435–485 nm); CFP: cyan fluorescent protein filter (Ex: 420–445 nm, Em: 460–510 nm); GFP: green fluorescent protein filter (Ex: 455–485 nm, Em: 500–545 nm); CY3: cyanine 3 filter (Ex: 528–553 nm, Em: 590–650 nm). Source: From Tao et al. [32]/with permission of Royal Society of Chemistry.

Moreover, the emission position and displacement are calculated to be linear with the number of water molecules (Figure 11.2c) [31], enabling the use of photoluminescence from FF nanotubes to detect the number of water molecules or humidity levels.

Due to the significant contribution of water molecules to the composition of W*w* (L-tryptophan-D-tryptophan) crystal, an alternative doping strategy was employed using deuterium oxide (D_2O) instead of regular water (H_2O) as the solvent [32]. This strategy involves introducing neutrons, which preserves the molecular backbone of the W*w* crystal with only a slight change in symmetry. In Figure 11.2d, it can be observed that the emission maximum of the original W*w* side-chain indole rings shows a significant red-edge excitation shift (REES) due to the highly polar environment surrounding it [36, 37], including a wide range of hydrogen-bonded water molecules and backbone terminals. The relaxation or reorientation of the molecules in such polar media is slower than that required for fluorescence lifetime [38], resulting in the W*w* combination emitting different colors from blue to red under different excitations, as shown in Figure 11.2e (upper row). The redshift was measured to be 10 nm (from 384 to 394 nm), with a linear slope of 0.1 nm nm^{-1}. A plausible explanation for this phenomenon is that during the relaxation of the highly restricted water molecules in relation to the excited indole rings, various intermediate states (I state) are formed between the initial excited state (Franck–Condon state, FC state) and the final relaxed state (R state). This minimizes the interaction energy [36] and leads to the REES phenomenon [38]. Neutrons are introduced through the exchange of H_2O and D_2O. Both doping strategies prompt the motion of the hydrogen-bonded water molecules, attenuating the REES effect and facilitating charge hopping.

11.2.3 The Effect of Temperature

Temperature is another critical factor affecting the photoluminescence of minimalistic peptide assemblies [39–41]. When FF and dileucine (LL) nanotubes are heated to 140–180 °C, they undergo irreversible transformation into ultrathin amyloid-like nanofibrils due to the conversion of linear dipeptide molecules into their cyclic dipeptide [42]. This process results in new secondary β-sheet structures and new optical properties, namely blue photoluminescence. This phenomenon has been observed in other dipeptide-based nanostructures [23, 43], suggesting that heating is a general method for producing peptide nanostructures with blue luminescence.

Conversely, low temperatures can also affect the photoluminescence of minimalistic peptide nanostructures. At room temperature, FF microtubes emit a near-UV broad band centered at 290 nm [8, 39]. However, when the temperature drops below −73 °C, a strong near-UV/blue/green emission of 350–500 nm occurs [40]. The photoluminescence intensity and lifetime of FF nanotubes also decrease significantly when the temperature rises from 5 to 65 °C (Figure 11.3a, b). Nonradiative recombination occurs in excitons located at the defect site, inducing the thermal quenching

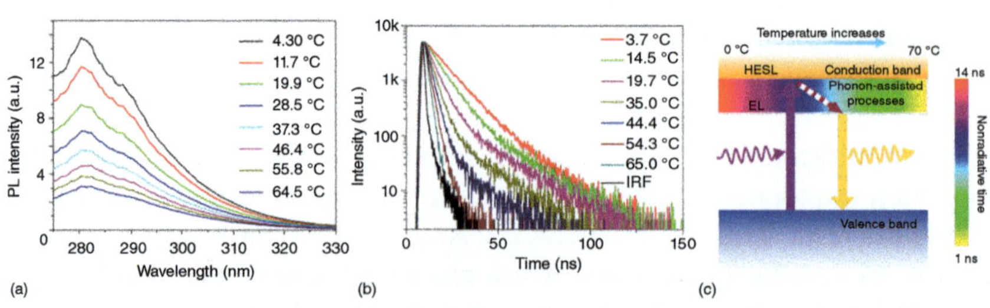

Figure 11.3 (a) Temperature-dependent PL and (b) TRPL spectra of the FF nanotubes. (c) Schematic representation of the mechanism underlying the temperature dependence of the photoluminescence lifetime. Source: From Gan et al. [39]/with permission of American Chemical Society.

of the photoluminescence [25]. High temperatures can activate the thermally assisted energy transfer process in FF nanotubes, accelerating nonradiative events and ultimately reducing photoluminescence lifetime and intensity (Figure 11.3c) [39]. By monitoring the photoluminescence intensity and lifetime of FF nanotubes, rapid changes in local temperature and absolute temperature between 5 and 65 °C can be detected *in situ*. This temperature-sensing capability is independent of the ionic (K^+) concentration or pH (6.0–8.0) in a physiological microenvironment, suggesting the potential use in bioimaging and monitoring temperature fluctuations *in vivo* [39].

11.2.4 The Effect of Chemical Conjugations

The chemical conjugation of functional groups provides another effective way to prepare minimalistic peptide nanostructures with adjustable and functional optical properties [44, 45]. Among the biomaterials with the properties of QD, modified FF has made a great contribution to the development of self-assembled bioinspired QDs. For example, the QC in a region of peptide nanocrystalline was formed within Boc-FF nanospheres (Figure 11.4a), which exhibit the optical absorption properties of QDs with pronounced luminescence of excitonic origin [26, 46, 49]. The QC phenomena can also be found in fluorenylmethyloxycarbonyl-modified FF (Figure 11.4b). The presence of a step-like absorption pattern [46, 50], along with a long-wavelength edge peak resulting from enhanced Coulomb interactions of excitons, is a telltale sign of the formation of 2-D quantum well confinement structures in nanofibrillar hydrogels [49]. Porphyrin molecules have historically been studied as photosensitizers (PS) and have found extensive applications in various types of

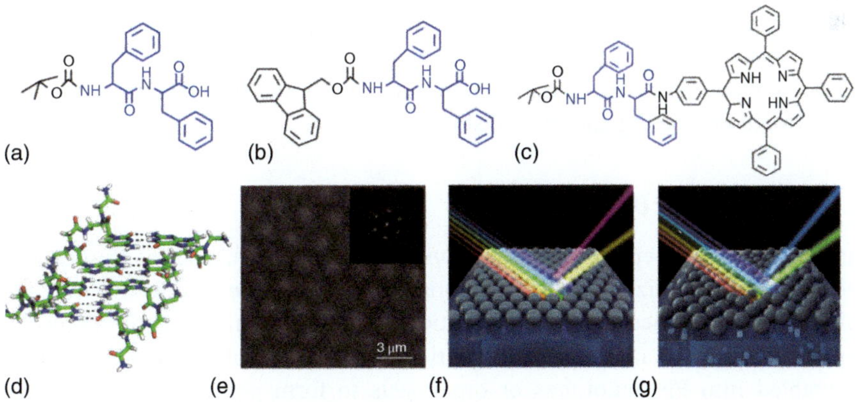

Figure 11.4 (a–c) Chemical structures of various FF derivatives: (a) Boc-FF, (b) Fmoc-FF, and (c) monoaminoporphyrin-FF-Boc. Source: From Chen et al. (2015) [46]/Figure 11.1/with permission of Elsevier. (d) Each di-PNA-GC molecule forms hydrogen bonds with a neighboring unit between the cytosine and guanine residues during crystallization. Source: From Berger et al. (2015) [47]/Figure 11.2b/with permission of Springer Nature. (e) Large area ordered arrays organized by the aromatic protected PNA building blocks self-assembling spheres show crystallized structures with hexagonal symmetry (f and g) Schematic representation of the PNA sphere array (f) before and (g) following the addition of NaCl, demonstrating the geometrical alternation leading to the finely modulated reflectance color. Source: From Berger et al. [48]/with permission of John Wiley and Sons.

dye-sensitized solar cells [51]. Some photosensitizer molecules can be also conjugated to FF to meet photoelectronic performance requirements. One example is monoaminoporphyrin-FF-Boc (Figure 11.4c), after modification with porphyrin, which can self-assemble into fluorescence-enhanced and electronically conservative porous vesicles [46, 52]. In addition, the modification of side chains has been proven to effectively enhance and control the optical properties of minimalist peptides. For instance, peptide nucleic acids (PNAs) combine the chemical versatility of peptides with the specific molecular recognition advantages of nucleic acids [47]. Specifically, Watson–Crick base pairing and stacking (as depicted in Figure 11.4d) drive the self-assembly of diPNA-guanine-cytosine (di-PNA-GC) [47]. The integration of extensive and directional noncovalent interactions results in excitation-dependent fluorescence spanning from 440 to 684 nm, covering nearly the entire visible region. Notably, when these self-assemblies are placed onto a chip, the nanostructures respond to gate voltage in the same manner as a field-emission transistor, displaying lower current rates as voltage increases. When a device was constructed using these structures and alternating voltages (5 to −5 V) were applied, bright electroluminescence was emitted, indicating its potential for optical biosensing and light emission-based applications, including organic light-emitting diodes and imaging labels that are tunable via optical or electrical modulation.

Moreover, after various structural modifications, only an aromatic protected PNA was found to be also able to assemble into supramolecular architectures [53]. The heat-cooling process can form uniform spherical particles with a size of 1.7 μm [48], which is very rare in self-assembled biological systems and similar to inorganic particles such as silica. Large area ordered layers of these spheres could form at the water–air interfaces exhibiting vivid colors. The results of TEM images indicated that crystalline structures of the layers displaying hexagonal symmetry were formed by self-association of the spheres (Figure 11.4e). The addition of NaCl to a spherical solution can achieve visible color changes (as shown in Figure 11.4f, g) [48], simulating the skin color changes of chameleons (changes in the spacing of the guanine lattice between the resting and excited states).

11.2.5 The Effect of Doping

Doping minimalistic peptide assemblies with optical entities, such as fluorescent molecules, photosensitizers, lanthanide ions, and QDs, can be an effective method to enhance luminescence and extend functionality of the bioinspired supramolecular structures. For example, photosensitizers and lanthanide ions have been co-assembled into FF nanotubes or organogels to form photoactive biosystems (Figure 11.5a) [54, 57]. Peptide assemblies act as both host matrixes and antennas for lanthanide ions, resulting in enhanced photoluminescence due to energy transfer cascades from the peptides to lanthanide ions through the photosensitizers. As a result, FF nanotubes with various colors (Figure 11.5b) or short peptide organogels can be obtained [54], and their luminescence wavelengths can be easily tuned by changing the composition of the assemblies.

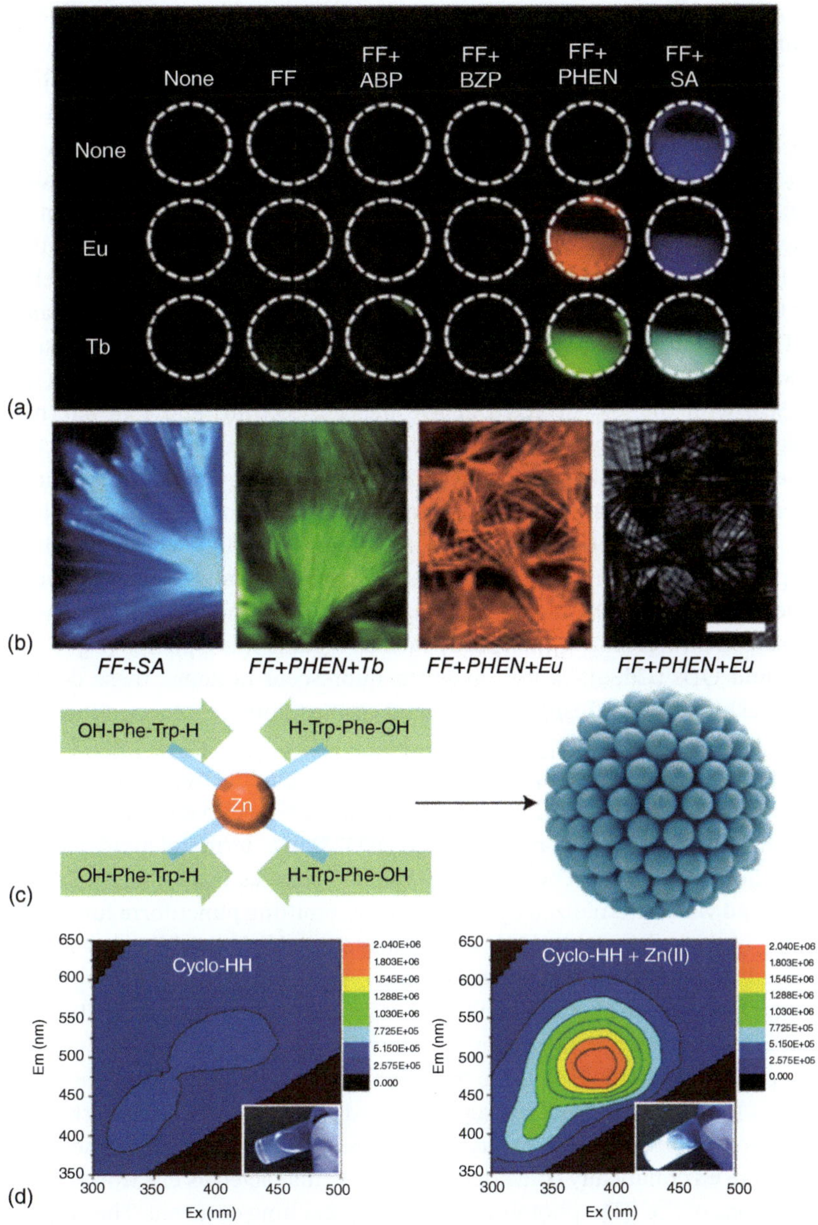

Figure 11.5 Influence of doping on the photoluminescence properties of short peptide self-assemblies. (a) Influence of FF nanotubes (NTs) and PS on the photoluminescence (PL) of lanthanide ions (Eu and Tb). The PL properties of lanthanide ions and PS were compared in the absence or presence of FF NTs (λ_{ex} = 254 nm). (b) Optical microscopy images of FF-NT complex with or without UV excitation (the last one). Scale bar = 50 μm. Source: From Ryu et al. (2009) [54]/Figure 11.1/with permission of John Wiley and Sons. (c) Through hydrogen bonding, π–π stacking interactions, and Zn(II) coordination, the dipeptide self-assembles to form spherical nanoparticles. Both H-Trp-Phe-OH and OH-Phe-Trp-H represent WF dipeptides. Source: From Fan, Z. et al. (2016) [55]/Figure 11.1c/with permission of Springer Nature. (d) Fluorescence contour profiles of cyclo-HH in the absence (left) or presence (right) of Zn(II). The insets show photographic pictures of the corresponding peptide solutions under UV light irradiation (365 nm). Source: From Tao et al. [56]/with permission of John Wiley and Sons.

Carboxyl groups of amino acids and some amino-acid side chains can coordinate with metal ions, such as the imidazole group of histidine and the indole group of tryptophan [58]. This coordination can further participate in the self-assembly of peptides and affect their photoluminescence properties. For example, Zn^{2+}-conjugated BFPms1, a green fluorescent protein variant [59], has been shown to exhibit enhanced luminescence. Based on this method, WF, an aromatic dipeptide, was designed to assemble into spherical nanoparticles that emit blue light at 423 nm when excited at 370 nm (Figure 11.5c) [55]. Moreover, aiming to increase the efficiency, Zn(II) was introduced to the aromatic cyclic dipeptide solutions, thereby attempting to decrease the electron transfer energy barrier and enhance the quantum yield. Fluorescence characterization demonstrated that the emission intensity of the cyclo-HH solution was significantly enhanced in the presence of Zn(II) (Figure 11.5d), with the λ_{Em} detected at 490 nm (Ex: 395 nm) and τ of 4.63 ns. Simultaneously, the photoluminescence efficiency was improved, with an enhanced quantum yield of 15% for cyclo-HH + Zn(II). The emission of the cyclo-HY solution was also enhanced to some extent in the presence of Zn(II), with the λ_{Em} detected at 490 nm (Ex: 380 nm), τ of 4.79 ns, and a QY of 9%, similar to cyclo-WW [56]. Doping these nanoparticles with Zn^{2+} significantly stabilized them and increased their quantum yield, leading to sustained and enhanced photoluminescence. Compared to other advanced fluorescent counterparts, such as organic fluorophores (e.g. rhodamine 6G) and QDs (CdSe), as well as green fluorescent proteins, these doped nanoparticles exhibited significant and sustained visible fluorescence as well as biocompatibility [55, 56].

Minimalistic peptide assemblies have been shown to be effective in tuning the photoluminescence properties of QDs [56]. For instance, 3-D colloidal spheres were prepared by encapsulating CdSeS nanocrystals (QD523) in C-terminal amidated FF (cationic dipeptide, CDP) gel. The resulting colloidal spheres displayed good biocompatibility and were internalized by cultured cells, emitting punctiform luminescent spots for biological imaging [56]. Minimalistic peptide assemblies have also been employed as ultrasensitive biosensors based on specific recognition/targeting probe binding fluorescence changes, such as photoluminescence quenching. To create a peptide hydrogel-based biosensing platform, enzymes (glucose oxidase or horseradish peroxidase) and QDs (CdTe or CdSe) were incorporated into Fmoc-FF self-assemblies [60]. The resulting hydrogels formed a 3-D nanofiber network that retained the photoluminescence property of the QDs. When the glucose or phenolic compounds were enzymatically converted to products that acted as electron acceptors on the surface of the QDs, photoluminescence quenching occurred. The degree of quench ability was found to correlate with the concentration of the analyte [60], demonstrating the potential of this short peptide hydrogel as a versatile platform for enzymatic detection of metabolic analytes. Moreover, this strategy could also be used to selectively detect neurotoxins in organophosphates and nitro-group compounds using photoluminescent FF nanotubes [61]. These findings suggest that short peptide assemblies have promising applications in biosensing and biological imaging.

11.3 Optical Waveguide Properties

11.3.1 Intrinsic Optical Waveguiding Properties

Quantum confined organic nanomaterials are potential candidates for the next generation of environmentally friendly optoelectronic devices. These bioinspired supramolecular architectures present the ability to manipulate and propagate light on a subwavelength scale [62–64]. Short peptides, due to their high photoluminescent properties, optical tunability, and excellent self-assembly propensity, are expected to be effective building blocks for future miniaturized optoelectronics manufacturing, enabling the generation and propagation of light [65]. The waveguiding property of peptide crystals has been confirmed by focusing the excited light on one end of a platelet and observing the guided photoluminescence emitted from the other end (arrow in Figure 11.6a–d) [66]. The external conjugation effect at the end is typical of waveguiding, and these peptide crystals act as active waveguides. The photoluminescence is generated within the peptide platelets due to the inherent properties of the cyclo-FF dipeptide. This discovery suggests that a peptide crystal with long-range order can function as an optical microcavity, enabling photoluminescence emission to propagate along the longitudinal axis. The photoluminescence emission of dipeptide crystals can be easily modulated by the addition of guest dyes during crystallization. For instance, the addition of fluorescent Nile Red (NR) into peptide platelets results in red emission upon excitation at 515–560 nm. When one end of the platelet is excited, the emitted red photoluminescence can propagate along the platelet and be coupled out at the other end [66]. Overall, these findings demonstrate that short peptides have great potential for use in the development of miniaturized optoelectronics that can effectively manipulate and propagate light.

11.3.2 Oriented Optical Waveguide Devices

The oriented alignment of organic materials is essential in enhancing their functions [69, 70]. Through the dip-coating process, ultralong, well-aligned single FF crystals can be formed, which are promising materials for large-scale optical devices [67]. This approach enables the continuous directional growth of homogeneous single crystals that can reach several centimeters in length on silicon wafers (Figure 11.6e) [67]. By controlling the process parameters, the structure of FF single crystals can be precisely manipulated to exhibit active optical waveguide properties. Additionally, a one-dimensional FF single crystal with a length of 4 cm can be obtained using a capillary [68]. When excited at one end of the ultralong single FF crystal, photoluminescence can be observed at the other end, demonstrating its active waveguide capabilities (Figure 11.6f) [68].

The development of ultralong-oriented short peptide crystals is expected to expand the potential applications of aligned organic nanomaterials in optical devices. These findings highlight the potential of dip coating and capillary techniques for growing well-aligned organic materials with superior optical properties.

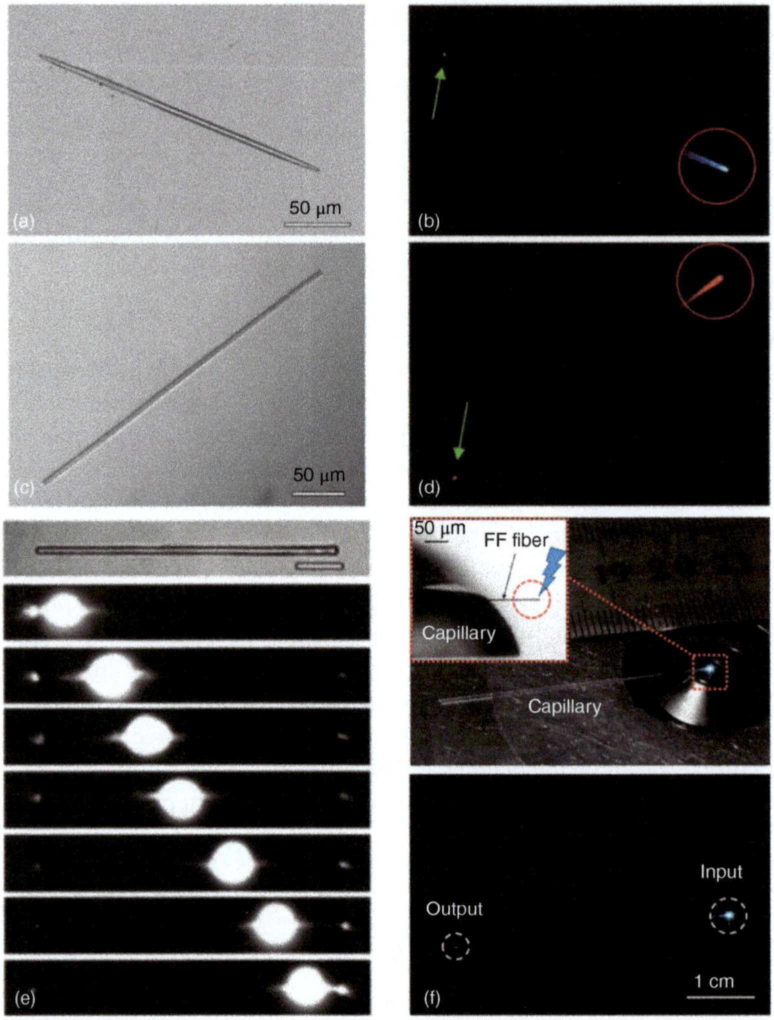

Figure 11.6 Oriented optical waveguiding properties of short peptide self-assemblies. (a–d) Bright-field images (a and c) and optical waveguiding images (b and d) of an individual crystal (a, b) and a crystal incorporating Nile Red dye (c, d) with local excitation at one end (red circle). The green arrow denotes the outcoupling of PL emission at the other end. Source: From Yan et al. [66]/with permission of John Wiley and Sons. (e) Photograph of an FF-RhB single crystal showing the transition of the excitation spot along the crystal upon excitation at 488 nm. The scale bar (right) is 10 mm. Source: From Sun et al. [67]/with permission of American Chemical Society. (f) Photographic image of an ultralong single FF-RhB crystal in capillary (upper panel) and the corresponding fluorescence image showing the optical waveguiding properties (lower panel). The circles mark the area of excitation (488 nm) and emission. Source: From Sun et al. [68]/with permission of American Chemical Society.

11.3.3 Second-harmonic Generation (SHG)

SHG is a physical effect that depends on the symmetry of crystals and the electronic properties of dielectric materials. It is observed only in crystals with no center of symmetry and is associated with ferroelectric phenomena [71] as well as linear electrooptical and piezoelectric effects [72]. The SHG effect is also used in optical microscopy, such as medical and biological research [73]. It can detect two-photon emission from biomaterials and biopolymers that lack a center of symmetry [74]. Additionally, this effect has been combined with quantitative indicators to diagnose various diseases [75]. Recently, second-order responses have also been observed in bioinspired aromatic FF-PNT with a hexagonal space group $P6_1$ [9]. In recent years, SHG has been observed in bioinspired aromatic FF-PNT with a hexagonal space group $P6_1$ [9], which has a noncentrosymmetric structure that satisfies the requirements of second-order nonlinear optical devices. Specifically, as shown in Figure 11.7, SHG was observed in the minimalistic peptide nanostructures L-phenylalanine dipeptide (FF) and L-phenylalanine tripeptide (FFF) [76]. The study involved measuring the light transmission of normally oriented FF-PNT at two different frequencies: the fundamental frequency at a wavelength of $\lambda = 800\,\text{nm}$ (Figure 11.7d) and the double SHG frequency at a wavelength of $\lambda = 400\,\text{nm}$

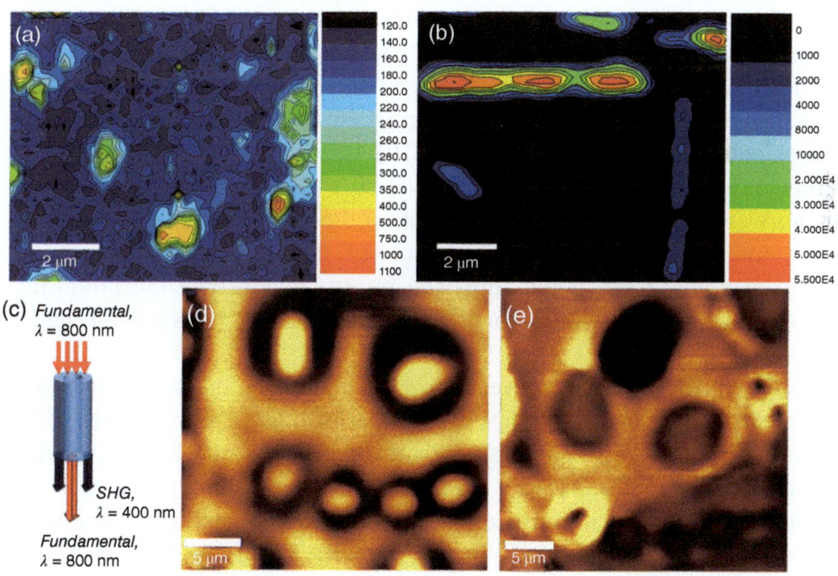

Figure 11.7 SHG imaging of (a) FFF-peptide nanospheres and (b) FFF-peptide nanobelts; (c) illustration of the fundamental and second-harmonic generated wave (red arrows-fundamental wave, blue arrows-SHG wave); (d) linear waveguiding of FF-peptide nanotubes; (e) nonlinear waveguiding of FF nanotubes. Source: From Handelman et al. [76]/ with permission of John Wiley and Sons.

(Figure 11.7e) [76]. When observed at the second-order nonlinear response frequency (a selective narrow filter operating at 400 nm was inserted before the lower detector), the position of the FF-PNT remained unchanged (Figure 11.7e), and the cladding of the FF-PNT emitted light at double frequency, indicating the potential use of FF PNT as either linear or nonlinear waveguides [76–78]. Overall, the SHG effect is a valuable tool for studying and manipulating the properties of dielectric materials, particularly in the context of biomedical research and nanotechnology.

SHG is a fundamental physical effect that depends on the symmetry of crystals and the electronic properties of dielectric materials. Bioinspired peptide nanostructures based on L-F dipeptide and tripeptide have been shown to exhibit SHG, which suggests the potential of minimalistic peptide assemblies for use in a new generation of nonlinear optical nanomaterials.

11.3.4 Electroluminescent Properties

The optical properties of feature assemblies have been studied extensively and have shown potential for use as emission material in electroluminescent prototypes [56, 79, 80]. The researchers combined a mixture of dried cyclo-HH + Zn(II) dots and polydimethylsiloxane with a 420-nm emissive indium gallium nitride chip to create the LED [56]. The cyclo-HH is an aromatic cyclodipeptide, and the Zn(II) dots are zinc ions coordinated to two organic compounds, creating structures called coordination complexes. As shown in Figure 11.8a, b, when the LED was operated under a voltage of 3.0 V, it emitted bright green light with an emission around 565 nm. And the chromaticity coordinates of the emitted light were (0.37, 0.40) as determined by Commission Internationale de L'Eclairage (CIE) measurements. Additionally, the color temperature of the light was measured to be 4415 K (Figure 11.8c). Further experiments were conducted to confirm the role of the Zn(II) dots in enhancing the green photoluminescence of the LED. A control LED that only contained cyclo-HH emitted intrinsic blue light of the chip. The results demonstrate the promising potential of using self-assembling aromatic cyclodipeptide structures in eco-friendly optoelectronics.

Further investigation by Chen revealed that CHH-Zn could also be used as a bioorganic luminescent material in optoelectronics [80]. A simple natural peptide-derived bioorganic LED (OLED) prototype was prepared using CHH-Zn-blended poly(N-vinyl carbazole) (PVK) as the emissive layer (Figure 11.8d). The resulting OLED demonstrated bright, uniform, and defect-free surface green electroluminescence emission. The maximum EQE was 0.25%, and the corresponding current efficiency was 0.58 cd A^{-1} (the corresponding L_{max} was 1385 cd m^{-2}). Due to the stable fluorescence of CHH-Zn, the emission spectrum of the bio-OLED showed no temporal degradation under the corresponding operating conditions, further demonstrating the feasibility of using self-assembled peptides to fabricate biointegrated chips with eco-friendly and customized optoelectronic properties. These findings illustrate the frontiers of using biologically inspired organic chemistry to deliver smart materials with the potential for better interfaces between sustainable optoelectronics and biomedical applications [79, 80].

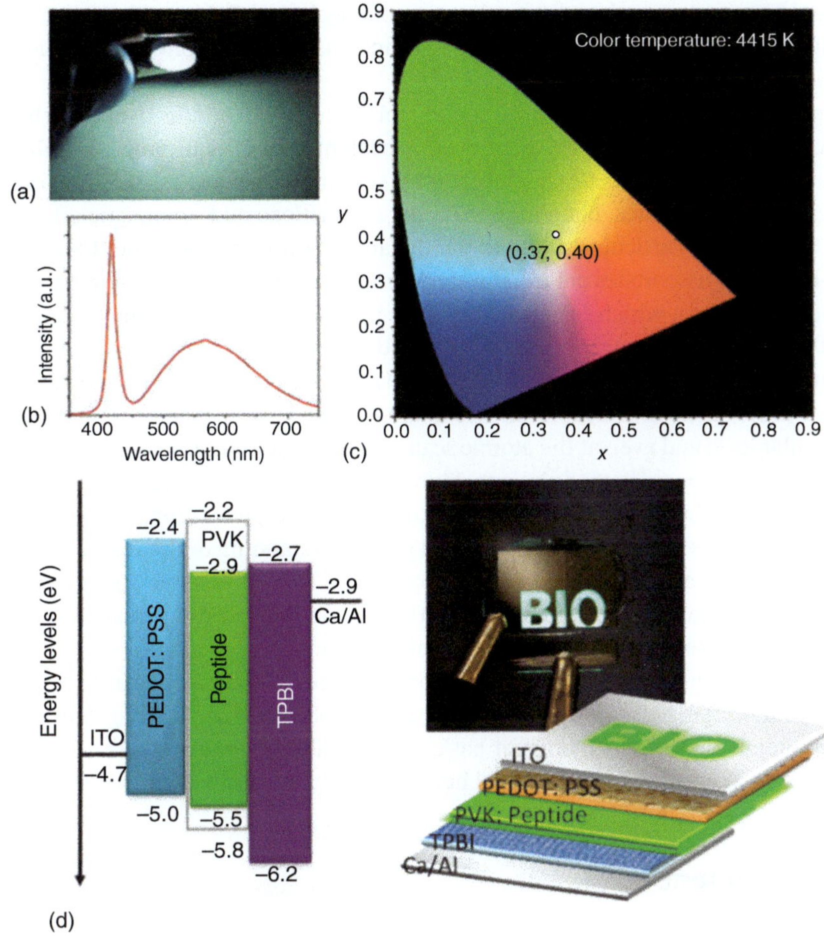

Figure 11.8 Luminescent applications of the CHH-Zn self-assemblies. (a) Photographic picture depicting the prototype, emitting bright green light upon excitation at 420 nm. (b) Emission spectrum of the LED operated under a voltage of 3.0 V. (c) CIE coordinates of the operating LED and the corresponding color temperature. Source: From Tao et al. [56]/ with permission of John Wiley and Sons. (d) OLED structure, energy diagram, and operation photograph. Source: From Chen et al. (2020) [80]/with permission of American Chemical Society.

11.4 Conclusions and Perspectives

In this chapter, the recent developments of photoactive self-assembling short peptides are reviewed, and their promising application directions for future researches are discussed. Self-assembling functional micro/nanomaterials can be easily manufactured using minimalistic peptide building blocks, and a wide range of architectures ranging from nano- to microscale can be prepared. The morphologies and optical properties of the bioinspired architectures can be finely tuned. After decades of research efforts, the biocompatible and unique optical properties of short peptide self-assemblies can be employed to various nanobiotechnical fields, such as manufacturing of bioorganic optoelectronic devices and construction of photo-controlled drug delivery systems.

Minimalistic peptide-based supramolecular architectures can be effectively manipulated and improved through covalent or noncovalent strategies to prepare multifunctional self-assemblies. Co-assembly with other biomolecules can endow customer-designed features, benefiting from the interactions and synergistic effects of each component. Particularly, due to the dynamic nature of noncovalent bonds, peptide-based self-assemblies are inherently appropriate for constructing responsive and adaptive systems, which is highly desirable for potential applications in smart drug delivery and precise cancer theragnostic fields. The stability of the architectures (through the formation of chemical bonds) is also essential to maintain their functions over the long term. A deeper understanding of the nature of self-assembly, at the molecular level and even at the atomic scale, will provide clues for new strategies of structural and properties modulations. Moreover, studies of this class of optoelectronic supramolecular materials could not only provide next-generation, biologically inspired alternatives to optics but also expand the understanding of the role of polypeptides or proteins optics in physiology and pathology. This may enable direct, unlabeled, and real-time monitoring and perception of metabolic activities as well as analysis and possible intervention of biological systems.

With these goals in mind, photoactive minimalistic peptide systems can be attractive candidates for bridging chemistry, materials science, biology, optics, and engineering, providing promising technical tools for basic biology, biomedicine, and nanobiology research. These suggestions are presented here and are by no means exhaustive.

Acknowledgment

This work was supported by the National Key R&D Program of China within the China-Israel Cooperative Scientific Research (2022YFE0100800) (Israeli No. 3-18130), National Natural Science Foundation of China (52175551), Leading Innovative and Entrepreneur Team Introduction Program of Zhejiang Province (2022R01001), Zhejiang University Global Partnership Fund, and Open Foundation of the State Key Laboratory of Fluid Power and Mechatronic Systems. The authors thank the members of all the laboratories for helpful discussions.

References

1 Chan, F.T., Pinotsi, D., Gabriele, S. et al. (2014). Structure-specific intrinsic fluorescence of protein amyloids used to study their kinetics of aggregation. In: *Bio-Nanoimaging: Protein Misfolding and Aggregation* (ed. V.N. Uversky and Y.L. Lyubchenko), 147–155. Elsevier.

2 Huang, C., Wagner-Valladolid, S., Stephens, A.D. et al. (2019). Intrinsically aggregation-prone proteins form amyloid-like aggregates and contribute to tissue aging in Caenorhabditis elegans. *Elife* 8: e43059.

3 Sharpe, S., Simonetti, K., Yau, J. et al. (2011). Solid-state NMR characterization of autofluorescent fibrils formed by the elastin-derived peptide GVGVAGVG. *Biomacromolecules* 12 (5): 1546–1555.

4 Tcherkasskaya, O. (2007). Photo-activity induced by amyloidogenesis. *Protein Sci.* 16 (4): 561–571.

5 Chan, F.T., Schierle, G.S., Kumita, J.R. et al. (2013). Protein amyloids develop an intrinsic fluorescence signature during aggregation. *Analyst* 138 (7): 2156–2162.

6 Tadepalli, S., Slocik, J.M., Gupta, M.K. et al. (2017). Bio-optics and bio-inspired optical materials. *Chem. Rev.* 117 (20): 12705–12763.

7 Pinotsi, D., Buel, A.K., Dobson, C.M. et al. (2013). A label-free, quantitative assay of amyloid fibril growth based on intrinsic fluorescence. *ChemBioChem.* 14 (7): 846–850.

8 Amdursky, N., Molotskii, M., Aronov, D. et al. (2009). Blue luminescence based on quantum confinement at peptide nanotubes. *Nano. Lett.* 9 (9): 3111–3115.

9 Rosenman, G., Beker, P., Koren, I. et al. (2011). Bioinspired peptide nanotubes: deposition technology, basic physics and nanotechnology applications. *J. Pept. Sci.* 17 (2): 75–87.

10 Knowles, T.P., Oppenheim, T.W., Buell, A.K. et al. (2010). Nanostructured films from hierarchical self-assembly of amyloidogenic proteins. *Nat. Nanotechnol.* 5 (3): 204–207.

11 Wang, H., Feng, Z., and Xu, B. (2017). Bioinspired assembly of small molecules in cell milieu. *Chem. Soc. Rev.* 46 (9): 2421–2436.

12 Wong, G.C., Lin, A., Tang, J.X. et al. (2003). Lamellar phase of stacked two-dimensional rafts of actin filaments. *Phys. Rev. Lett.* 91 (1): 018103.

13 Adler-Abramovich, L., Arnon, Z.A., Sui, X.M. et al. (2018). Bioinspired flexible and tough layered peptide crystals. *Adv. Mater.* 30 (5): 1704551.

14 Tao, K., Makam, P., Aizen, R. et al. (2017). Self-assembling peptide semiconductors. *Science* 358 (6365): eaam9756.

15 Tao, K., Fan, Z., Sun, L. et al. (2018). Quantum confined peptide assemblies with tunable visible to near-infrared spectral range. *Nat. Commun.* 9 (1): 3217.

16 Gardel, M.L., Shin, J.H., MacKintosh, F.C. et al. (2004). Elastic behavior of cross-linked and bundled actin networks. *Science* 304 (5675): 1301–1305.

17 Duboz, J.Y., Gupta, J.A., Byloss, M. et al. (2002). Intersubband transitions in InGaNAs/GaAs quantum wells. *Appl. Phys. Lett.* 81 (10): 1836–1838.

18 Alivisatos, A.P. (1996). Semiconductor, clusters, nanocrystals, and quantum dots. *Science* 271 (5251): 933–937.

19 Richardson, J.S. (1981). The anatomy and taxonomy of protein structure. *Adv. Protein. Chem.* 34: 167–339.

20 Yao, H.B., Fang, H.Y., Wang, X.H. et al. (2011). Hierarchical assembly of micro–/nano-building blocks: bio-inspired rigid structural functional materials. *Chem. Soc. Rev.* 40 (7): 3764–3785.

21 Akdim, B., Pachter, R., and Naik, R.R. (2015). Self-assembled peptide nanotubes as electronic materials: an evaluation from first-principles calculations. *Appl. Phys. Lett.* 106 (18): 183707.

22 Pelletier, O., Pokidysheva, E., Hirst, L.S. et al. (2003). Structure of actin cross-linked with α-actinin: a network of bundles. *Phys. Rev. Lett.* 91 (14): 148102.

23 Lee, J.S., Yoon, I., Kim, J. et al. (2011). Self-assembly of semiconducting photoluminescent peptide nanowires in the vapor phase. *Angew. Chem. Int. Edit.* 50 (5): 1164.

24 Hauser, C.A. and Zhang, S. (2010). Peptides as biological semiconductors. *Nature* 468 (7323): 516–517.

25 Amdursky, N., Gazit, E., and Rosenman, G. (2010). Quantum confinement in self-assembled bioinspired peptide hydrogels. *Adv. Mater.* 22 (21): 2311–2315.

26 Amdursky, N., Molotskii, M., and Gazit, E.t al. (2010). Elementary building blocks of self-assembled peptide nanotubes. *J. Am. Chem. Soc.* 132 (44): 15632–15636.

27 Bhattacharya, P., Ghosh, S., and Stiff-Roberts, A.D. (2004). Quantum dot opto-electronic devices. *Annu. Rev. Mater. Res.* 34: 1–40.

28 Qian, F., Li, Y., Grade, S. et al. (2008). *Nat. Mater.* 7: 701.

29 Tao, K., Xue, B., Li, Q. et al. (2019). Stable and optoelectronic dipeptide assemblies for power harvesting. *Mater. Today* 30: 10–16.

30 Tao, K., Hu, W., Xue, B. et al. (2019). Bioinspired stable and photoluminescent assemblies for power generation. *Adv. Mater.* 31 (12): 1807481.

31 Wang, M., Xiong, S., Wu, X. et al. (2011). Effects of water molecules on photoluminescence from hierarchical peptide nanotubes and water probing capability. *Small* 7 (19): 2801–2807.

32 Tao, K., O'Donnell, J., Yuan, H. et al. (2020). Accelerated charge transfer in water-layered peptide assemblies. *Energ Environ. Sci.* 13 (1): 96–101.

33 Kim, J., Han, T.H., Kim, Y.I. et al. (2010). Role of water in directing diphenylalanine assembly into nanotubes and nanowires. *Adv. Mater.* 22 (5): 583.

34 Görbitz, C.H. (2006). The structure of nanotubes formed by diphenylalanine, the core recognition motif of Alzheimer's β-amyloid polypeptide. *Chem. Commun.* 22: 2332–2334.

35 Guo, C., Luo, Y., Zhou, R. et al. (2012). Probing the self-assembly mechanism of diphenylalanine-based peptide nanovesicles and nanotubes. *ACS Nano.* 6 (5): 3907–3918.

36 Lakowicz, J.R. and Keating-Nakamoto, S. (1984). Red-edge excitation of fluorescence and dynamic properties of proteins and membranes. *Biochemistry* 23 (13): 3013–3021.

37 Chattopadhyay, A. and Haldar, S. (2014). Dynamic insight into protein structure utilizing red edge excitation shift. *Acc. Chem. Res.* 47 (1): 12–19.

38 Kabir, M.L., Wang, F., and Clayton, A.H.A. (2021). Red-edge excitation shift spectroscopy (Rees): application to hidden bound states of ligands in protein-ligand complexes. *Int. J. Mol. Sci.* 22 (5): 2582.

39 Gan, Z., Wu, X., Zhang, J. et al. (2013). In situ thermal imaging and absolute temperature monitoring by luminescent diphenylalanine nanotubes. *Biomacromolecules* 14 (6): 2112–2116.

40 Nikitin, T., Kopyl, S., Shur, V.Y. et al. (2016). Low-temperature photoluminescence in self-assembled diphenylalanine microtubes. *Phys. Lett. A.* 380 (18–19): 1658–1662.

41 Semin, S., Van, E., A., Cattaneo, L. et al. (2015). Strong thermo-induced single and two-photon green luminescence in self-organized peptide microtubes. *Small* 11 (9–10): 1156–1160.

42 Handelman, A., Natan, A., and Rosenman, G. (2014). Structural and optical properties of short peptides: nanotubes-to-nanofibers phase transformation. *J. Pept. Sci.* 20 (7): 487–493.

43 Reches, M. and Gazit, E. (2003). Casting metal nanowires within discrete self-assembled peptide nanotubes. *Science* 300 (5619): 625–627.

44 Sun, H.L., Chen, Y., Han, X. et al. (2017). Tunable supramolecular assembly and photoswitchable conversion of cyclodextrin/diphenylalanine-based 1D and 2D nanostructures. *Angew. Chem. Int. Edit.* 56 (25): 7062–7065.

45 Bhowmik, S., Jadhav, R.G., and Das, A.K. (2018). Nanoporous conducting covalent organic polymer (cop) nanostructures as metal-free high performance visible-light photocatalyst for water treatment and enhanced CO_2 capture. *J. Phys. Chem. C.* 122 (1): 274–284.

46 Chen, C., Liu, K., Li, J. et al. (2015). Functional architectures based on self-assembly of bio-inspired dipeptides: structure modulation and its photoelectronic applications. *Adv. Colloid. Interfac.* 225: 177–193.

47 Berger, O., Adler-Abramovich, L., Levy-Sakin, M. et al. (2015). Light-emitting self-assembled peptide nucleic acids exhibit both stacking interactions and Watson-crick base pairing. *Nat. Nanotechnol.* 10 (4): 353–360.

48 Berger, O., Yoskovitz, E., Adler-Abramovich, L. et al. (2016). Spectral transition in bio-inspired self-assembled peptide nucleic acid photonic crystals. *Adv. Mater.* 28 (11): 2195–2200.

49 Amdursky, N., Molotskii, M., Gazit, E. et al. (2009). Self-assembled bioinspired quantum dots: optical properties. *Appl. Phys. Lett.* 94 (26): 261907.

50 Tao, K., Yoskovitz, E., Adler-Abramovich, L. et al. (2015). Optical property modulation of Fmoc group by pH-dependent self-assembly. *Rsc. Adv.* 5 (90): 73914–73918.

51 Wang, Y., Chen, B., Wu, W. et al. (2014). Efficient solar cells sensitized by porphyrins with an extended conjugation framework and a carbazole donor: from molecular design to cosensitization. *Angew. Chem.* 126 (40): 10955–10959.

52 Charalambidis, G., Kasotakis, E., Lazarides, T. et al. (2011). Self-assembly into spheres of a hybrid diphenylalanine-porphyrin: increased fluorescence lifetime and conserved electronic properties. *Chem. Eur. J.* 17 (26): 7213–7219.

53 Teyssier, J., Saenko, S.V., Van, D. et al. (2015). Photonic crystals cause active colour change in chameleons. *Nat. Commun.* 6 (1): 6368.

54 Ryu, J., Lim, S.Y., and Park, C.B. (2009). Photoluminescent peptide nanotubes. *Adv. Mater.* 21 (16): 1577–1581.

55 Fan, Z., Sun, L., Huang, Y. et al. (2016). Bioinspired fluorescent dipeptide nanoparticles for targeted cancer cell imaging and real-time monitoring of drug release. *Nat. Nanotechnol.* 11 (4): 388–394.

56 Tao, K., Chen, Y., Orr, A.A. et al. (2020). Enhanced fluorescence for bioassembly by environment-switching doping of metal ions. *Adv. Funct. Mater.* 30 (10): 1909614.
57 Liu, X., Zhu, P., Fei, J. et al. (2015). Synthesis of peptide-based hybrid nanobelts with enhanced color emission by heat treatment or water induction. *Chem. Eur. J.* 21 (26): 9461–9467.
58 Faller, P., Hureau, C., and Berthoumieu, O. (2013). Role of metal ions in the self-assembly of the Alzheimer's amyloid-β peptide. *Inorg. Chem.* 52 (21): 12193–12206.
59 Barondeau, D.P., Kassmann, C.J., Tainer, J.A. et al. (2002). Structural chemistry of a green fluorescent protein Zn biosensor. *J. Am. Chem. Soc.* 124 (14): 3522–3524.
60 Kim, J.H., Lim, S.Y., Nam, D.H. et al. (2011). Self-assembled, photoluminescent peptide hydrogel as a versatile platform for enzyme-based optical biosensors. *Biosens. Bioelectron.* 26 (5): 1860–1865.
61 Kim, J.H., Ryu, J., and Park, C.B. (2011). Selective detection of neurotoxin by photoluminescent peptide nanotubes. *Small* 7 (6): 718–722.
62 Zhao, Y.S., Fu, H., Peng, A. et al. (2010). Construction and optoelectronic properties of organic one-dimensional nanostructures. *Acc. Chem. Res.* 43 (3): 409–418.
63 Zhao, Y.S., Peng, A., Fu, H. et al. (2008). Nanowire waveguides and ultraviolet lasers based on small organic molecules. *Adv. Mater.* 20 (9): 1661–1665.
64 Muller, E.B., Flagg, P.B., DG, D. et al. (2007). Resonance fluorescence from a coherently driven semiconductor quantum dot in a cavity. *Phys. Rev. Lett.* 99 (18): 187402.
65 Zhang, C., Zou, C.L., Yan, Y. et al. (2011). Two-photon pumped lasing in single-crystal organic nanowire exciton polariton resonators. *J. Am. Chem. Soc.* 133 (19): 7276–7279.
66 Yan, X., Su, Y., Li, J. et al. (2011). Uniaxially oriented peptide crystals for active optical waveguiding. *Angew. Chem.* 123 (47): 11382–11387.
67 Sun, B., Li, Q., Riegler, H. et al. (2017). Self-assembly of ultralong aligned dipeptide single crystals. *ACS Nano.* 11 (10): 10489–10494.
68 Sun, B., Riegler, H., Dai, L. et al. (2018). Directed self-assembly of dipeptide single crystal in a capillary. *ACS Nano.* 12 (2): 1934–1939.
69 Briseno, A.L., Mannsfeld, S.C., Ling, M. et al. (2006). Patterning organic single-crystal transistor arrays. *Nature* 444 (7121): 913–917.
70 Kim, K., Rho, Y., Kim, Y. et al. (2016). A lattice-strained organic single-crystal nanowire array fabricated via solution-phase nanograting-assisted pattern transfer for use in high-mobility organic field-effect transistors. *Adv. Mater.* 28 (16): 3209–3215.
71 Boyd, R.W. (2020). *Nonlinear Optics*. Academic Press.
72 Lang, S.B. (2000). Piezoelectricity, pyroelectricity and ferroelectricity in biomaterials: speculation on their biological significance. *IEEE. T. Dielect. El. In.* 7 (4): 466–473.
73 Chen, X., Nadiarynkh, O., Plotnikov, S. et al. (2012). Second harmonic generation microscopy for quantitative analysis of collagen fibrillar structure. *Nat. Protoc.* 7 (4): 654–669.

74 Haupert, L.M. and Simpson, G.J. (2009). Chirality in nonlinear optics. *Annu. Rev. Phys. Chem.* 60: 345–365.

75 Brown, D.J., Morishige, N., Neekhra, A. et al. (2007). Application of second harmonic imaging microscopy to assess structural changes in optic nerve head structure ex vivo. *J. Biomed. Opt.* 12 (2): 024029-024029-5.

76 Handelman, A., Lavrov, S., Kudryavtsev, A. et al. (2013). Nonlinear optical bioinspired peptide nanostructures. *Adv. Opt. Mater.* 1 (11): 875–884.

77 Heredia, A., Bdikin, I., Kopyl, S. et al. (2010). Temperature-driven phase transformation in self-assembled diphenylalanine peptide nanotubes. *J. Phys. D Appl. Phys.* 43 (46): 462001.

78 Amdursky, N., Beker, P., Koren, I. et al. (2011). Structural transition in peptide nanotubes. *Biomacromolecules* 12 (4): 1349–1354.

79 Chen, Y., Tao, K., Ji, W. et al. (2022). Histidine as a key modulator of molecular self-assembly: peptide-based supramolecular materials inspired by biological systems. *Mater. Today.* 60: 106–127.

80 Chen, Y., Orr, A.A., Tao, K. et al. (2020). High-efficiency fluorescence through bioinspired supramolecular self-assembly. *ACS Nano.* 14 (3): 2798–2807.

12

Peptide-based Coacervates: A Biomimetic Protocell

Manzar Abbas

Department of Chemistry, Khalifa University of Science and Technology, P.O. Box 127788, Abu Dhabi, United Arab Emirates
Advanced Materials Chemistry Center, Khalifa University of Science and Technology, P.O. Box 127788, Abu Dhabi, United Arab Emirates

12.1 Introduction

There are certain defined hallmarks to understanding the features of living systems, for example compartmentalization, replication, energy transduction, growth and division, metabolism, and adoption [1, 2]. The complexity of life-like properties by the mixture of molecules could have been a longstanding challenge to understand. Based on the physical and chemical properties of biomimetic living systems, compartmentalization is considered the fundamental hallmark because of its physical similarity to a living cell where most of the metabolic reactions are held and have defined boundaries of a living system [3, 4]. Along with the importance of this hallmark, other hallmarks have their own advantages, but compartmentalization could have been considered as a starting point for the investigation into the origin of life [5]. Therefore, several studies have been reported on potential design rules and synthesizing rational protocells, which are considered basic compartments from which living cells can emerge. The ideal protocell models could be formed if compartments are made of abiotic molecules and show biomimicry properties of living cells [6, 7].

Over the past decade, several protocell models formed from different building blocks have been reported. Because of similar properties of modern cells, liposomes formed from phospholipids or fatty acids have been considered attractive candidates for protocellular compartments [8, 9]. Other than lipids, several different molecules such as polymers, peptides, proteins, and inorganic colloidal particles have the ability to form vesicles – compartments [10]. However, several limitations, for example the prebiotic irrelevance of some of the components, difficulty to form spontaneously, and permeability for nutrients, are real challenges. In 1929, Dutch scientist Hendrik G. Bungenberg de Jong used the term coacervate in his research on

Peptide Self-Assembly and Engineering: Fundamentals, Structures, and Applications, First Edition.
Edited by Xuehai Yan.

lyophilic colloids [11]. Later, in the 1930s, Oparin used coacervates to get the insights into origins of life as an alternative type of primitive cell that could have been formed within membraneless compartments. The coacervation or droplet formation happens spontaneously, and liquid–liquid phase separation (LLPS) drives this physical occurrence, where two phases get separated: liquid condensed phase – rich in coacervates forming molecules and an aqueous solution – lacking the coacervates forming molecules [12]. Coacervation is a spontaneous physical process, which is similar to the separation of oil and water, and the main difference between phase-separated coacervate systems and oil-in-water emulsions is that water is also present inside coacervates as a continuous phase [13]. As coacervates are membraneless organelles (MLOs), they can easily encapsulate solutes from outside the aqueous phase due to noncovalent interactions of molecules with coexisting phases and ultimately cause sequestration of micro- and macromolecules [14].

The apparent physical properties such as spontaneous formation, dynamic exchange of solute molecules, and the ability of selective accumulation make coacervates suitable compartments for protocell studies [15]. The increased concentration of coacervating motifs in the condensed phase and the chemically unique environment inside the coacervate droplets could play a role to increase the rate of reaction inside the droplet and thus make coacervates with properties similar to closed environment catalysts [16]. Nevertheless, the absence of a membrane outside the coacervate droplets shows a disadvantage because without membrane coacervates either simple or complex are tended to coalesce due to the least stability. Furthermore, coacervate formation needs long and highly charged building blocks such as polyelectrolytes and nucleic acids, which have complex chemical structures and are bigger in size. Therefore, it is a great challenge to identify a small and simple peptide residue in peptide coacervates – protocell model that can show phase separation under ample conditions and most importantly could provide a balance between their abilities to fuse and encapsulation of the solutes.

Over the past two decades, several polyelectrolyte molecules of synthetic and natural polymers such as proteins, nucleotides, polypeptides, and their chemically modified derivatives have been reported for the formation of coacervate droplets [17]. Interestingly, it is discovered that dynamic intracellular compartments in living cells are created by the natural physical process of phase separation. For example, disorder proteins with low-complexity regions showed phase separation [18]. These findings have developed tremendous interest in coacervates as protocells and could lead to new insights into designing short peptide residues which are simple and able to show phase-separating properties. Short peptides and their functionalized derivatives are particularly promising candidates for making coacervate protocells because of their simple chemical structure and the multiple roles of functional groups in different amino-acid residues and their side chains [14]. Resultantly, their structures are able to incorporate multiple moieties which carry the intermolecular interactions causing the phase separation.

In this chapter, we discuss the LLPS of peptides, an interesting property of peptide-based coacervates, and their central role in protocell formation with the context of the origin of life. After the brief introduction of coacervates, we focus on the fundamental driving forces for simple and complex coacervate systems comprising one or two peptide components where one component can be a peptide and other can be a nonpeptide, while simple coacervates are formed from a single peptide component. Similar to disorder proteins, peptide residues show LLPS in solution to make liquid droplets; thus, classification of the chapter is based on simple and complex coacervates forming peptides and their derivatives along with driving forces needed for the formation of coacervate droplets for protocell models. We also present the role of active coacervates for protocell formation along with a brief explanation of the encapsulation of dyes and pigments and their chemical reaction role in the synthetic cells.

12.2 Driving Forces for Coacervation and Types of Phase Separation

LLPS is usually driven through noncovalent interactions that are fundamental in the self-assembly process of short peptides to make ordered structures such as nanoparticles and nanofibers. This physical property governs the phase separation and leads the molecules to form liquid-like droplets – coacervates – spontaneously. The types of noncovalent interactions involved in phase separation to make simple and complex coacervates are briefly described, and the respective schematic representation is shown in Figure 12.1.

Charge–charge interactions: Charge–charge interactions are also named as electrostatic interactions between two oppositely charged molecules, and typical complex coacervates are formed from these interactions. Positive and negative charges on the respective polyelectrolytes either natural or synthetic polymers or proteins and nucleotides originate from the functional groups such as carboxylic acids and amine group present in the chemical structures of coacervates forming molecules as shown in Figure 12.1. The strength of the interaction is proportional to the charges and inversely proportional to the square of the distance between them.

Dipole–dipole interactions or H-bonds: These interactions are present between polar molecules that have dipole moments and have the ability to form complex coacervates. The positive polarized end of one molecule attracts the negative polarized end of another molecule/structure and vice versa to make order structures or cause phase separation. This type of interaction is between two atoms, where one atom can be an electronegative atom (O, N, and F), and the other atom is a hydrogen atom (H). Hydrogen bonding is also present between polar molecules, especially the

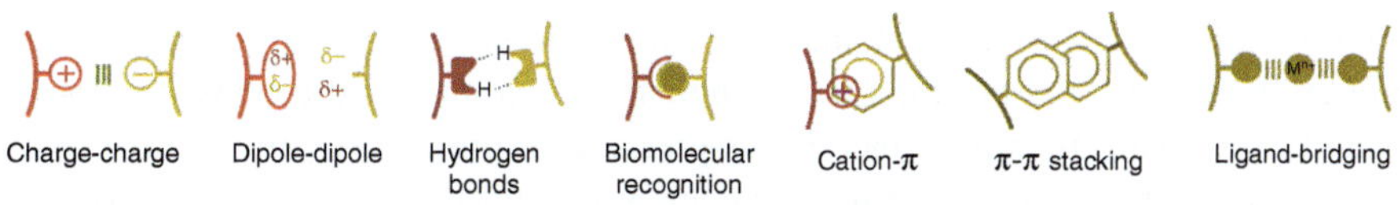

Figure 12.1 Types of molecular interactions that drive the LLPS of peptides to make simple and complex coacervates.

molecules that have electronegative atoms. This interaction is strong enough to cause the phase separation of biomolecules including peptides.

Cation–π interactions: These interactions are the attractive forces between positively charged species such as metal ions or ammonium ions and electron-rich aromatic rings. This interaction is found in the metal-binding sites in proteins, amino-acid side chains that are positively charged, and those that have aromatic side groups. This type of interaction triggers the metal-coordinated phase separation to make complex coacervates.

π–π stacking: It is the interactions between π electrons in p orbitals of aromatic rings that resultantly stack the aromatic rings to form ordered (assembled) and disordered (disordered) materials such as coacervates. Peptides that contain aromatic amino-acid residues (tryptophan, tyrosine, and phenylalanine) are usually phase separate due to these π–π stacking interactions. Other than their involvement in coacervation, these interactions can be found in DNA between the nitrogen bases that stack on top of one another, stabilizing the double-helix structure.

Bimolecular recognition: This involves the specific interactions between two biomolecules, which can be between proteins, nucleic acids, and carbohydrates. This specific interaction occurs with the help of other noncovalent interactions such as electrostatic interactions and hydrogen bonding. The arrangement and stereochemistry of each molecule make interactions specific. This is mainly found in enzyme–substrate interactions, antigen–antibody interactions, and even in designing target-specific drugs, somehow similar to host–guest interactions.

Ligand bridging: This interaction involves bridging the ligands such as carbonyl, chloride, and pyrazine, to establish a connection between two metal centers, and peptide/protein subunits or vice versa. Examples include metalloproteins (certain enzymes) in which there is a ligand bridge between the protein and the metal and nitride- or carbide-bridged phthalocyanine. Such kinds of ligand-bridging interactions are also central to LLPS to form complex coacervates.

The LLPS is of three types: (1) associative, (2) segregative, and (3) simple phase separation as shown in Figure 12.2a,b. In the associative phase separation, two molecules that are water soluble can be together in the same phase because of mutually attractive interactions. The condensed phase where the two molecules are gathered is named a coacervate, which is concentrated with both molecules and exhibits a considerable amount of solvent – more than 50% weight. The coexisting phase is depleted of both solutes and contains mostly solvent. A definitive example of this type of LLPS is the phase separation of two oppositely charged polymers. In the second type of phase separation, segregative, two molecules of peptides, polymers, and nucleotides that are soluble cannot mix because of the mutually repulsive interactions. Resultantly, both molecules can be separated into two phases, and each phase will be concentrated with one molecule. Usually, coexisting phases and occurrence of phase separation are described by a schematic diagram called a phase diagram (Figure 12.1c). Lastly, the third type, simple phase separation, is a relatively new type of LLPS, and it is the state where attractive forces of single molecule make it insoluble at solution temperature, pH, and concentration of that particular single compound. This phase separation also forms the concentrated phase, which is called

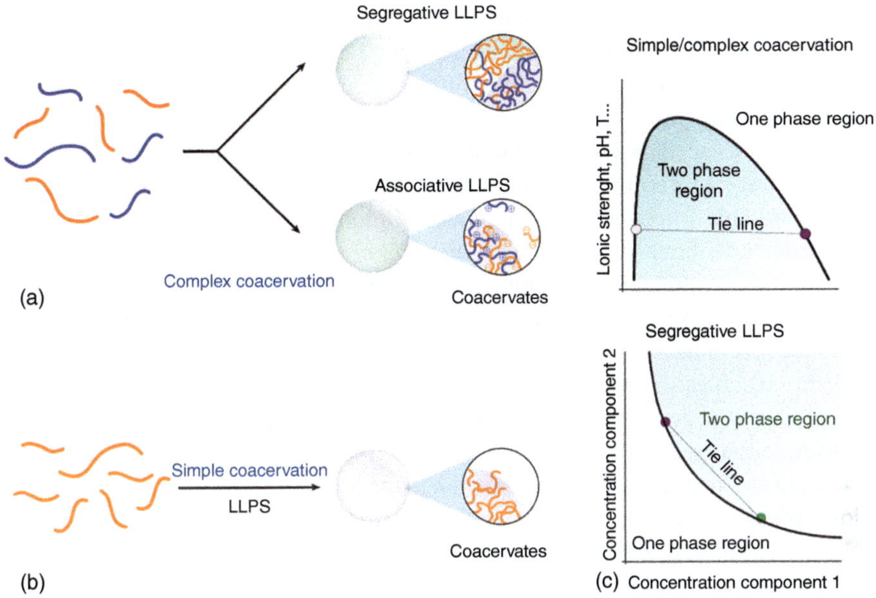

Figure 12.2 Schematic and phase diagrams to show the simple and complex coacervation of peptides: (a) types of LLPS – segregative and associative; (b) simple phase separation; and (c) phase diagram of simple and complex coacervates and segregative phase separation.

simple coacervate, having similar properties to the complex coacervates formed from associative phase separation. The proteins containing disordered domains also show simple phase separation to form MLOs and caused by charge–charge interactions, cation–pi, and pi–pi stacking (Figure 12.1). From a theoretical perspective, Flory–Huggins theory is used to model the simple coacervation in a similar way of segregation of polymer/nucleotides/peptides/proteins and solvent. Nonetheless, this phase separation has a similarity with associative phase separation, where molecules are enriched in a single phase and termed simple coacervation.

12.3 Peptide-based Complex Coacervates

Complex coacervates are formed from electrostatic interactions of two components. In this section, we consider the peptides as essential components with other motifs including nucleotides (3.1), peptides (3.2), and metal ions (3.3) to form complex coacervates.

12.3.1 Peptide/Nucleotide-based Coacervates

Similar to the co-assembly of two components, nucleotides and peptides as electro-opposite motifs can show phase separation to form complex coacervates. As it is evident from the literature that disordered proteins and nucleotides are central components of MLOs, compounds with the same chemical features can create biomimetic MLOs or protocells at an *in vitro* level. The first study on poly-L-lysine and adenosine triphosphate (ATP) and other nucleotides was reported by Mann and his team in 2011, and phase separation was observed into coacervate droplets from the solution [19]. Coacervates were stable at higher temperatures up to 90 °C and also in a certain range of pH like between 2 and 10. Interestingly, coacervates compartments were able to concentrate the client photoactive molecules, enzymes, and nanoparticles without changing the functions of guest molecules, which makes these systems prominent candidates for making protocells and reactors for biocatalysis. Similar peptide- and mono/polynucleotide-based systems were shown for enzymatic reactions in a dynamic controlled fashion. The phosphorylation of phosphoserine residues in arginine-based peptides and dephosphorylation of ADP to ATP could lead to the formation and deformation of coacervates [20]. This is important to note that single phosphorylation in the coacervate systems that comprised the poly-U and peptide $(RRASL)_3$ can change the ability of peptides for phase separation completely and lead to dissolve coacervate, and in the same way, single dephosphorylation in ATP/poly-L-lysine ends up in the deformation of coacervates. This study shows that the charge density has a central role in the stability of coacervates, and single phosphorylation of peptide residue decreases the net charge density of building blocks, which resultantly affects the stability of coacervates.

Though the building blocks of peptides and nucleotides or nucleic acids are quite simple, the coacervation can make complex arrangements in the condensed phase of components. Spruijt and his team showed recently that coacervates of RNA or ssDNA or nucleotides and different cationic polypeptides can form multiphase coacervates [21]. Such systems provide insights into biomimetic protocells as a container with subcompartments can give a hint to separate the noncompatible molecules and reaction pathways.

The sequence of peptides involved in coacervation increases the complexity and that also helps to get insights into the multitude of interactions causing the phase separation. For example, Das and his team found that peptide derivative Fmoc-VVVRRKK-NH_2, three hydrophobic amino-acid residues along with hydrophobic Fmoc group and four cationic residues, cannot form coacervates with ATP molecules polyanion [22]. The multitude of effects of the amphiphilic side chain of the peptide and high charge density region of peptide sequence and nucleotides contribute to forming the nanofibers (Figure 12.3a). Sequence modification of peptide derivative by scrambling the position of amino acids into alternative hydrophobic and hydrophilic manner (Fmoc-KVRVRVK-NH_2) can ultimately change the amphiphilic properties due to suppression in the local charge density. Apparently, this scrambled peptide motif was able to form complex coacervates with ATP.

Interestingly, these peptides also show enzymatic responsive behavior when using ATPase and dissembled into a transparent solution (Figure 12.3b). This suggests that certain charge spacing between amino-acid residues in peptide sequence is crucial for phase separation. Additionally, the adjacent uncharged residues can influence the overall charge and length of the peptide sequence, which is essential for the phase-separation process, and polycations of arginine and lysine – homopeptides

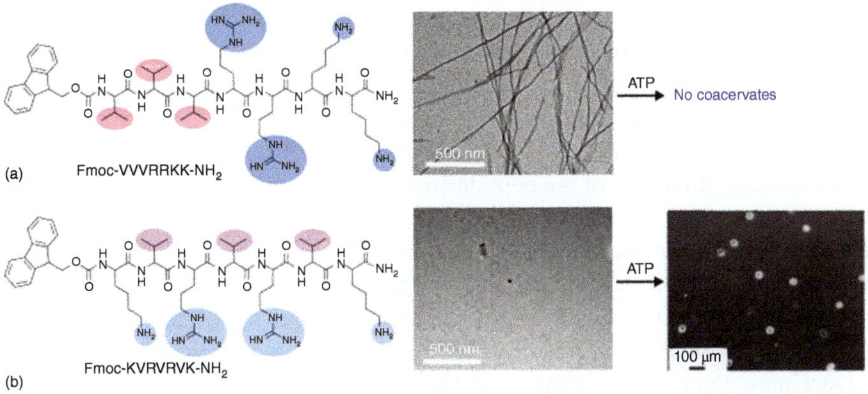

Figure 12.3 Chemical structures and respective assembly trail to show the structure–function relationship of peptide derivatives to make peptide condensates. Source: Reproduced with permission from Ref. (22) Royal Society of Chemistry, copyrights 2019.

with less than 10 amino-acid residues – cannot phase separate with ATP [23]. Contrary to more number of residues, only four cationic amino acids with alternative hydrophobic amino acids are enough to show phase separation with ATP. Similar to poly-L-lysine/ATP phase-separating system, these Fmoc-KVRVRVK-NH$_2$/ATP-based complex coacervates can take up the macromolecules such as proteins – cytochrome *c* (CytC) – and are also able to catalyze the redox reaction in dissipative conditions. These properties of the control assembly, encapsulation of catalytic proteins, and role in redox chemistry make complex coacervates form this system promising protocell models.

12.3.2 Peptide/Peptide-based Coacervates

Complex coacervates are liquid droplets formed from two oppositely charged peptide motifs, and the electrostatic interactions play a central role in the phase-separation phenomenon. Such kinds of complex coacervates of peptide–peptide interactions, for example polylysine (positive charge) and polyglutamate (negative charge), and charge–charge interactions are major driving forces for phase separation to construct liquid droplets. This phase separation of homopeptides is not only dependent on charge interactions, but another type of noncovalent interactions could be involved, depending on the chemical design of peptide motifs. A classic system of peptide–peptide interactions was reported first by Perry et al., where they identified that mixing of homopeptides either all L isomers or all D isomers of the same and opposite chirality formed solid aggregates rather than liquid droplets coacervates. With this information, they concluded that the formation of coacervates needs a minimum of one component with mixed enantiomeric configuration in the polypeptide [23]. This difference in morphology can be reasoned to the involvement of other interactions such as hydrogen bonding along with charge–charge interactions that lead to aggregation because of the formation of beta-sheet structures. The mixed isomers in the peptide sequence change the pattern of the peptide backbone which is needed to form beta sheets, stabilized by intermolecular hydrogen bonding, and this change can cause the LLPS. Conversely, this effect was not observed in short peptides – 10 amino-acid residues of glutamate, aspartate, arginine, and lysine could form coacervates [24]. However, poly-L-glutamate with 20-mers length forms aggregates in combinations (Figure 12.4). This attributes to the fact that a certain length of the peptide sequence is essential for the formation of stable beta sheets.

The propensity of forming liquid droplets via LLPS is dependent on (i) stereochemistry, (ii) length of polypeptide, (iii) charge density, and (iv) chemical properties of functional groups. Because of certain charge balances and interactions, poly-L-aspartate and poly-L-glutamate can form coacervates with poly-L-lysine and poly-L-arginine. For example, Keating and team investigated that a minimum of 10 amino-acid residues are required for the formation of peptide/peptide coacervates. Nevertheless, poly-L-arginine shows stronger interactions as compared to poly-L-glutamate, and poly-L-glutamate shows stronger interactions in comparison to poly-L-aspartate in mixtures for coacervation. This stronger interaction propensity of arginine amino acids with negatively charged nucleotides and peptides is due to charge–charge interactions that stem from the fact of its higher pK_a; it also shows a

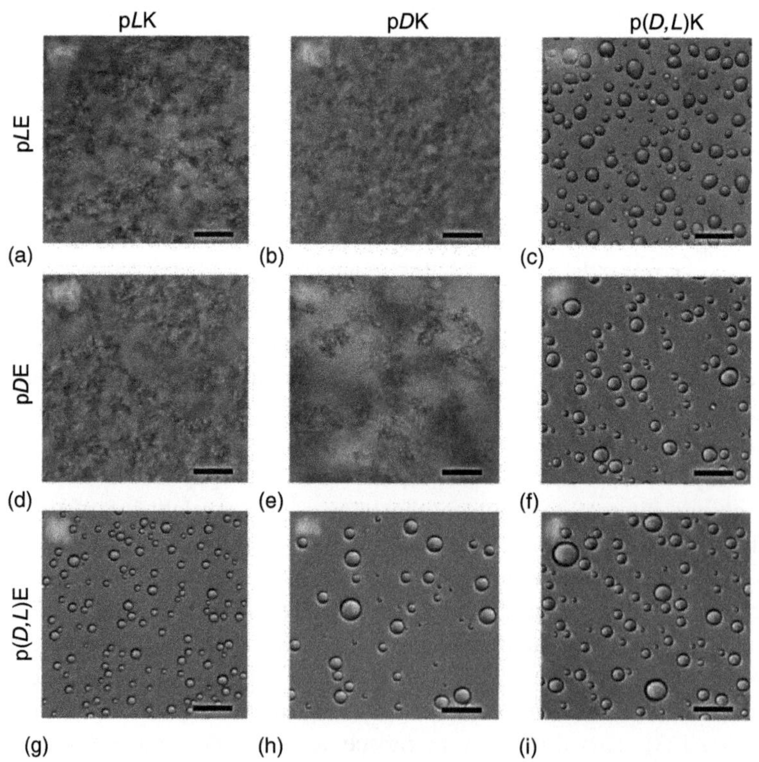

Figure 12.4 Optical microscope images of coacervates formed from different combinations of charged amino acids with chirality effect. Source: Adapted with permission from ref. [23]/ Springer Nature / CC BY 4.0.

tendency to form hydrogen bonds with phosphates, etc. and also cation–pi interactions with aromatic systems [25].

12.3.3 Peptide/Metal-based Coacervates

In living cells, metal ions are central to many functions as cofactors, catalysts, and coordination interactions with proteins and enzymes. The metalloenzymes have active sites to show metal and ligand interactions and play a role in protein and nucleotide folding. As metal ions have the tendency to bind with proteins, peptides and metal ions interactions can have a potential role in self-assembly and phase separation to make biomaterials. Such interactions between metal ions and peptides are usually with an amino group at the N terminal, heteroatom functional groups such as nitrogen and sulfur, and carboxylic acid at the C-terminal of the peptide sequence. Though there are several reports for supramolecular peptide nanomaterials formation using peptide and metal ions, recently, researchers are exploring the metal ions on phase separation to form liquid droplets. Increasing the concentration of divalent metal ions such as Mg^{2+} or Ca^{2+} in arginine-rich peptides and RNA mixture triggers the reentrant phase transition. Interestingly, if charge–charge interactions between peptide and RNA are minimized, then metal ions concentration leads to the dissolution of coacervates. It is important to note that electrostatic repulsive forces between negatively charged RNA arise and that resultantly due to base pairing and stacking interactions could drive the phase separation of homotypic coacervates of RNA–RNA [26].

Along with metal ions, small inorganic anions can also initiate phase separation of peptide residues through bridging interactions with positively charged peptide residues. For example, Boeynaems et al. [27]. show that divalent anions such as phosphate and carbonate can lead to the phase separation of arginine–proline-containing residues (PR_{20}) in the presence of the crowding agent polyethylene glycol (PEG). Similarly, the same research group investigated that bigger counterions with negative charges such as citrate, phosphate, sulfate, and nitrate (monoanionic) show ion pairing with positively charged arginine-rich proteins and are able to form simple coacervates from mussel-foot-derived proteins (fp-3F) without any external phase-separating agents such as crowders [28]. These charge–charge interactions can be screened by the salt and resultantly prevent the phase separation of particular peptide residue or protein. Nonetheless, even at higher concentrations of sulfate, it was unable to prevent the phase separation because of the bridging effect between the divalent and peptide.

Besides the metal ions and inorganic anions, inorganic metal clusters are able to form complex coacervates with short peptides. For example, Xiangyi et al. [29] used inorganic polyoxometalates (POM) – negatively charged with tripeptides – positively charged to form coacervates through condensation. The electrostatic interactions were considered the main driving force to induce the coacervation, but other interactions between the two components were also involved. The hypothesis was confirmed when they observed that the difference in coacervation by changing the sequence of tripeptides GVK at pH 6.5 did not form coacervates with SiW_{11}, and

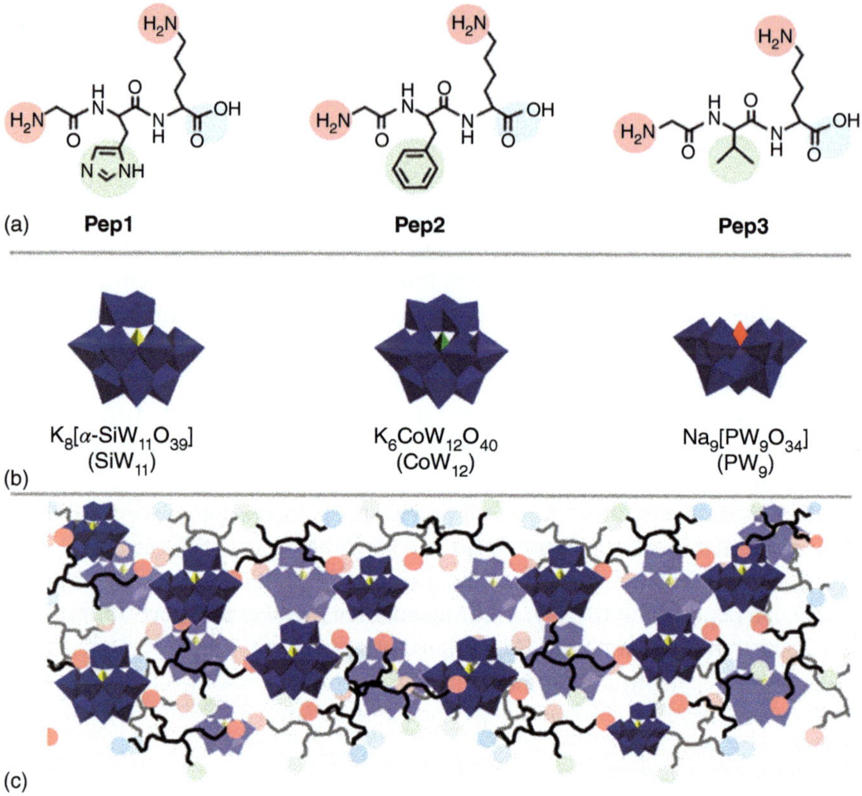

Figure 12.5 Chemical structures of tripeptides and polyoxometalates (POM) that form complex coacervates through electrostatic interactions. Source: Reproduced with permission from Ref. (29) American Chemical Society, @copyright 2019.

GHK and GFK formed complex coacervates. This difference comes from the other interactions of hydrogen bonding between histidine and POMs and the hydrophobic local environment that enhanced the charge–charge interactions between the two components (Figure 12.5a–c). Other than tuning the phase behavior in these systems, physiochemical properties can also be tuned especially in the histidine-containing peptide that shows different properties at different pHs due to the protonation and deprotonation effects of a histidine residue.

12.4 Peptide-based Simple Coacervates

Contrary to complex coacervates, simple coacervates could have formed from one component which is responsible to trigger the LLPS and thus resultantly make liquid droplets – compartments. This reveals that the dense phase coacervate is formed from the one component and aqueous phase. Similar to complex coacervates, the phase separation in simple coacervate is also driven by a multitude of weak intermolecular forces, stemming from the chemical structures of peptides. However, there are no concrete design roles for simple coacervates from single components of peptides. We discuss the simple coacervates formed from small peptide derivatives (4.1) and bioinspired peptides (4.2).

12.4.1 Designer Small Peptide Derivatives

Usually, to understand the protocells, simple as possible molecular building blocks are required that can form coacervates, alone or in combination with other motifs. Typical phase-separating small molecular motifs could be designed from low-complexity domains of phase-separating proteins, *de novo* design from molecular insights and different modeling outcomes. For example, Chilkoti and his team designed the minimal motif – octapeptide repeat units inspired by the disordered protein resilin from *Drosophila* and synthesized 64 variants with sequence length difference and different points of mutations [30]. They concluded that weak molecular interactions such as cation–π and hydrophobic drive the phase separation in laboratory-based designed disordered molecules. The aromaticity and length of amino-acid residues or simply the molecular weight of proteins are directly connected to the upper critical solution temperature (UCST) property of designer molecules – a temperature above which no coacervation happens and resultantly used to measure the strength of intermolecular interactions. Increasing the aromaticity and length of the designer molecule can increase the UCST. However, these designer molecules are not short peptides because of their sequence having more than 160 amino acids and 17 kDa size, and the simplicity of octapeptide repeats provides the basic information to design the shorter peptides for coacervate protocell models.

In order to find short peptide building blocks for phase separation, researchers have started using the peptide motifs which have shown self-assembling properties to make ordered structures. For example, the simple diphenylalanine (FF) motif – a well-known dipeptide – forms amyloid-like assemblies ranging from hydrogel to various morphologies of nanostructures. The derivative of FF dipeptide, *N*-carboxybenzyl protected (Z-FF), forms the metastable coacervates which ultimately transformed into nanofibers on aging [31]. The addition of silver ions to coacervates enhances the stability for more than 10 hours, and the stability of such droplets by variation in sequence could be promising for protocell models.

Recently, we developed the sticker–spacer model of short peptides derivatives with hydrophobic residues of FF and hydrophilic spacer cystamine, and interestingly, this water-soluble peptide derivative was able to phase separate by just tuning the pH [16]. The polar nature of the sticker and spacer has central role to tune the phase separation, and amazingly, strong hydrophobic effect in the sticker could lead to aggregation. The chemistry of sticker and spacer makes the peptide derivatives water soluble, and upon increasing the pH, they show LLPS to make liquid droplets – coacervates – spontaneously in a similar fashion to intrinsically disordered proteins (Figure 12.6). The redox chemistry functional present in spacer and amine functionality at the terminus of peptide derivative also play a role in phase separation and make the coacervates more responsive toward oxidation and reduction chemistry. Interestingly, to show the ability of these simple coacervates as protocell models, several light-sensitive pigments and macromolecules such as nucleotides and molecular beacons were sequestered and uptaken because of the hydrophobic regions of stickers that most probably exist inside the core of the coacervates. More importantly, these coacervates were able to host the anabolic organic reactions and catalyze the addition reactions that make the coacervates biocatalysis chambers and mimic the metabolic reactions happening in living cells. Simple coacervates of peptide derivatives have significantly enhanced the rate of reaction of addition reactions within coacervate as compared to control reactions in aqueous media alone with the same pH conditions.

Several other designer peptide molecules have been recently reported where tyrosine amino acid has played a role to make the formation of simple coacervates in a specific pH window and is prone to oxidation of tyrosine residues that resultantly forms different microstructures from coacervates [32]. Similarly, Leshem et al. [33] reported that the peptide composition of designer molecules dictates the propensity of LLPS, the dynamic nature of coacervates, and encapsulation efficiency, and interestingly, arginine and aromatic residues are central to make droplets and control the dynamic nature of coacervates.

12.4.2 Mussel Foot Peptide Derivatives

The proteins and peptides derived from underwater organisms are natural biomolecules to make coacervates and have been used as bioadhesive agents for underwater applications. One example is mussels foot protein (mfp) which is able to coacervate in aqueous media and interestingly has the ability to attach with wet surfaces.

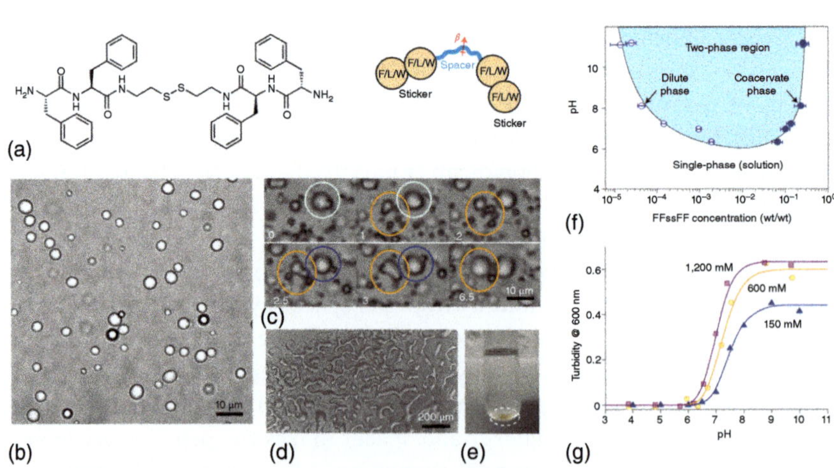

Figure 12.6 Chemical structure and phase-separation properties of sticker–spacer–sticker model designer peptides: (a) chemical structure; (b–e) microscope image, fusion ability, wetting, and coalescence, respectively; (f) phase diagram; and (g) responsive behavior of peptide coacervates. Source: Reproduced with permission from Ref. (16) Springer Nature Publisher, @copyright 2021.

This mfp, also referred as mfp-3S (45 amino acids), shows LLPS with the help of multitude of inter- and intramolecular interactions including electrostatic, charge-aromatic regions, and hydrophobic forces with apolar regions of the molecule. Wei and team designed a shorter, even simple self-coacervating peptide than mfp-3S named as mfp-3s-pep (25 amino acids – GYDGYNWPYGYNGYRYGWNKGWNGY), and coacervates were formed at the same conditions of external factors pH and salt as used for full-length protein [34]. The shorter peptide sequence is way easy to synthesize than the full-length protein and could be used as bioadhesive in aqueous environments. This short peptide mimic provides basic molecular insights into peptide-based simple coacervates. Liquid–liquid phase (LLPS) of mfp-3s-pep depends on ionic interactions and pH; salt can minimize the repulsive electrostatic interactions between cations and is thus able to promote cation–π interactions (Figure 12.7) [34, 35]. Besides that, the range of LLPS can be reduced by changing 3,4-dihydroxy-L-phenylalanine (DOPA) by phenylalanine residues in the design, which makes the peptide more hydrophilic and least propensity for LLPS. However, this enhanced the adhesive characteristics on metal ions because of hydrogen bond formation with the DOPA motif. So far, before the discovery of FF-based phase-separating

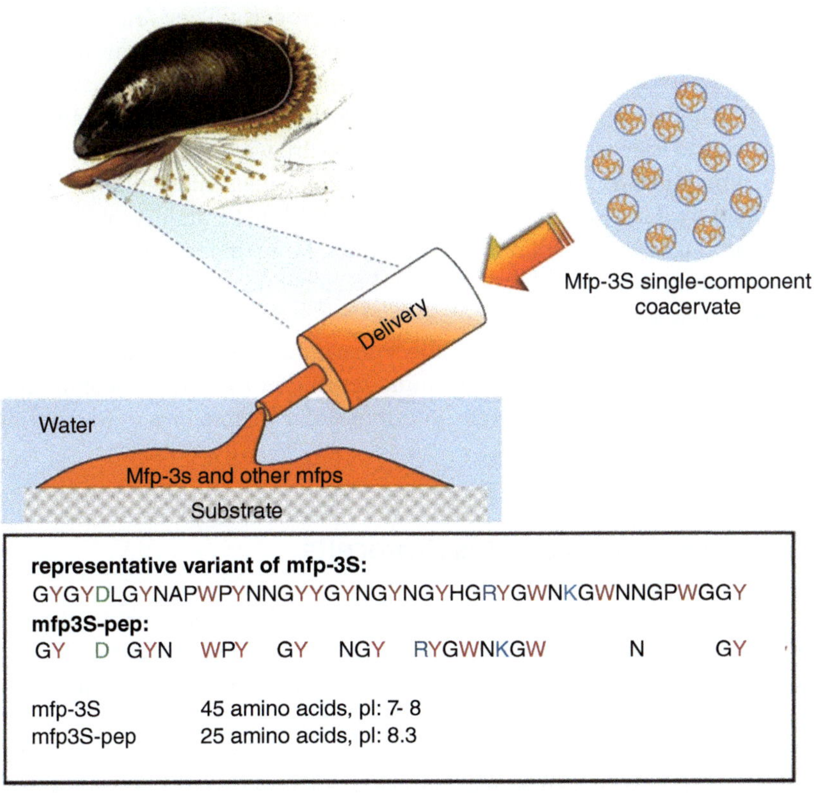

Figure 12.7 Schematic illustration of bioinspired peptide sequence for simple coacervation. Source: Reproduced with permission from Ref. (34) Wiley-VCH, @copyright 2016.

molecules, mussel foot-derived peptides are simple peptide derivatives that undergo coacervation, and that is why, such short molecules can be central to designing simple peptide protocells.

Other than mussel-foot protein derivative, jumbo squid is an interesting marine organism that can exploit protein-based coacervates. The secretion of histidine-rich proteins from the hard beak of the marine organism is named histidine-rich beak proteins (HBPs), which are capable to form coacervates [36]. Interestingly, these coacervates assemble at the growing beak and resultantly end up into a hard β-sheet-rich structure. This protein has a resemblance with elastin and elastin-like peptides in terms of coacervation, which undergoes phase separation to form coacervates due to the presence of pentapeptide repeats VPGXG; X is any amino acid except proline. While HBPs contain $GHGX_1X_2$ repeating motifs in the sequence, X_1 represents proline, valine, or leucine, and X_2 represents tyrosine. To understand the interactions that were involved in the LLPS, two molecules that are analogues of HBPs with several repeat units of $GHGX_1X_2$ were developed. It was found that hydrophobic interaction is the central driving force in the self-coacervation like salt facilitates the coacervation in mfp-3 s-pep by minimizing the effect of charge [37].

Further HBP-1 sequence was investigated by Gabryelczyk et al. for getting insights into the factors that influence the phase-separation property. They found that the GHGLY part in the sequence caused the phase separations, and a minimum of four repeats of this sequence are required for droplet formation. The amino acids tyrosine and histidine in the sequence are crucial for LLPS because on increasing the pH, histidine amino acid is deprotonated and forms the hydrogen bonding with tyrosine amino acids, which makes the coacervate phase more stable. Interestingly, on further increasing the pH, tyrosine amino acid gets deprotonated and thus weakens the interactions and causes the dissolution of coacervates. More importantly, they found that peptides having GAGFA instead of GAGLH sequence in the middle of all repeats cannot phase separate to form coacervates but end up as compact and dense hydrogel [38]. This difference in the assembly behavior can be reasoned to increase hydrophobic interactions due to phenylalanine residue in the GAGFA-containing peptides, which resultantly enhances the intermolecular interactions of the overall molecule.

12.5 Biomimetic Coacervates Protocells

Researchers have identified fundamental hallmarks to make artificial or biomimetic primitive cells that are close to living cells in their functions and structure. We discuss those hallmarks briefly in this section in light of the structure–function relationship of peptide motifs.

12.5.1 Compartmentalization and Sequestration

The main structure of a living cell exhibits the compartment-like morphology, and living materials including nucleotides and macromolecules are enclosed inside that compartment. There are several types of compartments including polymersomes, liposomes, proteinosomes, and coacervates designed to study the modern cells which could be considered as chemical ancestors of living cells [12]. This compartmentalization is the first step toward the journey to make artificial cells, which we mainly discussed in Sections 3 and 4. However, we discuss the sequestration of guest molecules inside the coacervates which make them attractive compartments due to unavailability of membrane on the periphery; resultantly, a wide range of micro–macromolecules get sequestered inside the coacervates. A central reason why client micro- and macromolecules get sequestered inside is the decreased polar environment of coacervates. Usually, the polarity inside the coacervates is lower in comparison to the surrounding environment, or we can say that the hydrophobic forces are strong within the core that provides the opportunity to hydrophobic guest molecules including thioflavin-T, Nile Red, and many more to be accumulated [16] (Figure 12.6). Interestingly, when the concentration of client molecules reaches the concentration of coacervates forming molecules, then partition coefficient cannot be described by concentration effect, and then, partitioning molecules become part of associative phase separation that ends up to form hybrid coacervate. Other than hydrophobic effect, charge–charge interactions of basic and acidic amino acids and hydrogen bonding are also important factors that contribute in this physical phenomenon of sequestration. Usually, partitioning coefficients determine the ability of any external guest molecules for sequestration.

When guest and host molecules have same intermolecular interactions, big molecules can also encapsulate into coacervates on the cost of small coacervate-forming molecules due to the increase in entropy of the system that released from coacervate phase to dilute phase [39]. For example, coacervates of polyamine/ATP have partitioning coefficients around 10^5, and the local concentration of ATP is about 500-fold higher than the dilute phase [40]. This entropy-driven effect was also observed in complex coacervates of polyamines and short peptides of decamers of lysine or arginine [41].

12.5.2 Catalytic Nature

Catalysis within coacervates is primarily linked with the encapsulation and sequestration of small and large molecules including enzymes and small organic reactants. Simple or complex coacervates provide the environment for a reaction presumably happening inside the coacervates, which are completely different from the surrounding environment in physicochemical aspects. This environment changes the energy states of client molecules and their products; resultantly, the conformation of reactants and products can be changed. In other words, the suitable interactions in coacervate core and encapsulated species could also stabilize the compartments and ultimately could enhance the rate of reaction [39]. There is one other way the intermediate complex or the product within the compartments can be stabilized, through the addition of crowding agents [42]. Nonetheless, the chemical properties of coacervate droplets could have played a central role in determining the reactivity than the crowding effect, and interestingly, transcription has been possible in the core of coacervates formed from poly-L-lysine/CMDex systems [43], where the rate of gene expression rate was lower and yield was also reduced in comparison to PEG-induced coacervates.

Recently, we observed a collective impact of concentration on condensation reactions of small organic molecules within coacervates. Dipeptide derivatives with polar spacer enhanced the rate of two addition reactions of aldol condensation and hydrazone formation significantly with 50- and 12-fold higher rates as compared to control reactions [16]. In this designer system, FF motifs are hydrophobic in nature and were responsible to enrich the reactants 2–5 times higher due to hydrophobic interactions, and this effect lowers the energy barrier up to 6 kJ/mol. There are several examples where coacervates of synthetic polymers, nucleotides, and Ru-polyoxometalate catalysts [44] have been used for enhancing the rate of reaction, but peptide-based coacervate systems are not well reported until today.

12.6 Conclusion and Future Perspectives

Peptide-based protocells are promising candidates to understand and get the insights to recapitulate the features of a minimal biological cell. For this purpose, compartment-type microstructures including coacervates could help to provide life-like properties that stem from the simple molecules. Simple or complex coacervates formed from peptides spontaneously have become attractive candidates for biomimetic protocells because of their physical features that mimic modern living cells. These coacervates formed from LLPS show advantages over their counterpart protocellular microstructures such as the ability to sequester and enriched the solutes for chemical reactions. Rather than using bigger synthetic and natural polymers, proteins, now minimal building blocks of short peptides and small ionic species, form liquid droplets through minute changes in pH, salt, and temperature. Due to their physicochemical properties including growth and potential to divide and replicate, peptide-based coacervates could have emerged as biomimetic protocells.

Noncovalent interactions given in Figure 12.1 including electrostatic interactions, hydrogen bonding, and hydrophobic interactions that are responsible for coacervate formation also make these compartments promising candidates for synthetic cells. Electrostatic interactions that neutralize charges, hydrogen bonding, and π–π stacking together make the cores of coacervate droplets more hydrophobic, which cause the encapsulation of guest macromolecules. Recently, one step closer to real cells are to host the small molecules and enhance the rate of reaction that somehow mimic the metabolic reactions. However, cell division is way difficult to understand in coacervate systems, but this could be possible by tailoring different small molecules, particularly short peptide-based systems. Besides this, we need to make coacervate systems that could have life-like properties including growth and division. In general, current advancements in peptide-based coacervates have shown promising physicochemical properties that short peptides can make protocells that could be best biomimetic natural cells.

Acknowledgment

MA greatly acknowledges the financial support from Khalifa University, project number 8474000462.

References

1 Yewdall, N.A., Mason, A.F., and Van Hest, J.C. (2018). The hallmarks of living systems: towards creating artificial cells. *Interface Focus.* 8 (5): 20180023.

2 Slootbeek, A.D., van Haren, M.H., Smokers, I.B., and Spruijt, E. (2022). Growth, replication and division enable evolution of coacervate protocells. *Chem. Commun.* 58 (80): 11183–11200.

3 Radzvilavicius, A.L. and Blackstone, N.W. (2018). The evolution of individuality revisited. *Biol. Rev.* 93 (3): 1620–1633.

4 Wang, X., Wu, S., Tang, T.-Y.D., and Tian, L. (2022). Engineering strategies for sustainable synthetic cells. *Trends Chem.* 4 (12): 1106–1120.

5 Gaut, N.J. and Adamala, K.P. (2021). Reconstituting natural cell elements in synthetic cells. *Adv. Biol.* 5 (3): 2000188.

6 Ruiz-Mirazo, K., Briones, C., and de la Escosura, A. (2014). Prebiotic systems chemistry: new perspectives for the origins of life. *Chem. Rev.* 114 (1): 285–366.

7 Schreiber, A., Huber, M.C., and Schiller, S.M. (2019). Prebiotic protocell model based on dynamic protein membranes accommodating anabolic reactions. *Langmuir* 35 (29): 9593–9610.

8 Misuraca L. Protomembranes at the Origin of Life: Université Grenoble Alpes; 2021.

9 Monnard, P.-A., DeClue, M., and Ziock, H.-J. (2008). Organic nano-compartments as biomimetic reactors and protocells. *Current Nanosci.* 4 (1): 71–87.

10 Li, M., Huang, X., Tang, T.D., and Mann, S. (2014). Synthetic cellularity based on non-lipid micro-compartments and protocell models. *Curr. Opin. Chem. Biol.* 22: 1–11.

11 de Jong, H. (1932). Die Koazervation und ihre Bedeutung für die Biologie. *Protoplasma.* 15 (1): 110–173.

12 Haren, M.H.I., Nakashima, K.K., and Spruijt, E. (2020). Coacervate-based protocells: integration of life-like properties in a droplet. *J. Syst. Chem.* 8: 107–120.

13 Dardelle, G. and Erni, P. (2014). Three-phase interactions and interfacial transport phenomena in coacervate/oil/water systems. *Adv. Colloid Interface Sci.* 206: 79–91.

14 Abbas, M., Lipiński, W.P., Wang, J., and Spruijt, E. (2021). Peptide-based coacervates as biomimetic protocells. *Chem. Soc. Rev.* 50 (6): 3690–3705.

15 Martin, N. (2019). Dynamic synthetic cells based on liquid–liquid phase separation. *ChemBioChem.* 20 (20): 2553–2568.

16 Abbas, M., Lipiński, W.P., Nakashima, K.K. et al. (2021). A short peptide synthon for liquid–liquid phase separation. *Nature Chemistry.* 13 (11): 1046–1054.

17 Spoelstra, W.K., Deshpande, S., and Dekker, C. (2018). Tailoring the appearance: what will synthetic cells look like? *Curr Opin Biotechnol.* 51: 47–56.

18 Hughes, M.P., Sawaya, M.R., Boyer, D.R. et al. (2018). Atomic structures of low-complexity protein segments reveal kinked b sheets that assemble networks. *Science* 359 (6376): 698–701.

19 Koga, S., Williams, D.S., Perriman, A.W., and Mann, S. (2011). Peptide–nucleotide microdroplets as a step towards a membrane-free protocell model. *Nat. Chem.* 3 (9): 720–724.

20 Aumiller, W.M. Jr. and Keating, C.D. (2016). Phosphorylation-mediated RNA/peptide complex coacervation as a model for intracellular liquid organelles. *Nat. Chem.* 8 (2): 129–137.

21 Mountain, G.A. and Keating, C.D. (2019). Formation of multiphase complex coacervates and partitioning of biomolecules within them. *Biomacromolecules* 21 (2): 630–640.

22 Saha, B., Chatterjee, A., Reja, A., and Das, D. (2019). Condensates of short peptides and ATP for the temporal regulation of cytochrome c activity. *Chem. Commun.* 55 (94): 14194–14197.

23 Perry, S.L., Leon, L., Hoffmann, K.Q. et al. (2015). Chirality-selected phase behaviour in ionic polypeptide complexes. *Nat. Commun.* 6 (1): 6052.

24 Cakmak, F.P., Choi, S., Meyer, M.O. et al. (2020). Prebiotically-relevant low polyion multivalency can improve functionality of membraneless compartments. *Nat. Commun.* 11 (1): 5949.

25 Kaur T, Raju M, Alshareedah I, Davis RB, Potoyan DA, Banerjee PR. 2020.

26 Onuchic, P.L., Milin, A.N., Alshareedah, I. et al. (2019). Divalent cations can control a switch-like behavior in heterotypic and homotypic RNA coacervates. *Sci. Rep.* 9 (1): 12161.

27 Boeynaems, S., Bogaert, E., Kovacs, D. et al. (2017). Phase separation of C9orf72 dipeptide repeats perturbs stress granule dynamics. *Mol. Cell.* 65 (6): 1044–1055. e5.

28 Yang, B., Jin, S., Park, Y. et al. (2018). Coacervation of interfacial adhesive proteins for initial mussel adhesion to a wet surface. *Small* 14 (52): 1803377.

29 Li, X., Zheng, T., Liu, X. et al. (2019). Coassembly of short peptide and polyoxometalate into complex coacervate adapted for pH and metal ion-triggered underwater adhesion. *Langmuir* 35 (14): 4995–5003.

30 Dzuricky, M., Rogers, B.A., Shahid, A. et al. (2020). De novo engineering of intracellular condensates using artificial disordered proteins. *Nat. Chem.* 12 (9): 814–825.

31 Yuan, C., Levin, A., Chen, W. et al. (2019). Nucleation and growth of amino acid and peptide supramolecular polymers through liquid–liquid phase separation. *Angew. Chem. Int. Ed.* 58 (50): 18116–18123.

32 Abbas, M., Law, J.O., Grellscheid, S.N. et al. (2022). Peptide-based coacervate-core vesicles with semipermeable membranes. *Adv. Mater.* 34 (34): 2202913.

33 Baruch Leshem, A., Sloan-Dennison, S., Massarano, T. et al. (2023). Biomolecular condensates formed by designer minimalistic peptides. *Nat. Commun.* 14 (1): 421.

34 Wei, W., Petrone, L., Tan, Y. et al. (2016). An underwater surface-drying peptide inspired by a mussel adhesive protein. *Adv. Funct. Mater.* 26 (20): 3496–3507.

35 Kim, S., Yoo, H.Y., Huang, J. et al. (2017). Salt triggers the simple coacervation of an underwater adhesive when cations meet aromatic π electrons in seawater. *ACS Nano.* 11 (7): 6764–6772.

36 Tan, Y., Hoon, S., Guerette, P.A. et al. (2015). Infiltration of chitin by protein coacervates defines the squid beak mechanical gradient. *Nat. Chem. Biol.* 11 (7): 488–495.

37 Cai, H., Gabryelczyk, B., Manimekalai, M.S. et al. (2017). Self-coacervation of modular squid beak proteins–a comparative study. *Soft Matter.* 13 (42): 7740–7752.

38 Gabryelczyk, B., Cai, H., Shi, X. et al. (2019). Hydrogen bond guidance and aromatic stacking drive liquid-liquid phase separation of intrinsically disordered histidine-rich peptides. *Nat. Commun.* 10 (1): 5465.

39 Nakashima, K.K., Vibhute, M.A., and Spruijt, E. (2019). Biomolecular chemistry in liquid phase separated compartments. *Front. Mol. Biosci.* 6: 21.

40 Frankel, E.A., Bevilacqua, P.C., and Keating, C.D. (2016). Polyamine/nucleotide coacervates provide strong compartmentalization of Mg2+, nucleotides, and RNA. *Langmuir* 32 (8): 2041–2049.

41 Poudyal, R.R., Guth-Metzler, R.M., Veenis, A.J. et al. (2019). Template-directed RNA polymerization and enhanced ribozyme catalysis inside membraneless compartments formed by coacervates. *Nat. Commun.* 10 (1): 490.

42 André, A.A. and Spruijt, E. (2020). Liquid–liquid phase separation in crowded environments. *Int. J. Mol. Sci.* 21 (16): 5908.

43 Tang, T.-Y.D., Van Swaay, D., DeMello, A. et al. (2015). In vitro gene expression within membrane-free coacervate protocells. *Chem. Commun.* 51 (57): 11429–11432.

44 Gobbo, P., Tian, L., Pavan Kumar, B.S. et al. (2020). Catalytic processing in ruthenium-based polyoxometalate coacervate protocells. *Nat. Commun.* 11 (1): 41.

13

Water-induced Peptide Self-assembly and Its Function

Wei Fan[1], *Chengqian Yuan*[1], *and Xuehai Yan*[1,2,3]

[1] *State Key Laboratory of Biochemical Engineering, Institute of Process Engineering, Chinese Academy of Sciences, Beijing, China*
[2] *School of Chemical Engineering, University of Chinese Academy of Sciences, Beijing, P. R. China*
[3] *Center for Mesoscience, Institute of Process Engineering, Chinese Academy of Sciences, Beijing, P. R. China*

13.1 Introduction

Organisms have formed many complex and exquisite ordered structures in the process of millions of years of evolution and realized the diversification of biological functions [1, 2]. Many ordered structures in organisms are formed by the synergistic self-assembly of biomolecules through multiple weak interactions between molecules [3]. Modern biophysical chemistry and cell biology studies have shown that in the process of biomolecular assembly, water is not only used as a solvent but more importantly as a structural water to participate in the assembly of biomolecules, maintain the structure of biomolecules, and achieve life functions [4, 5].

Water structure and its dynamic behavior are fundamentally related to the special properties of biological molecules such as proteins and nucleic acids. Water molecules are actually involved in biomolecular interactions, determining the key sources of macromolecular conformation and association. They form a very wide range of structures, such as linear and reticular cluster water structures, which contribute to structural stability and information transfer in cells. In addition, water molecules can promote proton transfer, mediate electrostatic interactions, and serve biological functions [6, 7].

During the process of peptide self-assembly, water molecules play an essential role by reducing the energy barrier of monomer peptide/protein molecular structure [8–10]. Hydration helps release water molecules and facilitates the formation of salt bridges that stabilize the monomer's structure. The release of bound water between the layers encourages the formation of precursors. Additionally, the ordering of water molecular structure promotes the amphiphilic peptide self-assembly

Peptide Self-Assembly and Engineering: Fundamentals, Structures, and Applications, First Edition.
Edited by Xuehai Yan.

process. As a result, water molecules significantly contribute to inducing peptide self-assembly. Utilizing the interaction between water molecules and peptide molecules, researchers have designed a range of water-responsive functional materials and crystalline peptide materials for wearable devices, soft robots, and other fields.

This chapter concentrates on the structures of liquid, solid, and interfacial water molecules and their impact on peptide and protein microstructure, behavior, and function. Moreover, we provide a summary of the role of dynamic water molecule behavior in inducing peptide self-assembly as well as the design of water-responsive materials based on the interaction between water molecules and peptide-assembled crystals.

13.2 The Structure of Water

13.2.1 Radial Distribution Function of Water

The structure of liquid water can be described by the radial distribution function (rdf) g(r), which is related to its density [11]. However, since water has an oscillatory radial distribution function, a single function g(r) cannot fully characterize its structure. Therefore, a series of partial rdfs gXY(r) must be defined to express the probability of finding atom Y at a distance r from the center of atom X. For instance, gOO(r) represents the radial distribution of oxygen atoms in other H_2O molecules around any given molecule. In Figure 13.1, gOO(r) generates a separation peak that exceeds the average molecular diameter (or radius of gyration) σ, with a peak value of about 1.4σ [12]. These results indicate that water molecules are separated by hydrogen bonds, which impose geometric constraints on their close contact. Hydrogen bonds have a clear linear trait, with hydrogen atoms located approximately between two oxygen atoms' axes to maximize orbital overlap and increase bonding strength.

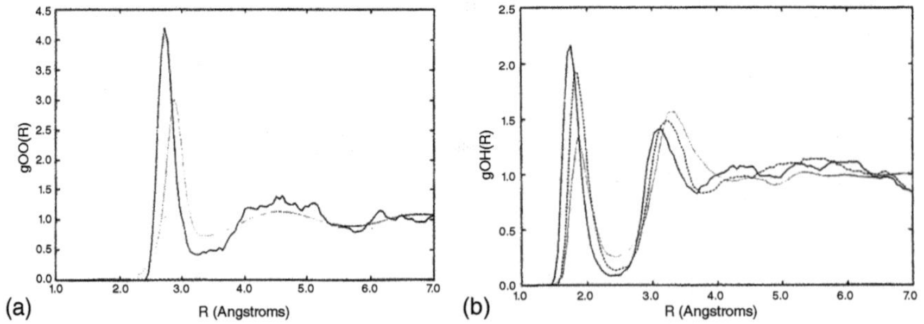

Figure 13.1 (a) Oxygen–oxygen radial distribution functions, DFT/SPC (normal line) and experimental (dotted line). (b) Oxygen–hydrogen radial distribution functions, DFT hydrogen–TIP3P oxygen (normal line), DFT oxygen–TIP3P hydrogen (dashed line), and experimental (dotted line). *Source:* Adapted with permission from Ref. [12]; © 1996, Wiley-VCH.

13.2.2 Liquid and Solid Water Cluster Structures

In recent decades, significant advancements have been made in the accurate description of the structure of water [13]. Despite the establishment of several force field models and the development of powerful *ab initio* molecular dynamics methods, key properties of liquid water, such as heat capacity, density, and dielectric constant, cannot be accurately calculated [14]. Although the hydrogen bond network within liquid water has been elucidated, and its fluctuation and rearrangement dynamics have been studied, there are currently no experimental techniques available that can provide detailed molecular-level information about its structure.

In 1989, Busarow and colleagues [15] captured the first far-infrared spectrum of gaseous water clusters near $22\,cm^{-1}$ (455 μm), providing a precise characterization of the dimer's ground state. Subsequently, Zwart et al. [16] expanded on this data to other quantum states. With the development of technology and the expansion of spectrometer frequency, Puliano and Saykally [17] first measured the intermolecular vibrational–rotational–tunneling (VRT) spectra of water clusters in 1992 and detected the torsion of D_2O trimer near $89.5\,cm^{-1}$. In Berkeley, the follow-up work of Liu et al. [18] produced a broader trimer spectrum and assigned the transitions in detail for the first time. Cruzan et al. [19] found the VRT spectrum of tetramers, and Liu et al. further detected pentamers [20] and hexamers [21]. Thus, the molecular structure of liquid water presents a variety of cluster structures [22]. As illustrated in Figure 13.2, as the number of water molecules increases, the hydrogen-bonding mode between the water molecules also significantly changes, forming rings or cages, among others [23].

Compared to the molecular structure of liquid water, the configuration of solid water molecules is also intricate. There are over 20 types of solid water, and these molecules are often subject to phase transitions in specific environments. For example, Tulk et al. [24] reported that at 100 K, increasing pressure causes the normal form

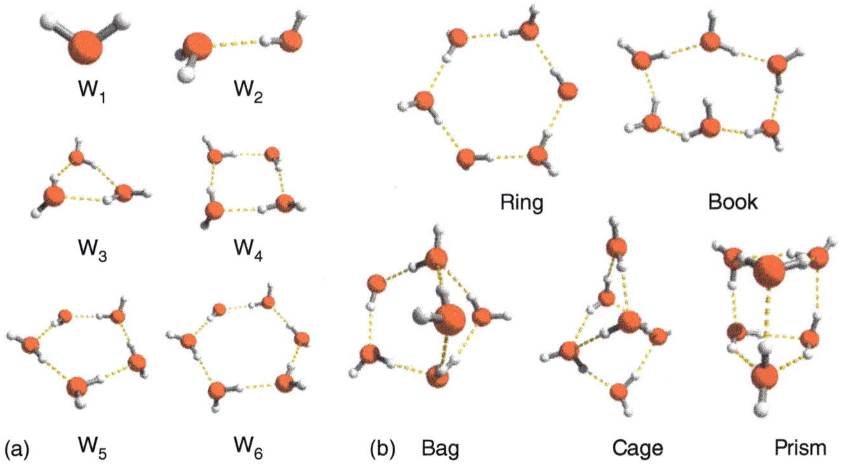

Figure 13.2 (a) Different cluster structures of liquid water. (b) Types of hexamolecular water clusters. *Source:* Adapted with permission from Ref. [23]; © 2001, Wiley-VCH.

of ice (ice Ih) to convert sequentially into other crystalline forms (ice IX', ice XV', and ice VIII') after resting for an hour at each pressure, rather than forming an amorphous solid as observed in previously reported compression experiments. In short, the distinctive properties of water, whether solid or liquid, make it highly susceptible to environmental factors because of the hydrogen bonding among its molecules.

13.2.3 Interface-ordered Water

In 1959, Kauzmann proposed that the mutual attraction of hydrophobic solutes in water is a crucial driving force for protein folding and aggregation [25]. Furthermore, the hydration of water in hydrophobic entities is more orderly and rigid [25]. Later, Frank and Evans' "classic iceberg" diagram explained hydrophobic attraction using the concept of "entropy" [26]. When two hydrophobic surfaces are close, the release of some "ordered water" drives hydrophobic attraction. The concept of "ordered water"is an attractive one, as seen in polywater [27]. Solutes or surfaces can disturb water molecules, leading to the formation of more or less ordered molecular structures of water, providing a convenient theoretical basis for explaining the unique physical properties of water and the cooperative behavior of biomolecules in water.

The wetting phenomenon of "water nonwetting water monolayer" is remarkable [28, 29]. On hydrophobic surfaces, an organized molecular layer of water acts as a hydrophobic coating, impeding further wetting. This unique wetting phenomenon is prevalent on metal surfaces at very low temperatures [30], graphene [31], and may even occur on biomolecule surfaces [32]. Femtosecond time-resolved sum frequency generation probes have been used to study the ultrafast dynamics of ordered water molecules on polytetrafluoroethylene surfaces, revealing structures including cyclic tetramers in the molecular water layer [33]. Raman spectroscopy is often utilized to study the O—H bonding properties and the chemical environment of interfacial water [34]. Plasma-enhanced Raman spectroscopy is used to enhance the signal intensity of the O—H bond of interface water. *In situ* Raman spectroscopy directly distinguishes the two structural transformations of water on the surface of Au single-crystal electrode. As the potential becomes more negative, the structure of interfacial water molecules evolves from "parallel" to "one H-down" and then to "two H-down." The experiment also showed that the number of hydrogen bonds in the interface water changed twice.

The molecular structure and dynamics of water molecules in liquids, solids, and interfaces are intricate and diverse. Water molecules are significant participants that directly impact the functional expression of biomolecules, including proteins.

13.3 Water Determines Protein Structure and Function

13.3.1 Dynamic Hydration

Water molecules and polar groups on the surface of biomolecules form a layered water structure through hydrogen-bonding interactions, also known as a hydrated layer. Typical biomolecules such as protein [35], DNA double helix [36], and

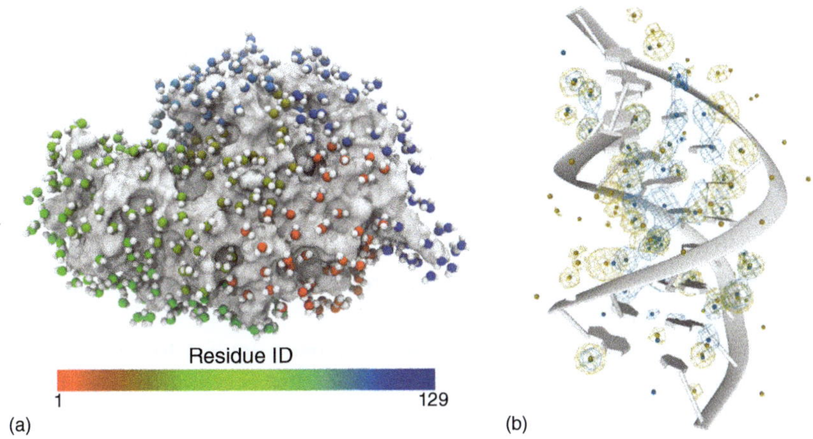

Figure 13.3 Typical biomolecular hydration layer: a protein and (b) DNA double strand. Part (a) *Source:* Ref. [35]/John Wiley & Sons/CC BY 4.0. Part (b) *Source:* Ref. [36]/John Wiley & Sons/CC BY 4.0.

phospholipid bilayer can form a hydrated layer [8, 9, 37] (Figure 13.3). The fluctuation of water molecules in the hydration layer affects the change in biological molecular structure and function expression. The dynamic process of biomolecular hydration has a very large span in space and time scales. In space, studies have shown that the thickness of the biomolecular hydration layer ranges from a single water molecule to a large pool; in time, the relaxation dynamics of water molecules are distributed on the femtosecond to microsecond time scale [38]. Studying the hydration process of biomolecules over such a wide range of time and space dimensions is extremely challenging. With the development of experimental techniques and computer simulation, the study of biomolecular hydration has shown diversity in terms of theory and experimental methods [10, 39]. X-ray and neutron scattering techniques can capture the structural information of biomolecules and their surface water molecules. In addition, ultrafast time-resolved vibration spectroscopy studies the dynamic changes of biomolecular hydration; combined with molecular simulation, it can provide a more accurate and visualized image of biomolecular hydration.

Compared with the relaxation time of small molecules in water, which is only a few picoseconds, the relaxation time of proteins reaches tens of picoseconds [40]. Water molecules cannot rotate under the influence of the protein's electrostatic field. In order to reorient, water molecules have to exchange with bulk water molecules. In other words, this slower kinetic process originates from the intrinsic slow kinetics of water, involving the exchange of water molecules between the bulk phase and the hydration layer. The movement of water on the surface of spherical proteins was systematically assessed using NMR techniques [41], and their reorientation and flipping were 3–5 times slower than bulk water. In addition, similar results have been reported in related theoretical simulation studies. Both the disordered geometrical hydrogen bonds and the local energy formed between water and polar groups on proteins lead to inhomogeneity and kinetic delay of water molecules in the hydration shell [38, 42].

Terahertz spectroscopy [10, 43] can directly detect changes in the low-frequency spectrum of solvated proteins, capturing the picosecond hydrogen bond dynamics of proteins and water molecules. The 1–10 THz frequency range corresponds to the low-frequency mode and plays a vital role in controlling conformational changes, triggering biochemical reactions, and energy transport through large amplitude modes (skeleton). Terahertz light extends the dielectric region from nanosecond to subpicosecond motion, allowing direct detection of free, translational, intermolecular, and collective motion of hydrated water and the large motion of biomolecules [10]. As such, terahertz spectroscopy provides a new means of directly detecting the coupling of protein hydration kinetics.

The extended jump model of water has recently been applied to describe the dynamic process of water-hydrogen bond reorientation near a wide range of solutes [43]. As illustrated in Figure 13.4, bulk water molecules experience a sudden large angle transition when the water hydroxyl group exchanges the hydrogen bond acceptor. This model is particularly suited for analyzing the dynamics of the hydration layer. The hopping model is affected by three key effects of the biomolecular interface: hydrophobic groups, hydrogen bond donors, and hydrogen bond acceptors [44]. Hydrophobic groups exclude new water–hydrogen bond acceptors, thus impeding their proximity through volume effects. Hydrogen bonds form with different strengths through hydrogen bond donors that act around the water oxygen with angular jumps, although the torque generated by these bonds on hydroxyl reorientation can be disregarded. Hydrogen bond acceptors can determine the hydrogen bond dynamics of water by excluding the volume effect and the strength of hydrogen bonds formed with water simultaneously.

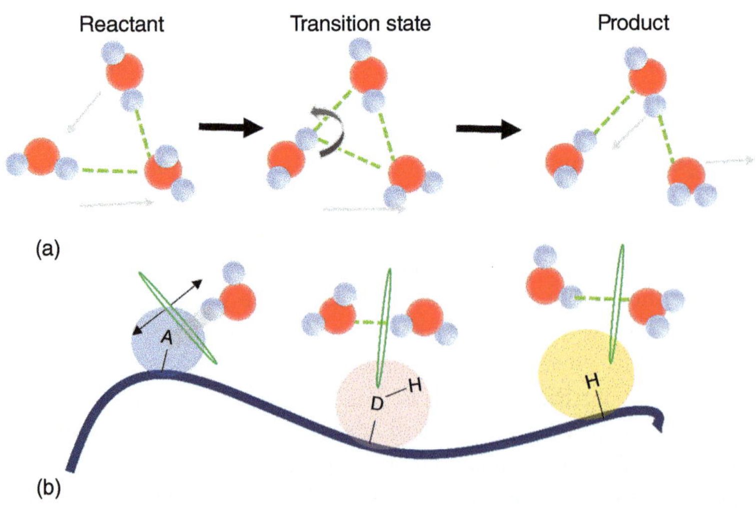

Figure 13.4 (a) Molecular hopping mechanism. (b) Schematic diagram of three types of sites containing protein interface hydrophobicity, hydrogen bond donor, and hydrogen bond acceptor and their disturbance types to hydrodynamics. Part (b) *Source:* Adapted with permission from Ref. [44]; © 2006, ACS.

In summary, changes in the hydrogen bond dynamics of water can induce alterations in the structure of biomolecules, which may significantly impact their functional expression. To gain a better understanding of how water behavior affects the structure and function of biomolecules, the upcoming section will concentrate on the influence of water on the expression of biomolecule functions.

13.3.2 Water Participating in Protein Proton Transfer

The internal bound water of protein can actively participate in the expression of biological functions and the transmission of information. The water molecules are combined to form long hydrogen bond chains through hydrogen bonds, which are often called "water wires" [7]. Protons can move rapidly in the water environment. After adding a proton at one end of the water wires, the hydrogen bond that flips down along the waterline releases an equivalent proton at the other end. This special process is usually called the Grotthuss hopping mechanism [45], which often occurs in the proton transport process of biomolecules. Ordered water molecules can connect proton donor sites and acceptor sites to promote rapid proton transfer. There is a strong correlation between the proton transfer efficiency and the hydrogen bonding of water molecules. The stronger the hydrogen bonding of water molecules, the faster the proton transfer rate. The behavior pattern of water molecules and the exact structure of water line are the key factors affecting proton transfer.

Water commonly participates in the proton transfer of protein molecules. For instance, PsbO, a subunit protein of photosystem II (PSII), has a hydrophilic cavity made up of polar amino-acid residues [5]. This cavity facilitates rapid proton transfer, while low-mobility water molecules covering the surface of the PsbO protein form an extended water–carboxylate network, promoting docking of PsbO and PSII complexes. The hydration shell of the composite consists of 1300 water molecules, offering multiple hydrogen bond channels for proton transfer. As demonstrated in Figure 13.5, water molecules efficiently participate in proton transfer, leading to protein structure isomerization [6]. Similarly, DNA repair

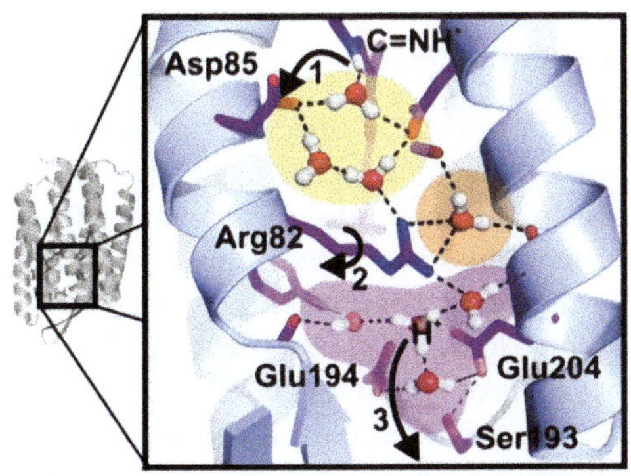

Figure 13.5 Proton release mechanism of membrane proteins during structural transition. *Source:* Adapted with permission from Ref. [6]; © 2010, Wiley-VCH.

process exhibits analogous water molecular dynamic behaviors. The DNA repair enzyme MutY removes DNA bases damaged by hydroxyl radical oxidation [4], whereby the protonation of an adenine nitrogen atom determines the rate-determining step of the cleavage process. The protein's hydrophobic and hydrophilic residues play a "water trap" role, forming five highly ordered water clusters skillfully assisted by other water clusters. Hence, water molecules actively take part in the proton transfer process of proteins, playing an essential role in protein functions' expression and information delivery.

13.3.3 Water Inducing Protein Allosteric Effect

Subtle changes in the folding structure of highly structured proteins can affect the expression of protein functions. This structural change can occur under different conditions, such as ligand binding and protein hydrolysis cleavage. This change in protein structure caused by external environmental disturbances is called allosteric phenomenon. For example, the binding of oxygen in one subunit of hemoglobin changes the affinity of oxygen in another subunit [46, 47]. The binding sites of the protein communicate with the distal active sites through allosteric phenomena. The hydration of proteins is the basis of structural stability and flexibility, and water molecules can cause nonnegligible disturbances to protein structure. The coupling of water molecules and proteins is an important factor in the allosteric phenomenon of proteins.

At present, the allosteric mechanism of water–protein coupling has been very clear. Dongping Zhong et al. [48] used intrinsic tryptophan as a localized optical probe for single-point-specific scanning of protein surfaces. At femtosecond time resolution, 16 mutants were detected to observe the relaxation process of the water molecular network. The local collective water network relaxation completes the initial ultrafast kinetic process ($\tau 1$) within a few picoseconds. At longer times ($\tau 2$), the water network undergoes structural rearrangements within layers, strongly coupling local protein fluctuations and dynamic exchange of bulk water. Water network relaxation is strongly correlated with local protein chemistry and structural rigidity.

13.3.4 Antifreeze Protein Inhibits Water Crystallization

Antifreeze proteins (AFPs) possess bound water with a high degree of spatial extension order, regulating ice crystal nucleation and growth via ice association [49]. The fish antifreeze protein Maxi comprises a hydrophobic channel composed of a four-helix bundle that entraps over 400 water molecules in a cage-like network dominated by five-membered rings in crystallography [50]. This ordered network extends outward through the gap between the spirals, forming an ordered layer of water molecules on the outer surface, which helps Maxi bind to ice crystals and impede their growth [50]. The structure of the protein hydration layer is distinct between highly active insect AFP and type III AFP, resulting in their recognition difference on different planes (see Figure 13.6) [49]. The highly active insect AFP exhibits a semicage structure, while the type III AFP has an ice-like structure.

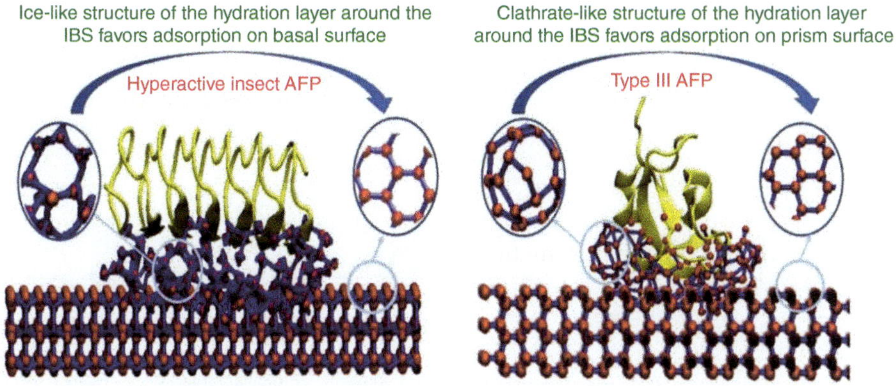

Figure 13.6 Different hydration layer structures around highly active antifreeze protein (AFP) and type III AFP and their different roles in different plane recognition. *Source:* Adapted with permission from Ref. [49]; © 2021, Wiley-VCH.

13.4 Water-induced Peptide Self-assembly and Its Function

13.4.1 The Role of Water in Protein Fibrosis

Proteins form amyloid fibrils by misfolding, which can induce a variety of neurodegenerative diseases, such as Alzheimer's disease and Parkinson's disease [51, 52]. The process of protein fibrosis is often accompanied by misfolding of natural structures, generating ordered β-sheet structures [53, 54]. The physical and chemical mechanisms based on the peptide/protein self-assembly process are extremely complex, involving protein structure changes on ultrafast time scales, and the synergy of multiple noncovalent bonds and hydrophobic interactions. In particular, the noncovalent interaction between water molecules and proteins is considered to be a key factor in amyloid fibrosis. However, it is generally believed that protein–protein interaction is the main driving force for protein aggregation, and the surrounding water is only a bystander. In fact, water molecules can actively participate in the process of protein fibrosis as active substances.

Water molecules play a crucial role in the process of protein fibrillation. For example, solid-state nuclear magnetic resonance and cryoelectron microscopy studies have revealed the presence of water channels formed by the hydration of polar side chains in transthyretin fibers [55]. X-ray fiber diffraction analysis has shown that the Aβ amyloid fiber at the trimer interface is "hollow" and filled with water [56]. During protein monomer transformation into fibers, N*-rich misfolded structures containing an abundance of water are preferentially created. Collisions with oligomers possessing diverse proportions of water molecules lead to the amplification of protein-rich droplet. The oligomers gradually develop nuclei as they grow in size, ultimately maturing into fibers exhibiting various shapes (Figure 13.7). Consequently, the shape and growth kinetics of fibers depend significantly on the creation process of N* aggregates and the kinetic behavior of water molecules [58, 59].

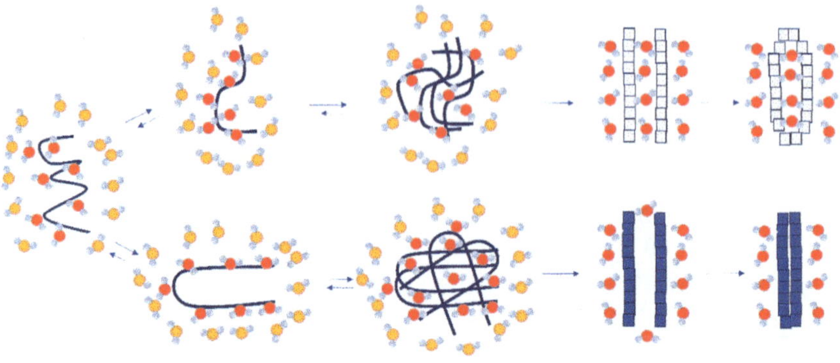

Figure 13.7 Protein aggregation mechanism of polymorphic fibers. *Source:* Adapted with permission from Ref. [57]; © 2012, ACS.

Experimental and molecular dynamics simulations of Aβ peptides and other peptides have demonstrated that peptide aggregates can adopt numerous conformations at different energy barriers, with the easily aggregated N* conformation typically occupying the lower free energy barrier [57]. The energy barrier of aggregation generally falls into two categories. Firstly, Aβ peptide fibers may partially unfold in their native state or partially fold in an unfolded state, necessitating the traversal of the free energy barrier and producing transient populations of the N* conformation of high assembly capacity. Secondly, in the context of mammalian prion protein aggregation, the N* structural ensemble has lower free energy than the eigenstate ensemble, signifying the metastability of the folding state. In either case, water-mediated interactions are responsible for generating a free energy barrier between the ground and N* states.

Water molecules can initiate the formation of salt bridges in protein molecules and affect the fibrillation process. In the structural model of Aβ1–40 fibrils based on solid-state NMR, it was found that the D23–K28 salt bridge played a crucial role in the early stage of Aβ peptide self-aggregation. A single water molecule forms hydrogen bonds with both D23 and K28 side chains [57]. Upon expulsion of the involved water molecules, a stable intramolecular salt bridge can form, necessitating overcoming a considerable desolvation barrier. Furthermore, the hydrogen bond lifetime of solvated D23 and K28 is nearly three times greater than that of bulk water, indicating stronger hydrogen bond interactions between these side chains and water molecules compared to those between water molecules. It has also been shown that the desolvation of D23 and K28 is an activation process. Protein molecules can form salt bridges by means of the desolvation process of water molecules, thereby reducing the barrier to monomer structure. In fact, experiments demonstrate that the polymerization rate of a monomer containing the salt bridge between D23 and K28 is 1000 times faster than that of a wild-type monomer.

Water molecules play an important role in the binding kinetics of two sheets formed by peptides rich in polar amino-acid side chains [57]. Throughout the binding process, beginning with completely dissolved pores, water molecules are affected by hydrogen bond interactions of polar amino acids, resulting in an ordered water line between the precursors. If there is displacement between the two layers, confined water molecules will exist in a disordered structure. Finally, the release of water molecules between the two layers leads to the precursor's formation. Interestingly, the ordered and disordered waters between the layers exhibit completely different behaviors during the release process, with ordered water extending orderly, and disordered water escaping from the gap on both sides of the precursor.

In summary, water molecules play multiple vital roles in peptide/protein fibrillation. In the early stages of fibrillation, hydrogen bond interactions between water molecules construct the free energy barrier between the ground state and the N* state. Additionally, the release of water molecules between polar amino acids establishes a salt bridge between residues, forming a transition state structure with stronger polymerization ability. In the final stage of fiber formation, the release of water molecules directly leads to the precursor's formation.

13.4.2 The Impact of Water Molecule Behavior on the Peptide Self-assembly Process

Amphiphilic peptide molecules possess the structural characteristics of amphiphilic surfactants and the bioactive functionality of peptides, making them promising for applications in biomedicine and biosensors [60, 61]. The self-assembly process of these molecules is influenced by external environmental factors such as pH, temperature, and charge. Various types of amphiphilic peptide molecules can assemble into different shapes, including sheets, spheres, and rods, depending on environmental stimuli [62]. Nanofibers created through the self-assembly of peptide amphiphiles (PAs) hold potential as light-harvesting fiber scaffolds [63]. Therefore, understanding the mechanism of fiber formation is crucial.

While it is widely recognized that water plays a critical role in the self-assembly of PAs, the active function of water molecules is often overlooked. Using multilevel coupling atomization and coarse-grained methods, molecular dynamics (MD) simulations can capture accurate atomic images of the early stages of PAs self-assembly into nanofibers and micelles, thereby revealing the crucial role of water molecules in promoting PAs self-assembly. Simulation results confirm that the structural order of the water molecule near the hydrophilic amino acid is significantly enhanced, promoting the balance to shift toward the fiber phase and stimulating structural order [64]. However, rupturing the ordered cages surrounding amphiphilic peptide molecules causes the hydrophobic structure to collapse and form micelles. In addition, measuring the infrared OH bond stretching vibration confirms the existence of an ice-like structure. This indicates that the solvated water in the formed peptide network becomes increasingly organized over time, highlighting the role of water in the self-assembly process. As a result, the local ordering and solvation effects of water drive the initial fiber and ensure the stability of the constructed fiber network.

Recently, it was found that water molecules play a critical role in triggering and facilitating the nucleation and aggregation processes of peptide self-assembly [65].

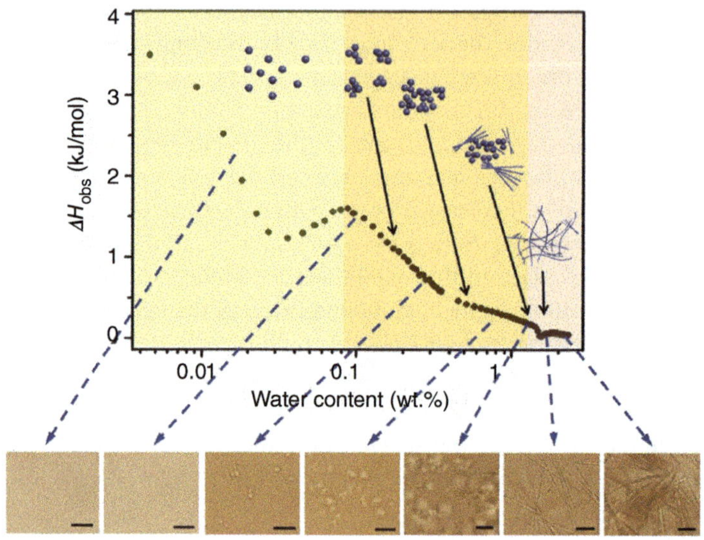

Figure 13.8 The FF peptide assembly process exhibits significant changes in morphology and thermodynamic energy. The scale bar in the optical microscope image is 10 μm. *Source:* Adapted with permission from Ref. [65]; © 2017, Wiley-VCH.

They shape morphology and significantly influence the system's thermodynamic characteristics during assembly. Variations in water content lead to different trends in latent heat changes during nucleation and aggregation processes. Research shows that the latent heat value undergoes a sudden drop at approximately 0.04 wt% water content (see Figure 13.8). This discovery provides novel perspectives on the role of water molecule in mediating biomolecule self-assembly.

13.4.3 Water Molecules Participating in and Stabilizing the Peptide Crystal Structure

Water molecules are common in peptide crystal. For instance, water plays a crucial role in the crystallization of aromatic dipeptides [66]. In the cyclo-phenylalanine–tryptophan (cyclo-FW) crystal, water molecules connect monomers from one main chain diketopiperazine ring to another side chain indole ring to form hydrogen bonds. In the cyclo-tryptophan–tryptophan (cyclo-WW) crystal, adjacent main chain diketopiperazine rings are connected by two pairs of water molecules forming three hydrogen bonds. Additionally, a water molecule on each side connects to the side chain indole ring to form two hydrogen bonds, creating a quasi-rectangular hydrophilic channel. The considerable differences between these structures indicate that the supramolecular structure and macroscopic properties of peptides can be altered by changing their sequence. Notably, the number ratio of water molecules to peptide in the crystal lattice can be finely tuned from zero (cyclo-diphenylalanine [FF], cyclo-glycine-tryptophan [GW]) to one (cyclo-FW, cyclo-W[d]W). Moreover, the number of water molecules can vary among 2 (linear-FF, linear-GW), 3 (cyclo-WW), and 4 water molecules (linear-W[d]W). Water molecules also influence the

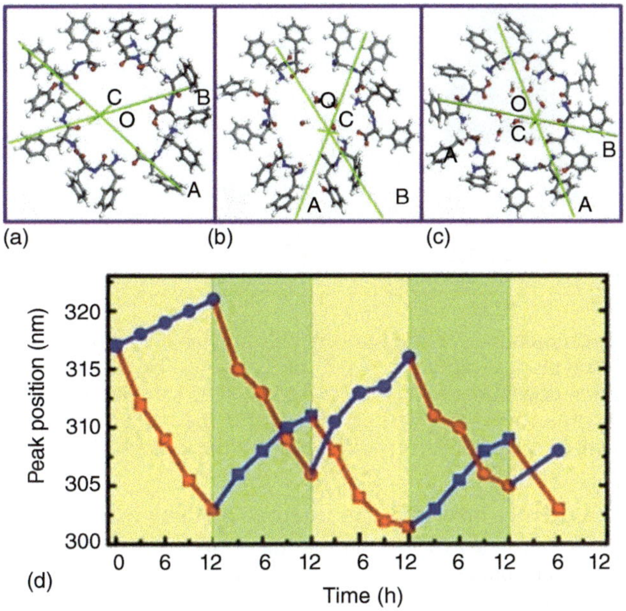

Figure 13.9 The water-containing supercell structure consists of six FF molecules, where (a) 0, (b) 4, and (c) 8 water molecules are located at the core of the channel. Red, gray, blue, and white spheres represent oxygen, carbon, nitrogen, and hydrogen atoms, respectively (d). The photoluminescence peak position changes with drying and water treatment. *Source:* Adapted with permission from Ref. [67]; © 2011, Wiley-VCH.

band gap of subnanocrystal structures formed via FF peptide self-assembly and can trigger a distinct UV photoluminescence (PL) peak. The position of this peak is directly proportional to the number of water molecules in PNTs (Figure 13.9). The introduction or removal of water molecules can lead to a red or blue shift of the UV PL peak. Finally, density functional theory calculations verify that introducing water molecules to PNTs causes the splitting of the valence band peak, corresponding to the shifting and splitting of the observed UV PL peak [67]. In summary, regulating the number of water molecules and the form of hydrogen bonding can effectively control the micro–nanostructures and properties of crystals.

13.4.4 Water-induced Mechanical Deformation of Self-assembled Materials

Humidity-responsive materials demonstrate notable structural responses to changes in relative humidity, providing opportunities for collecting and utilizing evaporative energy. They can serve as effective actuators driven by relative humidity for various applications, including soft robots and wearable devices [68, 69]. In nature, humidity response behavior has a crucial role in key biological functions, such as the opening and closing of the pineal gland, or the advancement of the wheat awn toward the ground. Some natural humidity-responsive materials have been applied outside their biological environment, demonstrating robust driving capabilities.

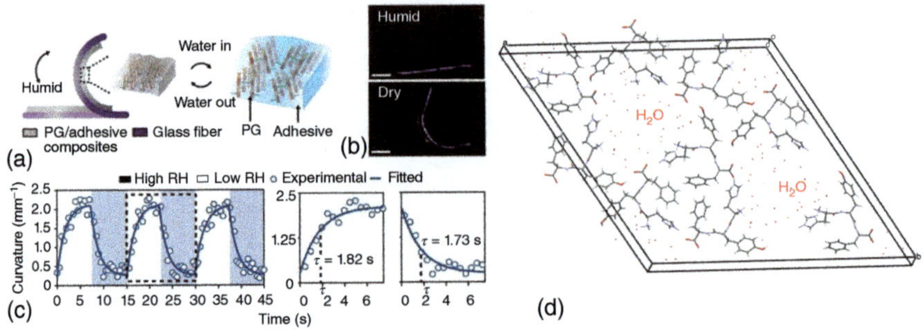

Figure 13.10 (a, b) Peptidoglycan/adhesive composite materials are humidity driven, bending under dry conditions and elongating under humid conditions. (c) The relaxation times for bending and elongation of peptidoglycan/adhesive composite materials are 1.82 seconds and 1.73 seconds, respectively. (d) Water cavities exist in the His-Tyr-Phe single crystal. Water-responsive tripeptide crystals. *Source:* Ref. [69]/John Wiley & Sons/CC BY 4.0.

Peptidoglycan/adhesive composite materials can undergo bending action under humidity control, as shown in Figure 13.10 [69]. These materials exhibit relaxation times of 1.82 and 1.73 seconds for bending and elongation, respectively. Figure 13.10d demonstrates that His-Tyr-Phe tripeptide crystals can serve as a water-responsive material due to the presence of water molecule cavities in their single crystal structure [68]. Altering the amount of water molecules can result in structural deformation in the single crystal.

At the fundamental level, the water responsiveness of materials is an energy storage process that entails the absorption and desorption of water molecules. This process is driven by the chemical potential difference of water. Hydrogen bonds and water–surface interactions are essential components of the energy storage process. These components enable the conversion of molecular interactions into physical structural deformation. Thus, the adsorption and desorption of water molecules provide the impetus for the physical deformation of water-responsive materials.

13.5 Conclusions

In conclusion, this chapter highlighted the key participant role that water molecules play in biological processes, including protein allostery, stability, information transfer, and self-assembly dynamics. Water molecules facilitate the formation of easily aggregated intermediates through hydration. Subsequently, their release promotes the formation of precursor fibers. The dynamic behavior and order of water molecules directly induce the self-assembly of peptides. By utilizing hydrogen-bonding interactions between water and peptide molecules, the synthesis of water-responsive peptide self-assembly materials presents potential applications in the field of molecular motors.

Acknowledgment

This work was financially supported by the National Natural Science Foundation of China (project nos. 22202207, 22172172, 22232006, and 22025207).

References

1 Knowles, T.P., Oppenheim, T.W., Buell, A.K. et al. (2010). Nanostructured films from hierarchical self-assembly of amyloidogenic proteins. *Nat. Nanotechnol.* 5: 204–207.

2 Yao, H.-B., Fang, H.Y., Wang, X.H., and Yu, S.H. (2011). Hierarchical assembly of micro-/nano-building blocks: bio-inspired rigid structural functional materials. *Chem. Soc. Rev.* 40: 3764–3785.

3 Yuan, C., Ji, W., Xing, R. et al. (2019). Hierarchically oriented organization in supramolecular peptide crystals. *Nat. Rev. Chem.* 3: 567–588.

4 Brunk, E., Arey, J.S., and Rothlisberger, U. (2012). Role of environment for catalysis of the DNA repair enzyme MutY. *J. Am. Chem. Soc.* 134: 8608–8616.

5 Umena, Y., Kawakami, K., Shen, J.R., and Kamiya, N. (2011). Crystal structure of oxygen-evolving photosystem II at a resolution of 1.9 Å. *Nature* 473: 55–60.

6 Wolf, S., Freier, E., Potschies, M. et al. (2010). Directional proton transfer in membrane proteins achieved through protonated protein-bound water molecules: a proton diode. *Angew. Chem. Int. Ed.* 49: 6889–6893.

7 Swanson, J.M., Maupin, C.M., Chen, H. et al. (2007). Proton solvation and transport in aqueous and biomolecular systems: insights from computer simulations. *J. Phys. Chem. B* 111: 4300–4314.

8 Duboue-Dijon, E., Fogarty, A.C., Hynes, J.T., and Laage, D. (2016). Dynamical disorder in the DNA hydration shell. *J. Am. Chem. Soc.* 138: 7610–7620.

9 Abel, S., Galamba, N., Karakas, E. et al. (2016). On the structural and dynamical properties of DOPC reverse micelles. *Langmuir* 32: 10610–10620.

10 Xu, Y. and Havenith, M. (2015). Perspective: watching low-frequency vibrations of water in biomolecular recognition by THz spectroscopy. *J. Chem. Phys.* 143: 170901.

11 Liu, J., He, X., Zhang, J.Z.H., and Qi, L.W. (2018). Hydrogen-bond structure dynamics in bulk water: insights from ab initio simulations with coupled cluster theory. *Chem. Sci.* 9: 2065–2073.

12 Tuñón, I., Martins-Costa, M., Millot, C. et al. (1996). A coupled density functional-molecular mechanics Monte Carlo simulation method: the water molecule in liquid water. *J. Comput. Chem.* 17: 19–29.

13 Stillinger, F. and H. (1980). Water revisited. *Science* 209: 451–457.

14 Chen, B., Xing, J., and Siepmann, J.I. (2000). Development of polarizable water force fields for phase equilibrium calculations. *J. Phys. Chem. B* 104: 2391.

15 Busarow, K.L., Cohen, R., Blake, G.A. et al. (1989). Measurement of the perpendicular rotation-tunneling spectrum of the water dimer by tunable far infrared laser spectroscopy in a planar supersonic jet. *J. Chem. Phys.* 90: 3937–3943.

16 Zwart, E., Ter Meulen, J., Meerts, W.L., and Coudert, L. (1991). The submillimeter rotation tunneling spectrum of the water dimer. *J. Mol. Spectrosc.* 147: 27–39.

17 Pugliano, N. and Saykally, R. (1992). Measurement of quantum tunneling between chiral isomers of the cyclic water trimer. *Science* 257: 1937–1940.

18 Liu, K., Elrod, M.J., Loeser, J.G. et al. (1994). Far-IR vibration–rotation–tunnelling spectroscopy of the water trimer. *Faraday Discuss.* 97: 35–41.

19 Cruzan, J., Braly, L., Liu, K. et al. (1996). Quantifying hydrogen bond cooperativity in water: VRT spectroscopy of the water tetramer. *Science* 271: 59–62.

20 Liu, K., Brown, M., Cruzan, J., and Saykally, R. (1996). Vibration-rotation tunneling spectra of the water pentamer: structure and dynamics. *Science* 271: 62–64.

21 Liu, K., Brown, M., Carter, C. et al. (1996). Characterization of a cage form of the water hexamer. *Nature* 381: 501–503.

22 Keutsch, F.N. and Saykally, R.J. (2001). Water clusters: untangling the mysteries of the liquid, one molecule at a time. *Proc. Natl. Acad. Sci. U.S.A.* 98: 10533–10540.

23 Ludwig, R. (2001). Water: from clusters to the bulk. *Angew. Chem. Int. Ed.* 40: 1808–1827.

24 Tse, J.S. (2019). *A Twist in the Tale of the Structure of Ice*. London: Nature Publishing Group UK.

25 Kauzmann, W. (1959). Some factors in the interpretation of protein denaturation. *Adv. Protein Chem.* 14: 1–63.

26 Glew, D. (1962). Aqueous solubility and the gas-hydrates. The methane-water system1. *J. Phys. Chem.* 66: 605–609.

27 Lippincott, E.R., Stromberg, R.R., Grant, W.H., and Cessac, G.L. (1969). Polywater: vibrational spectra indicate unique stable polymeric structure. *Science* 164: 1482–1487.

28 Limmer, D.T., Willard, A.P., Madden, P., and Chandler, D. (2013). Hydration of metal surfaces can be dynamically heterogeneous and hydrophobic. *Proc. Natl. Acad. Sci. U.S.A.* 110: 4200–4205.

29 Wang, C., Lu, H., Wang, Z. et al. (2009). Stable liquid water droplet on a water monolayer formed at room temperature on ionic model substrates. *Phys. Rev. Lett.* 103: 137801.

30 Kimmel, G.A., Petrik, N.G., Dohnálek, Z., and Kay, B.D. (2005). Crystalline ice growth on pt (111): observation of a hydrophobic water monolayer. *Phys. Rev. Lett.* 95: 166102.

31 Akaishi, A., Yonemaru, T., and Nakamura, J. (2017). Formation of water layers on graphene surfaces. *ACS Omega* 2: 2184–2190.

32 Ball, P. (2013). When water doesn't wet. *Nat. Mater.* 12: 289–289.

33 Zhang, J., Tan, J., Pei, R. et al. (2021). Ordered water layer on the macroscopically hydrophobic fluorinated polymer surface and its ultrafast vibrational dynamics. *J. Am. Chem. Soc.* 143: 13074–13081.

34 Li, C.Y., Le, J.B., Wang, Y.H. et al. (2019). In situ probing electrified interfacial water structures at atomically flat surfaces. *Nat. Mater.* 18: 697–701.

35 Kalayan, J., Chakravorty, A., Warwicker, J., and Henchman, R.H. (2023). Total free energy analysis of fully hydrated proteins. *Proteins Struct. Funct. Bioinf.* 91: 74–90.

36 Biedermannová, L., Mal, M., Nekardová, M., and Schneider, B. (2022). Knowledge-based prediction of DNA hydration using hydrated dinucleotides as building blocks. *Acta Cryst.* 78: 1032–1045.

37 Fogarty, A.C. and Laage, D. (2014). Water dynamics in protein hydration shells: the molecular origins of the dynamical perturbation. *J. Phys. Chem. B* 118: 7715–7729.

38 Makarov, V.A., Andrews, B.K., Smith, P.E., and Pettitt, B.M. (2000). Residence times of water molecules in the hydration sites of myoglobin. *Biophys. J.* 79: 2966–2974.

39 Nibali, V.C. and Havenith, M. (2014). New insights into the role of water in biological function: studying solvated biomolecules using terahertz absorption

spectroscopy in conjunction with molecular dynamics simulations. *J. Am. Chem. Soc.* 136: 12800–12807.
40 Zhong, D., Douhal, A., and Zewail, A.H. (2000). Femtosecond studies of protein–ligand hydrophobic binding and dynamics: human serum albumin. *Proc. Natl. Acad. Sci. U.S.A.* 97: 14056–14061.
41 Halle, B. (2004). Protein hydration dynamics in solution: a critical survey. *Phil. Trans. R. Soc. Lond. B* 359: 1207–1224.
42 Luise, A., Falconi, M., and Desideri, A. (2000). Molecular dynamics simulation of solvated azurin: correlation between surface solvent accessibility and water residence times. Proteins: struct., Funct. *Bioinformatics* 39: 56–67.
43 Laage, D. and Hynes, J.T. (2006). A molecular jump mechanism of water reorientation. *Science* 311: 832–835.
44 Laage, D., Elsaesser, T., and Hynes, J.T. (2017). Water dynamics in the hydration shells of biomolecules. *Chem. Rev.* 117: 10694–10725.
45 Agmon, N. (1995). The Grotthuss mechanism. *Chem. Phys. Lett.* 244: 456–462.
46 Henry, E.R., Jones, C.M., Hofrichter, J., and Eaton, W.A. (1997). Can a two-state MWC allosteric model explain hemoglobin kinetics? *Biochemistry* 36: 6511–6528.
47 Eaton, W.A., Henry, E., Hofrichter, J., and Mozzarelli, A. (1999). Is cooperative oxygen binding by hemoglobin really understood? *Nat. Struct. Biol.* 6: 351–358.
48 Zhang, L., Yang, Y., Kao, Y.T. et al. (2009). Protein hydration dynamics and molecular mechanism of coupled water–protein fluctuations. *J. Am. Chem. Soc.* 131: 10677–10691.
49 Pal, P., Chakraborty, S., and Jana, B. (2021). Differential hydration of ice-binding surface of globular and hyperactive antifreeze proteins. *Adv. Theory Simul.* 4: 2100090.
50 Sun, T., Lin, F.H., Campbell, R.L. et al. (2014). An antifreeze protein folds with an interior network of more than 400 semi-clathrate waters. *Science* 343: 795–798.
51 Shankar, G.M., Li, S., Mehta, T.H. et al. (2008). Amyloid-β protein dimers isolated directly from Alzheimer's brains impair synaptic plasticity and memory. *Nat. Med.* 14: 837–842.
52 Chiti, F. and Dobson, C.M. (2006). Protein misfolding, functional amyloid, and human disease. *Annu. Rev. Biochem.* 75: 333–366.
53 Fan, W., Xing, L., Chen, N. et al. (2019). Promotion effect of succinimide on amyloid fibrillation of hen egg-white lysozyme. *J. Phys. Chem. B.* 123: 8057–8064.
54 Jiang, D., Rauda, I., Han, S. et al. (2012). Aggregation pathways of the amyloid β (1–42) peptide depend on its colloidal stability and ordered β-sheet stacking. *Langmuir* 28: 12711–12721.
55 Fitzpatrick, A.W., Debelouchina, G.T., Bayro, M.J. et al. (2013). Atomic structure and hierarchical assembly of a cross-β amyloid fibril. *Proc. Natl. Acad. Sci. U.S.A.* 110: 5468–5473.
56 Mcdonald, M., Box, H., Bian, W. et al. (2012). Fiber diffraction data indicate a hollow core for the Alzheimer's aβ 3-fold symmetric fibril. *J. Mol. Biol.* 423: 454–461.
57 Thirumalai, D., Reddy, G., and Straub, J.E. (2012). Role of water in protein aggregation and amyloid polymorphism. *Acc. Chem. Res.* 45: 83–92.

58 Tarus, B., Straub, J.E., and Thirumalai, D. (2006). Dynamics of Asp23–Lys28 salt-bridge formation in Aβ10-35 monomers. *J. Am. Chem. Soc.* 128: 16159–16168.

59 Li, M.S., Reddy, G., Hu, C.-K. et al. (2010). Factors governing fibrillogenesis of polypeptide chains revealed by lattice models. *Phys. Rev. Lett.* 105: 218101.

60 Hartgerink, J.D., Beniash, E., and Stupp, S.I. (2002). Peptide-amphiphile nanofibers: a versatile scaffold for the preparation of self-assembling materials. *Proc. Natl. Acad. Sci. U.S.A.* 99: 5133–5138.

61 Hartgerink, J.D., Beniash, E., and Stupp, S.I. (2001). Self-assembly and mineralization of peptide-amphiphile nanofibers. *Science* 294: 1684–1688.

62 Hosseinkhani, H., Hosseinkhani, M., Khademhosseini, A. et al. (2006). Enhanced angiogenesis through controlled release of basic fibroblast growth factor from peptide amphiphile for tissue regeneration. *Biomaterials* 27: 5836–5844.

63 Fry, H.C., Garcia, J.M., Medina, M.J. et al. (2012). Self-assembly of highly ordered peptide amphiphile metalloporphyrin arrays. *J. Am. Chem. Soc.* 134: 14646–14649.

64 Deshmukh, S.A., Solomon, L.A., Kamath, G. et al. (2016). Water ordering controls the dynamic equilibrium of micelle–fibre formation in self-assembly of peptide amphiphiles. *Nat. Commun.* 7: 12367.

65 Wang, J., Yuan, C., Han, Y. et al. (2017). Trace water as prominent factor to induce peptide self-assembly: dynamic evolution and governing interactions in ionic liquids. *Small* 13: 1702175.

66 Tao, K., Xue, B., Li, Q. et al. (2019). Stable and optoelectronic dipeptide assemblies for power harvesting. *Mater. Today* 30: 10–16.

67 Wang, M., Xiong, S., Wu, X., and Chu, P.K. (2011). Effects of water molecules on photoluminescence from hierarchical peptide nanotubes and water probing capability. *Small* 7: 2801–2807.

68 Piotrowska, R., Hesketh, T., Wang, H. et al. (2021). Mechanistic insights of evaporation-induced actuation in supramolecular crystals. *Nat. Mater.* 20: 403–409.

69 Wang, H., Liu, Z.L., Lao, J. et al. (2022). High energy and power density peptidoglycan muscles through super-viscous nanoconfined water. *Adv. Sci.* 9: 2104697.